Mitochondria

CELLULAR ORGANELLES

Series Editor: PHILIP SIEKEVITZ
Rockefeller University, New York, New York

MITOCHONDRIA
Alexander Tzagoloff

Mitochondria

Alexander Tzagoloff
Columbia University, New York, New York

Plenum Press • New York and London

Library of Congress Cataloging in Publication Data

Tzagoloff, Alexander 1937 –
Mitochondria.

(Cellular organelles)
Includes bibliographies and index.
1. Mitochondria. I. Title. II. Series. [DNLM: 1. Mitochondria. QH 603.M5 T998m]
QH603.M5T94 574.87'342 81-23373
ISBN 0-306-40799-X AACR2
ISBN 0-306-40778-7 (pbk.)

A Division of Plenum Publishing Corporation
233 Spring Street, New York, N.Y. 10013

Printed in the United States of America

To Helen, Lydia, and Natasha

Preface

In writing this book, I found the choice of a suitable title to be a most vexing problem. Lehninger's excellent earlier monograph *The Mitochondrion* had already appropriated in the domain of library cards what appeared to be the most fitting description of the subject matter. Once the text was completed, however, it became obvious that pluralization was the simplest solution to this dilemma. Variations in the structure and function of this organelle and recent discoveries of the phylogenetic diversity in the organization and genetic content of its DNA all seemed to justify the idea that there are as many different mitochondria as there are mitochondriologists.

Even though my initial intention was to provide advanced undergraduate and graduate students of cell biology with supplemental reading material on topics usually dealt with in a cursory manner by most standard texts, inevitably the scope was broadened to attract the interest of more seasoned readers who might be familiar with some but not other areas of mitochondrial studies. Consistent with the original aim, literature citations have been kept to a minimum and are avoided in the main body of the text for purposes of readability.

Justice cannot be done to all the information that has emerged after a long period of experimental work in as active a field as the present one. Within the scope allowed by the practical consideration of manuscript size, I have therefore tried to emphasize those areas being most actively pursued at present. Historical developments have not been completely neglected since certain topics such as the electron transfer chain and oxidative phosphorylation are best understood when viewed against the background of earlier work. A substantial part of the discussion is devoted to cytochrome oxidase and the ATP synthetase, which have played an important role in the interpretation of mitochondrial energy metabolism and biogenesis of the inner membrane. Knowledge of the structure and protein compositions of these complex enzymes is also important in appreciating the biosynthetic and genetic autonomy of the organelle.

As a final note, I would like to express my debt to Dr. David E. Green, who first introduced me to the mysteries of the mitochondrion and whose enthusiasm and continued sense of interest in solving the outstanding problems of energy coupling have served as an example in my own endeavors. Thanks are also due to Dr. Philip Siekevitz, Dr. Gloria Coruzzi, and my wife Helen for their helpful comments; to Mrs. Edith Casper for magically transforming rough first drafts into clean copy; and to the staff of Plenum Press for their excellent editorial help.

Alexander Tzagoloff

New York

Contents

Chapter 3

Oxidative Pathways of Mitochondria

Chapter 4

The Electron Transfer Chain

Chapter 5

Cytochrome Oxidase: Model of a Membrane Enzyme

Chapter 6

Oxidative Phosphorylation

Chapter 7

The Mitochondrial Adenosine Triphosphatase

Chapter 8

Resolution and Reconstitution of Electron Transport and Oxidative Phosphorylation

Chapter 9

Mitochondrial Transport Systems

Chapter 10

Mitochondrial Biogenesis

Chapter 11

Mitochondrial Genetics

Mitochondria

1

Evolution of Mitochondrial Studies

The notion of the compartmentalization of cellular functions, often by means of different organelles, arose from the convergence of two independent lines of investigation, one concentrating on cytological observations of cell inclusions and the other on biochemical studies of various metabolic pathways. The convergence occurred during the period of 1940–1960 when many important technical advances were made in the fractionation and isolation of reasonably homogeneous subcellular components. This meant that it became possible in many instances to assign specific functions to particular organelles or compartments of the cell. Today, every student of biology is aware of the fact that the mitochondrion is a specialized organelle whose primary functions are the conservation of oxidatively derived energy and its utilization for ATP synthesis. Some of the earlier developments that led to this recognition and other more recent highlights of studies of mitochondrial structure and function will be traced briefly in this chapter.

I. Early Cytological Studies (1850–1900)

A large number of light microscopic studies, some published as early as 1850, indicated the presence of small granules in different types of cells. These were given a variety of names but were in fact what we know today to be mitochondria. One of the more systematic cytological studies was reported by Altman in 1890. Altman noted that the subcellular granules were quite similar in their shape and size to bacteria (Fig. 1.1). Based on these observations, he postulated that the granules (he named them bioblasts) were the basic units of cellular activity. He further speculated that bioblasts, like bacteria, were capable of an independent existence and that it was by virtue of their colonial association in the cytoplasm of the host cell that the latter acquired the properties of life. It is interesting that there is a resurgence of these ideas, albeit in a somewhat different form, in the endosymbiont theory of the evolutionary origin of mitochondria.

The term "mitochondrion," meaning threadlike granule, was introduced by Benda around the turn of the century. At about the same time, Michaelis

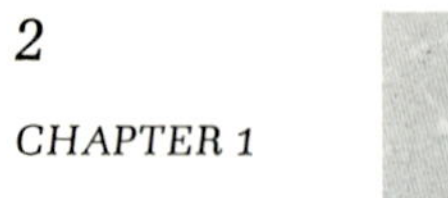

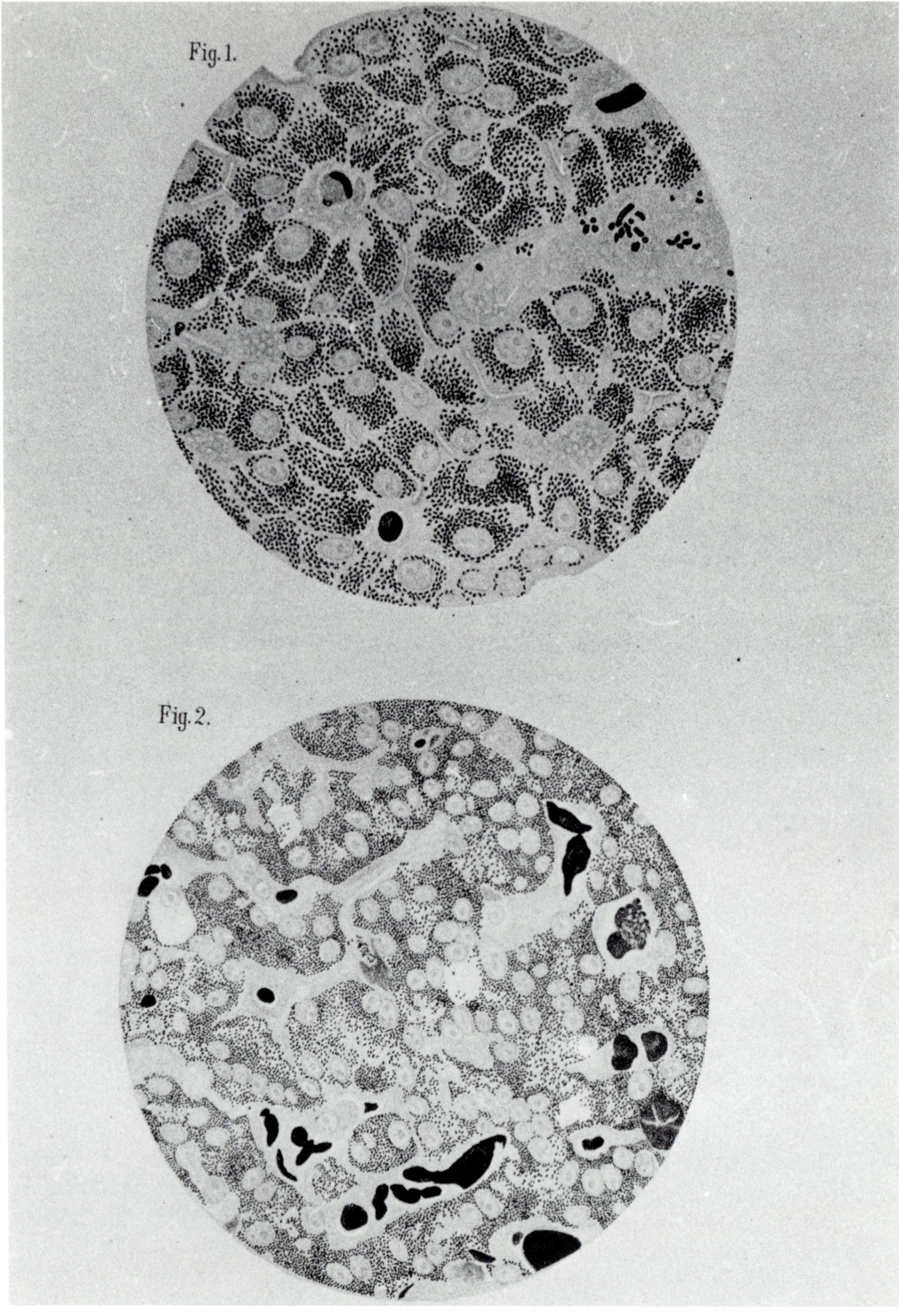

FIG. 1.1. Sections of frog liver fixed in osmium and stained with acid fuchsin (Altman, 1890).

developed a supravital technique for selectively staining mitochondria with the redox dye Janus Green. This was an important discovery because it provided a more definitive criterion for the cytological identification of the organelle and also because it was the first indication that mitochondria have the capacity to reduce a dye.

Although most cytologists in the period 1900–1930 recognized the mitochondrion to be a well-defined and ubiquitous cellular structure, there was no agreement as to its function. Some proposals made at that time were that mitochondria serve as centers of genetic information, protein synthesis, respiration, lipid synthesis, as well as a host of other assorted activities.

II. Early Studies of Biological Oxidation Reactions (1900–1940)

An extremely important question that became the concern of leading biochemists around 1900 was the problem of how substrates are oxidized in aerobic organisms. Two general mechanisms had been proposed, each based on what was known at the time from studies of model chemical reactions. According to Wieland, substrates were oxidized by virtue of a catalytic activation of the reduced compound which could then react with molecular oxygen as shown below.

$$SH_2 + \tfrac{1}{2}O_2 \rightarrow S + H_2O \qquad [1.1]$$

This interpretation arose from numerous examples in organic chemistry of reactions in which a reduced compound such as alcohol is first catalytically dehydrogenated, and the activated hydrogen then used to reduce some secondary acceptor such as oxygen.

$$RCH_2OH + \text{catalyst} \rightarrow RCH{=}O + \text{catalyst}(H_2) \qquad [1.2]$$

$$\text{catalyst}(H_2) + \tfrac{1}{2}O_2 \rightarrow \text{catalyst} + H_2O \qquad [1.3]$$

Wieland's ideas were challenged by Warburg who viewed the catalytic step to be an activation of oxygen. Here again, the evidence cited was largely based on known reactions such as the metal-catalyzed oxidation of certain organic compounds. In particular, Warburg championed the notion that iron was the chief catalyst in biological systems, since many of the properties of aerobic oxidation are mimicked by chemical reactions catalyzed by iron (e.g., inhibition by cyanide). Warburg devoted much of his research efforts to identifying and characterizing the natural iron catalyst which he termed "Atmungsferment." As will be seen, Warburg was successful in identifying the enzyme that reacts with oxygen and made many other fundamental contributions to our understanding of oxidation–reduction reactions. In the process, however, he had to abandon his original idea of oxygen activation.

In 1925, Keilin reported the first of an important series of investigations on the cytochrome system of pigments in aerobic cells. Actually, cytochromes had been discovered by MacMunn who, in 1886, proposed that they were heme-containing compounds that function in the transfer of oxygen. MacMunn's discovery occurred too early and had to await some 40 years as well as for the more complete studies of Keilin to be recognized and accepted.

The beauty of Keilin's experiments resided in their simplicity. Essentially, Keilin used a simple hand spectroscope to examine changes in the absorption properties of the cytochromes in a variety of cells under different metabolic conditions (Fig. 1.2). He concluded that there are three distinct pigments, which he called cytochromes *a*, *b*, and *c*, and that they underwent oxidation–reduction changes in a determined sequence such that cytochrome *b* was the furthest away from, and cytochrome *a* the closest to oxygen.

At about the time of Keilin's description of the cytochrome system, Warburg did an experiment that clearly showed that Atmungsferment was a hemoprotein. Warburg's experiment was to inhibit the terminal oxygen-utilizing reaction in yeast with carbon monoxide which he postulated competes with oxygen for the Atmungsferment. The Atmungsferment–carbon monoxide complex is light sensitive; i.e., carbon monoxide is dissociated from the enzyme by light. By correlating the efficiency with which carbon monoxide is dissociated

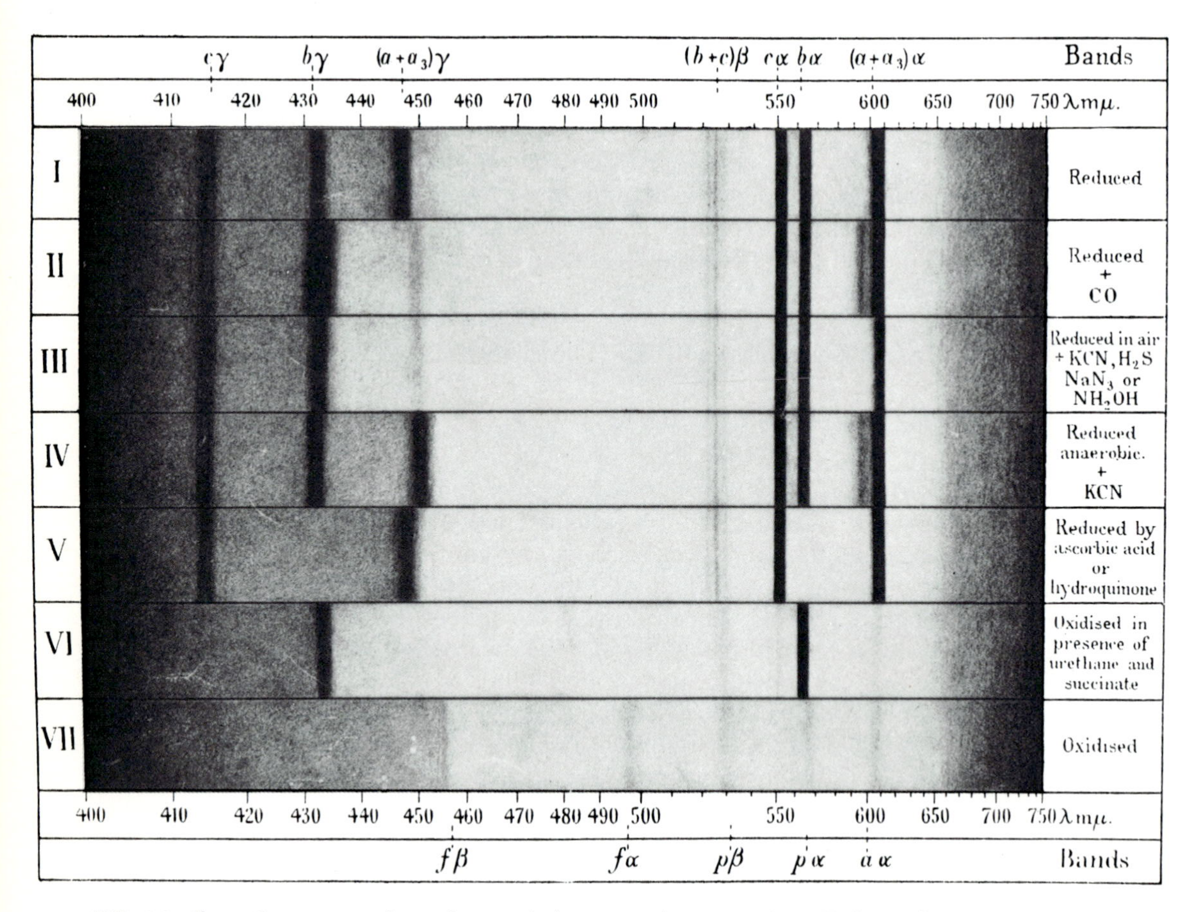

FIG. 1.2. Absorption spectra of cytochromes in heart muscle preparations (Keilin and Hartree, 1939).

at different wavelengths of light, Warburg obtained the action spectrum that is shown in Fig. 1.3. It was this result, showing the heme nature of the action spectrum, that convinced Warburg that Atmungsferment is a cytochrome. Although Keilin refused to accept this conclusion, he himself eventually provided its proof by showing that the Atmungsferment is identical to cytochrome a_3. Today we know that cytochrome a_3 is a catalytic component of cytochrome oxidase, the terminal enzyme of the respiratory chain.

In addition to his studies of the terminal oxidase, Warburg also characterized NAD^+ to be a hydrogen carrier in many oxidation–reduction reactions. Even more significant was his demonstration that flavoproteins act as intermediary carriers between NADH and the cytochrome system. Thus, by 1940, sufficient progress had been made to permit the following formulation of the electron transfer chain.

$$\text{NADH} \rightarrow \text{Flavoprotein} \rightarrow \text{Cytochromes } b \rightarrow c \rightarrow a \rightarrow a_3 \rightarrow O_2 \qquad [1.4]$$

Two other major landmarks in the field of energy metabolism occurred prior to 1940. The first was the description of the enzymatic reactions known as the tricarboxylic or citric acid cycle which lead to the complete conversion of pyruvate to CO_2 and H_2O. The second was the demonstration that the oxidation of various intermediates of the citric acid cycle is coupled to the generation of large amounts of ATP.

In 1935, Szent-Györgyi showed that the addition of different dicarboxylic acids to tissue homogenates stimulated oxygen consumption to a greater extent than could be explained by the oxidation of the added substrate alone. This suggested that the acids were acting in a catalytic fashion; i.e., they are regenerated from other substrates present in the homogenate and therefore participate in some kind of cyclic process. These observations were extended by Martius and Knoop who concluded that citrate is converted to succinate by way of isocitrate, oxalsuccinate, and α-ketoglutarate. Shortly thereafter, Krebs showed that citrate could be formed from pyruvate and oxalacetate, thus providing an explanation for the entry of pyruvate into the cycle (see Chapter 3, Section I).

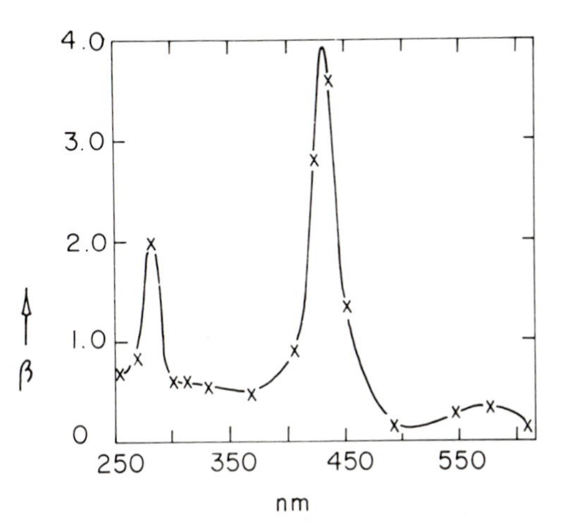

FIG. 1.3. Photochemical action spectrum of Atmungsferment of yeast (Warburg and Negelein, 1929).

Krebs worked out all of the intermediate steps of the citric acid cycle which constitutes one of the most elegant metabolic schemes in biochemistry.

By 1930, it was realized that the vast amount of energy released during the aerobic oxidation of carbohydrates must be trapped in some way for the use of the cell. Although ATP had been discovered in 1930, its importance as a repository of biochemically useful energy was not appreciated until the late 1930s when a number of reactions (e.g., oxidation of glyceraldehyde phosphate) of intermediary metabolism were found to be coupled to the synthesis of ATP. That decade also saw the discovery of oxidative phosphorylation. Kalckar demonstrated that, in aerobic systems, phosphorylation was dependent on the consumption of oxygen. At the same time, Belitser and Tsibakova made quantitative measurements of the oxygen-dependent conversion of phosphate to an organic form. These workers made the critical observation that the esterification of phosphate was stoichiometric with the amount of oxygen used. In their experiments, the ratio of phosphate esterified to oxygen utilized (P/O ratio) approximated 2. Based on these findings, Belitser and Tsibakova concluded that they were measuring the chemical conversion of energy released during the oxidation–reduction reactions mediated by the terminal respiratory pathway. This important conclusion was shown to be correct from subsequent work in the laboratories of Ochoa, Green, Lardy, Lipmann, and Lehninger.

III. Identification of Mitochondria as Centers of Energy Metabolism (1940–1950)

In 1932, Bensley and Hoerr reported the isolation from guinea pig liver of a granular fraction they identified as mitochondria. This report was greeted with skepticism for several reasons. The first had to do with the prevalent feeling among many cytologists of the day that the approach was bankrupt and that the real answers to cellular functions could only be attained in studies of the intact system. The second difficulty in Bensley's work was that some of the criteria previously used for the *in situ* identification of mitochondria were not fulfilled by the isolated fraction. For example, the isolated granules did not stain with Janus Green, a specific stain for mitochondria. Despite these ambiguities, Bensley's experiments were an important first step for subsequent work on cell fractionation and pointed to the feasibility of this type of approach.

The problem of fractionating subcellular components was taken up in 1940 by Claude who succeeded in separating what he called large and small subcellular granules by differential centrifugation of cell homogenates. Extensive studies on the enzymatic characteristics of these fractions led him to conclude that the large granules which he considered to be mitochondria had succinate oxidase and cytochrome oxidase activity. Neither of these activities were present in the fraction containing the small granules, indicating that the separation techniques were sufficiently resolving to yield functionally different subcellular fractions. Claude's success, however, was still marred by the fact that the suspending medium (saline) employed caused rather extensive damage of mitochondria, resulting in preparations cross contaminated with secretory granules. These problems were finally solved by Hogeboom, Schneider, and Palade in

1948. These workers incorporated a high concentration of sucrose into the isolation medium to osmotically stabilize the mitochondria. Hogeboom and Schneider, with their improved preparation, were able to show that up to 80% of the cellular succinate oxidase is isolated in the mitochondrial fraction.

The same year that Hogeboom, Schneider, and Palade reported the successful isolation of intact mitochondria, two other important studies were published. Green, Loomis, and Auerbach showed that the entire citric acid cycle system of enzymes was associated with a particulate fraction obtained from both liver and kidney homogenates. This fraction, which was first called the cyclophorase system, was shortly afterwards shown to consist of mitochondria. Kennedy and Lehninger observed that the enzymes of fatty acid oxidation sedimented with a particulate fraction present in liver homogenates. A year later, Lehninger, using the method of Hogeboom, Schneider, and Palade to prepare liver mitochondria, showed that both fatty acid oxidation and the citric acid cycle were localized in the mitochondrion.

IV. Gross and Fine Structure of Mitochondria

Improved methods for the fixation, embedding, and sectioning of biological materials allowed a description of the gross morphological features of mitochondria. In 1953, Palade published the results of an extensive study in which he examined mitochondrial profiles in thin sections of different mammalian tissue. Palade noted that in all the tissues studied the mitochondria consisted of an outer limiting membrane and a set of internal membrane pleats or invaginations which he called "cristae mitochondriales." The presence of two membranes and their relative structural organization have been confirmed by many workers to be generally true of mitochondria in different plant, animal, and lower eucaryotic cells.

Although there was considerable interest in the fine structure analysis of mitochondrial membranes, the thin-sectioning technique was of limited usefulness. In fixed and stained tissue, the inner and outer membranes of mitochondria have the same railroad-track appearance as other cellular membranes (Fig. 1.4). This *status quo* persisted until 1962 when Fernandez-Moran, using the then new technique of negative staining, found that the inner mitochondrial membrane is lined with spherical particles 90 Å in diameter. This discovery stimulated investigators to identify the function of the particles. The problem was resolved by Racker and colleagues who, in 1965, conclusively demonstrated that the inner membrane particles were physically identical to the F_1 ATPase. This enzyme had been shown to be part of the ATP synthetase complex which is involved in the coupling mechanism of oxidative phosphorylation.

Further progress in understanding the organization of the enzymes of the inner membrane came from studies in Green's laboratory on the ultrastructure of the respiratory complexes of the electron transfer chain. These enzymes were shown to be globular proteins with dimensions similar to particles seen in the basal part of the inner membrane. Green and his co-workers found that purified preparations of cytochrome oxidase as well as the other respiratory

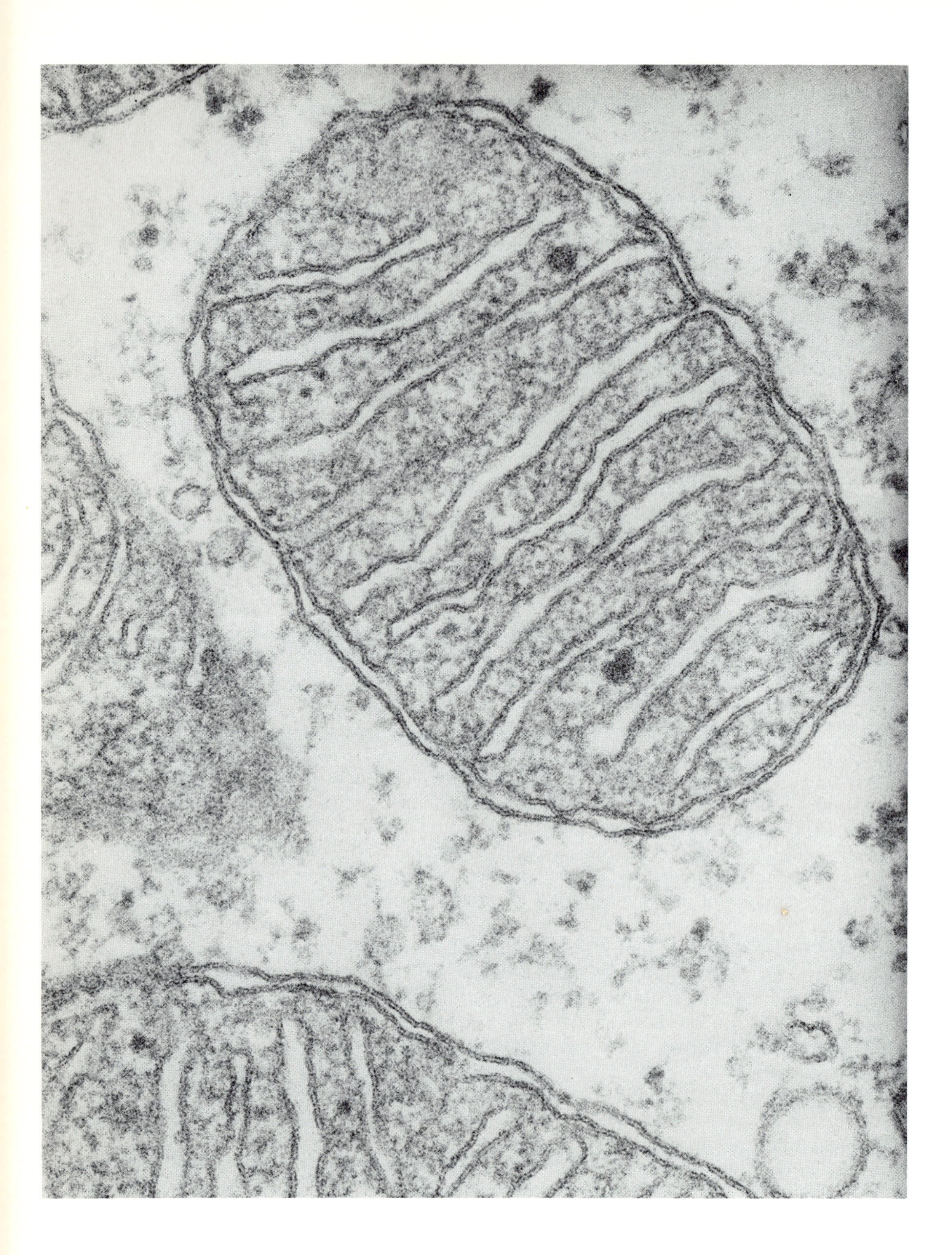

complexes had the interesting property of spontaneously associating into membranes whose ultrastructure was not unlike that of the inner membrane stripped of the ATPase particles. These studies provided fairly strong evidence that the electron transfer chain is composed of globular enzymes which are located in the membrane continuum. The capacity of a purified enzyme such as cytochrome oxidase to form a membrane, the only requirement being phospholipid, was also conceptually important, since it eliminated the need to invoke structural proteins in the organization of the inner membrane.

The interest in mitochondrial ultrastructure has continued to grow, and new physical and chemical methods are being used to probe fine details of the architecture of the inner membrane. Part of the current interest stems from the realization that the mechanisms of a number of functions carried out by membranes are intimately tied up with the structure and spatial arrangements of the component enzymes. Mitchell's notion of chemiosmotic coupling in oxidative phosphorylation, for example, demands certain topological arrangements of the respiratory complexes vis-a-vis each other and the ATP synthetase. A great deal of effort is therefore being devoted to studies of how the respiratory complexes are oriented in relation to the two faces of the inner membrane. Another area of interest is the quaternary structure of the enzymes themselves. Cytochrome oxidase, a multiprotein complex containing at least seven nonidentical subunits, is being examined from the standpoint of the function of the subunits and their spatial organization in the oligomeric enzyme.

V. Electron Transfer

Our knowledge of the electron transfer chain has been greatly enhanced in terms both of the oxidation–reduction components and of their role as catalytic subunits of respiratory complexes. Much of the new information stems from work in Chance's laboratory where sophisticated spectrophotometric methods were devised to analyze the visible components of the chain. These studies provided important kinetic data on the turnover rates of various carriers of the chain. Applications of other spectroscopic techniques such as electron spin resonance by Beinert were instrumental in showing that in addition to heme iron, the chain also contains different nonheme iron species which function in electron transfer. Another important member of the chain, coenzyme Q, was found by Crane to be present in mitochondrial and bacterial electron transfer systems. This lipoidal quinone is now firmly established to be an obligatory carrier in the oxidation of both NADH and succinate. Although new oxidation–reduction components have not been found during recent years, it is by no means certain that the electron transfer chain as it is pictured today is the final version.

←

FIG. 1.4. Electron micrograph of mitochondria in thin-sectioned pancreas centroacinar cells. The mitochondrial membranes are seen to have a bilayer structure. Magnification ×155,000. (Courtesy of Dr. G. Palade.)

All of the early investigations of mitochondrial electron transport were essentially aimed at defining the components of the chain and the sequence in which they function. Attempts at their isolation were largely unsuccessful because of a firm association of the proteins with the membrane and their lack of solubility in water. The exception to this was cytochrome *c* which had been purified by Keilin. These circumstances led to a rather abstract concept of the chain which was usually depicted as a sequence of the different oxidation–reduction carriers that had no real physical attributes. In the 1950s, however, investigators began to be successful in solubilizing enzymatically active portions of the chain. The work of Hatefi and his colleagues is especially noteworthy in this respect. These workers were able to purify four respiratory complexes from beef heart mitochondria, each containing a unique set of electron carriers and representing a segment of the overall chain. Hatefi also showed that under appropriate conditions, the oxidation of NADH and of succinate by oxygen could be reconstituted from the four respiratory complexes. Hatefi's studies formed the basis for a new way of thinking about the chain which could now be visualized as being a macromolecular assembly composed of four well-defined enzymes which carry out the entire electron transfer process. This interpretation was not generally accepted by many workers who felt that the procedures used to isolate the complexes caused artifactual fragmentation of the chain. The criticism, however, has been dispelled by the test of time, and the respiratory complexes are now recognized as real entities.

VI. Oxidative Phosphorylation

Although a tremendous amount of work has been done to determine how ADP and phosphate are esterified during oxidative phosphorylation, the number of experimental findings that have furthered our understanding of the process are relatively few. Soon after the discovery of the phenomenon, it was shown in several laboratories that there is a fixed stoichiometry of ATP formed to oxygen consumed. This was expressed as an efficiency of phosphorylation or P/O ratio which for NAD-linked substrates is 3 and for succinate, 2. The implication of these observed efficiencies was that there were different sites along the electron transfer chain where energy was conserved and used for ATP synthesis. The phosphorylation sites, or segments of the chain in which ATP was synthesized, were determined by two independent methods. Chance used a spectrophotometric method to study crossover points between electron transfer components (see Chapter 6, Section IVA) and determined the first site to be at the level of flavin and the second and third to be in the regions of cytochrome *b* and *a*, respectively. Other laboratories used site-specific assays for different segments of the chain to confirm this result.

The use of reagents that prevent ATP synthesis proved to be an extremely useful approach in dissecting the phosphorylation mechanism into two distinct stages. The earliest of such reagents to be used was dinitrophenol which belongs to the class of uncouplers of oxidative phosphorylation. These compounds were found to exert their effect by blocking some early event prior to the esterification of phosphate. Lardy discovered that in addition to uncouplers, it is also possible to inhibit oxidative phosphorylation with certain antibiotics such as oligomycin which act as inhibitors of the ATP synthetase. In contrast to uncouplers, inhibitors of oxidative phosphorylation prevent a terminal step involving the esterification of phosphate. The difference in the mode of action of uncouplers versus inhibitors was a crucial advance for two reasons. First, it demonstrated the existence of a nonphosphorylated intermediate or stored form of energy that can be used for a host of energy-demanding processes, e.g., ion transport, reversal of electron transfer. Second, it pointed to the essential difference between the coupling event, that is, the mechanism by which the energy of oxidation is first converted to a utilizable form, and the phosphorylation mechanism by which it is trapped in the form of ATP.

It was evident from the earliest studies with mitochondria that the phosphorylation mechanism is very labile and that the intactness of the organellar structure is essential for the preservation of a high efficiency of phosphorylation. Nonetheless, the realization that a resolution of the system into its component parts would be a powerful approach to understanding how it works was a stimulus for many researchers to persist in their efforts. A major breakthrough in these endeavors was made by Pullman and his colleagues who, in 1960, succeeded in isolating a soluble protein from beef heart mitochondria that was capable of restoring phosphorylation to submitochondrial particles in which this capacity was lost. This protein was a Mg-dependent ATPase which later was shown to be a catalytic component of the ATP synthetase complex. Resolution and reconstitution have continued to be a very useful means of studying the phosphorylation mechanism. In recent years, Racker's laboratory has been particularly successful in this approach and has been able to reconstitute oxidative phosphyloration in simple systems consisting of a respiratory complex and components of the ATP synthetase. These studies promise the real possibility that the essentiality of specific proteins can be tested and their functions determined.

In 1953, Slater formulated a scheme involving chemical intermediates to explain the mechanism of oxidative phosphorylation. The mechanism was based on what was known at the time about substrate level phosphorylation and entailed both nonphosphorylated and phosphorylated high-energy intermediates. Although the details of the scheme were revised in light of new information, it served as a working hypothesis for many years to workers in the field. During a decade of search on the part of many investigators, however, no substantive evidence for the existence of the postulated intermediates was obtained, and as a reaction to these abortive efforts, alternate mechanisms were proposed. One of the most innovative proposals was Mitchell's chemiosmotic hypothesis of energy coupling which dispensed with the need of any chemical intermediates during oxidative phosphorylation. Essentially, Mitchell suggested that during electron transfer, a hydrogen ion gradient is established across the inner membrane. The energy of the gradient is used by a reversible ATPase for the synthesis of ATP. Although the chemiosmotic hypothesis was not widely accepted at first, it has been gaining in popularity, and there is now a substantial body of experimental evidence from different quarters that supports the correctness of Mitchell's idea.

VII. Biogenesis and Genetics of Mitochondria

The mitochondrion was one of the first organelles to be isolated free of other cellular membranes and therefore has served as a useful model for studying various aspects of the structure, function, and, more recently, biogenesis of membranes. There were several key developments that helped to guide the direction of research on the problem of mitochondrial biogenesis. One of these was the finding by McLean and Simpson and later by Work and others that mitochondria possess a protein-synthesizing machinery that makes a limited number of proteins. This discovery led to investigations of the properties of the translational system as well as the identity and functional role of its products.

We know now that mitochondrial ribosomes resemble bacterial ribosomes but are quite different from the ribosomes found in the cytoplasm of eucaryotic cells. Linnane first showed that mitochondrial ribosomes are inhibited by antibiotics that inhibit bacterial but not plant or animal ribosomes. By use of selective inhibitors of either mitochondrial or cytoribosomal protein synthesis, it became possible to determine which of the mitochondrial proteins are derived from either of the two synthetic systems. It is now well established that certain enzymes of the mitochondrial inner membrane, including some of the respiratory complexes and the ATPase, are synthesized through the joint cooperation of the mitochondrial and cytoplasmic systems of protein synthesis. Much of the current work in this area centers around the problem of how proteins made in the cytoplasm are imported into the organelle and how they are integrated with their mitochondrially made partners into functional membrane enzyme complexes.

During the past 20 years, there has also been a burgeoning of studies on mitochondrial genetics so that this area of research can now be considered as a separate discipline. This field saw its beginning in the early 1950s when Ephrussi discovered a cytoplasmically inherited mutation in yeast that caused a deficiency of respiratory functions. Ephrussi's work indicated that mitochondrial morphogenesis is controlled by a nonchromosomal genetic element that he called the ρ factor. In 1964, Schatz reported the presence of DNA in yeast mitochondria, and a year later this was confirmed by Nass and her co-workers to be true of other mitochondria as well. The existence of mitochondrial DNA and its identification with the ρ factor was crucial to the development of mitochondrial genetics.

The genetic information stored in mitochondrial DNA has been a subject of considerable interest in recent years, both from the standpoint of the genetics of the organelle and of its biogenesis. The technology for studying the genetic properties of mitochondrial DNA was developed in the 1960s largely from the work done with yeast in the laboratory of Slonimski. Most of these studies were done with ρ^- mutants and with antibiotic-resistant mutants that had been isolated during that period. New types of mitochondrial mutants have been found by Tzagoloff and co-workers that define a new class of mitochondrial genes designated as mit genes. Mit genes have recently been shown to code for the same set of proteins that are translated on mitochondrial ribosomes.

Since mitochondrial DNA is relatively small, its coding capacity being of the same order of magnitude as that of some bacteriophages, it is safe to predict that it will be the first eucaryotic genome to be understood in detail.

Selected Readings

Altman, R. (1890) *Die Elementarorganismen und Ihre Beziehungen zur den Zellen,* Veit Co., Keipzig.

Annan, G., Banga, I., Blazsó, A., Bruckner, V., Laki, K., Straub, B., and Szent-Gyorgyi, A. (1935) Über die Bedeutung der Fumarsäure für die tierische Gewebeatmung, *Z. Physiol. Chem.* **244:**105.

Beinert, H., and Sands, R. H. (1959) On the function of iron in DPNH cytochrome *c* reductase, *Biochem. Biophys. Res. Commun.* **1:**171.

Belitser, V. A., and Tsibakova, E. T. (1939) The mechanism of phosphorylation associated with respiration, *Biokhimiia* **4:**516.

Benda, C. (1898) Weitere Mitteilungen über die Mitochondria, *Verh. Physiol. Ges. Berlin* 376–383.

Bensley, R. R., and Hoerr, N. L. (1934) Studies on cell structure: VI. The preparation and properties of mitochondria, *Anat. Rec.* **60:**251.

Chance, B., and Williams, G. R. (1955) A method for the localization of sites for oxidative phosphorylation, *Nature* **176:**63.

Chance, B., and Williams, G. R. (1956) The respiratory chain and oxidative phosphorylation, *Adv. Enzymol.* **17:**65.

Claude, A. (1946) Fractionation of mammalian liver cells by differential centrifugation: II. Experimental procedures and results, *J. Exp. Med.* **84:**61.

Coen, D., Deutsch, J., Netter, P., Petrochilo, E., and Slonimski, P. P. (1970) Mitochondrial genetics. I. Methodology and phenomenology, in *Control of Organelle Development,* Symposia of the Society for Experimental Biology (P. L. Miller, ed.), Vol. 24, Cambridge University Press, London, pp. 449–496.

Crane, F. L., Lester, R. L., Widmer, C., and Hatefi, Y. (1959) Studies on the electron transfer system. XVIII. Isolation of coenzyme Q from beef heart and beef heart mitochondria. *Biochim. Biophys. Acta.* **32:**73.

Cowdry, E. V. (1953) Historical background of research on mitochondria, *J. Histochem. Cytochem.* **1:**183.

Ephrussi, B., Hottinguer, H., and Chimenes, A. M. (1949) Action de l'acriflavine sur les levures. I. La mutation "petite colonie," *Ann. Inst. Pasteur* **76:**351.

Fernandez-Moran, H. (1962) Cell membrane ultrastructure. Low temperature electron microscopy and X-ray diffraction studies of lipoprotein components in lamellar systems, *Circulation* **26:**1039.

Green, D. E., Loomis, W. F., and Auerbach, V. H. (1948) Studies on the cyclophorase system. I. The complete oxidation of pyruvic acid to carbon dioxide and water, *J. Biol. Chem.* **172:**389.

Green, D. E., and Tzagoloff, A. (1966) The mitochondrial electron transfer chain, *Arch. Biochem. Biophys.* **116:**293.

Hatefi, Y., Haavik, A. G., and Griffiths, D. E. (1962) Studies of the electron transfer system. XL. Preparation and properties of mitochondrial DPNH–coenzyme Q reductase; XLI. Reduced coenzyme Q–cytochrome *c* reductase, *J. Biol. Chem.* **237:**1676, 1681.

Hobeboom, G. H., Schneider, W. C., and Palade, G. E. (1948) Cytochemical studies of mammalian tissue. I. Isolation of intact mitochondria from rat liver; some biochemical properties of mitochondria and submicroscopic particulate material, *J. Biol. Chem.* **172:**619.

Kalckar, H. (1937) Phosphorylation in kidney tissue, *Enzymologia* **2:**47.

Keilin, D. (1925) Cytochrome, a respiratory pigment common to animals, yeast and higher plants, *Proc. R. Soc. Lond.* [*Biol.*] **98:**312.

Keilin, D., and Hartree, E. F. (1939) Cytochrome and cytochrome oxidase, *Proc. R. Soc. Lond.* [*Biol.*] **127:**167.

Kennedy, E. P., and Lehninger, A. L. (1949) Oxidation of fatty acids and tricarboxylic acid cycle intermediates by isolated rat liver mitochondria, *J. Biol. Chem.* **179:**957, 964, 969.

Krebs, H. A., and Johnson, W. A. (1937) The role of citric acid in intermediary metabolism in animal tissue, *Enzymologia* **4**:148.

Lamb, A. J., Clark-Walker, G. D., and Linnane, A. W. (1968) The biogenesis of mitochondria. 4. The differentiation of mitochondrial and cytoplasmic protein synthesizing systems *in vitro* by antibiotics, *Biochim. Biophys. Acta* **161**:415.

Lardy, H. A., Johnson, D., and McMurray, W. C. (1958) Antibiotics as tools for metabolic studies. I. A survey of toxic antibiotics in respiratory, phosphorylative and glycolytic systems, *Arch. Biochem. Biophys.* **78**:587.

Lehninger, A. L. (1964) *The Mitochondrion*, W. A. Benjamin, New York.

MacMunn, C. A. (1887) Researches on myohaematin and the histohaematins, *Phil. Trans. R. Soc. Lond.* **177**:267.

Martius, C., and Knoop, F. (1937) Der physiologische Abbau der Citronsäure, *Z. Physiol. Chem.* **246**:1.

McLean, J. R., Cohn, G. L., Brandt, I. K., and Simpson, M. V. (1958) Incorporation of labeled amino acids into the protein of muscle and liver mitochondria, *J. Biol. Chem.* **233**:657.

Michaelis, L. (1900) Die vitale Färbung, eine Darstellungsmethode der Zellgrana, *Arch. Mikrosk. Anat.* **55**:558.

Mitchell, P. (1961) Coupling of phosphorylation to electron and hydrogen transfer by a chemiosmotic type of mechanism, *Nature* **191**:105.

Nass, M. M. K., Nass, S., and Afzelius, B. A. (1965) The general occurrence of mitochondrial DNA, *Exp. Cell Res.* **37**:190.

Ochoa, S. (1940) Nature of oxidative phosphorylation in brain tissue, *Nature* **146**:267.

Palade, G. E. (1953) An electron microscope study of the mitochondrial structure, *J. Histochem. Cytochem.* **1**:188.

Pullman, M. E., Penefsky, H. S., Datta, A., and Racker, E. (1960) Partial resolution of enzymes catalyzing oxidative phosphorylation. I. Purification and properties of soluble, dinitrophenol-stimulated adenosine triphosphatase, *J. Biol. Chem.* **235**:3322.

Racker, E., and Kandrach, A. (1973) Partial resolution of the enzymes catalyzing oxidative phosphorylation. XXXIX. Reconstitution of the third segment of oxidative phosphorylation, *J. Biol. Chem.* **248**:5841.

Racker, E., Tyler, D. D., Estabrook, R. W., Conover, T. E., Parsons, D. F., and Chance, B. (1965) Correlation between electron transport activity, ATPase and morphology of submitochondrial particles, in *Oxidases and Related Redox Systems* (T. E. King, H. S. Mason and M. Morrison, eds.), John Wiley & Sons, New York, pp. 1077–1094.

Roodyn, D. B., Reis, P. J., and Work, T. S. (1961) Protein synthesis in mitochondria, *Biochem. J.* **80**:9.

Schatz, G., Haslbrunner, E., and Tuppy, H. (1964) Deoxyribonucleic acid associated with yeast mitochondria, *Biochem. Biophys. Res. Commun.* **15**:127.

Slater, E. C. (1953) Mechanism of phosphorylation in the respiratory chain, *Nature* **172**:59.

Tzagoloff, A., Akai, A., Needleman, R. B., and Zulch, G. (1975) Assembly of the mitochondrial membrane system. Cytoplasmic mutants of *Saccharomyces cerevisiae* with lesions in enzymes of the respiratory chain and in the mitochondrial ATPase, *J. Biol. Chem.* **250**:8236.

Warburg, O. (1949) *Heavy Metal Prosthetic Groups and Enzyme Action*, Oxford University Press, London.

Warburg, O., and Negelein, E. (1929) Über das Absorptionsspektrum des Atmungsferment, *Biochem. Z.* **214**:64.

Wieland, H. (1932) *On the Mechanism of Oxidation*, Yale University Press, New Haven.

2

Mitochondrial Structure and Compartmentalization

I. Gross Morphology of Mitochondria

The average mitochondrion, whether in a mammalian or in a lower eucaryotic cell (e.g., yeast), has approximately the same dimensions as the bacterium *Escherichia coli.* It is most commonly observed as an oval particle, 1–2 μm long and 0.5–1 μm wide. Mitochondria, therefore, are sufficiently large to be seen in the light microscope, and the earliest descriptions of this organelle were based on observations of fixed and stained tissues. For a more detailed description of the internal organization of mitochondria, investigators had to make use of the more highly resolving electron microscope. These studies began with the pioneering work of Claude and Palade whose electron micrographs of mitochondria *in situ* were of such excellence that they compare favorably with micrographs seen in journals today.

Virtually all of the techniques developed for the electron microscopic visualization of subcellular structures have been applied to the study of mitochondria. Some of the techniques and the special features they reveal will be briefly described here and in the section on ultrastructure.

A. Fixed and Stained Thin Sections

When a piece of tissue or any biological material is placed in a solution of osmium tetroxide, most of the macromolecular components (proteins, lipids, nucleic acids, carbohydrates) become cross linked, both internally and to each other, and as a consequence are no longer free to diffuse. In other words, the arrangement of subcellular structures with respect to each other becomes frozen in space. The fixed specimen is then dehydrated and embedded in a material that penetrates the space formerly occupied by water. This is always a synthetic polymer that, on warming, polymerizes to form a solid matrix. At this point, it is possible to cut the embedded material with a glass or diamond knife into sections that can be made as thin as 300 Å. To increase the contrast between the embedding material which is electron transparent and the biological structures which are electron opaque but only poorly so, the sections are treated with a lead salt. The lead metal reacts with the proteins, lipids, and other cellular constituents and, being very electron opaque, helps to make them more visible.

In Fig. 2.1 are shown electron micrographs of various tissues and cells that have been sectioned and stained by this technique. The mitochondria of the different tissues are seen to be similar in their gross morphology. The common features are the presence of two separate membranes which define two different internal spaces. The peripheral membrane, known as the *outer membrane*, is a continuous bag that encloses the entire contents of the organelle. The second membrane is internal to the outer membrane and is also a topologically continous surface. This membrane is referred to as the *inner membrane*. Because of its larger surface area, the inner membrane forms a series of folds or invaginations that project into the interior space. The invaginations are known as the mitochondrial cristae and in cross-sectional views are seen to consist of closely apposed surfaces of the inner membrane.

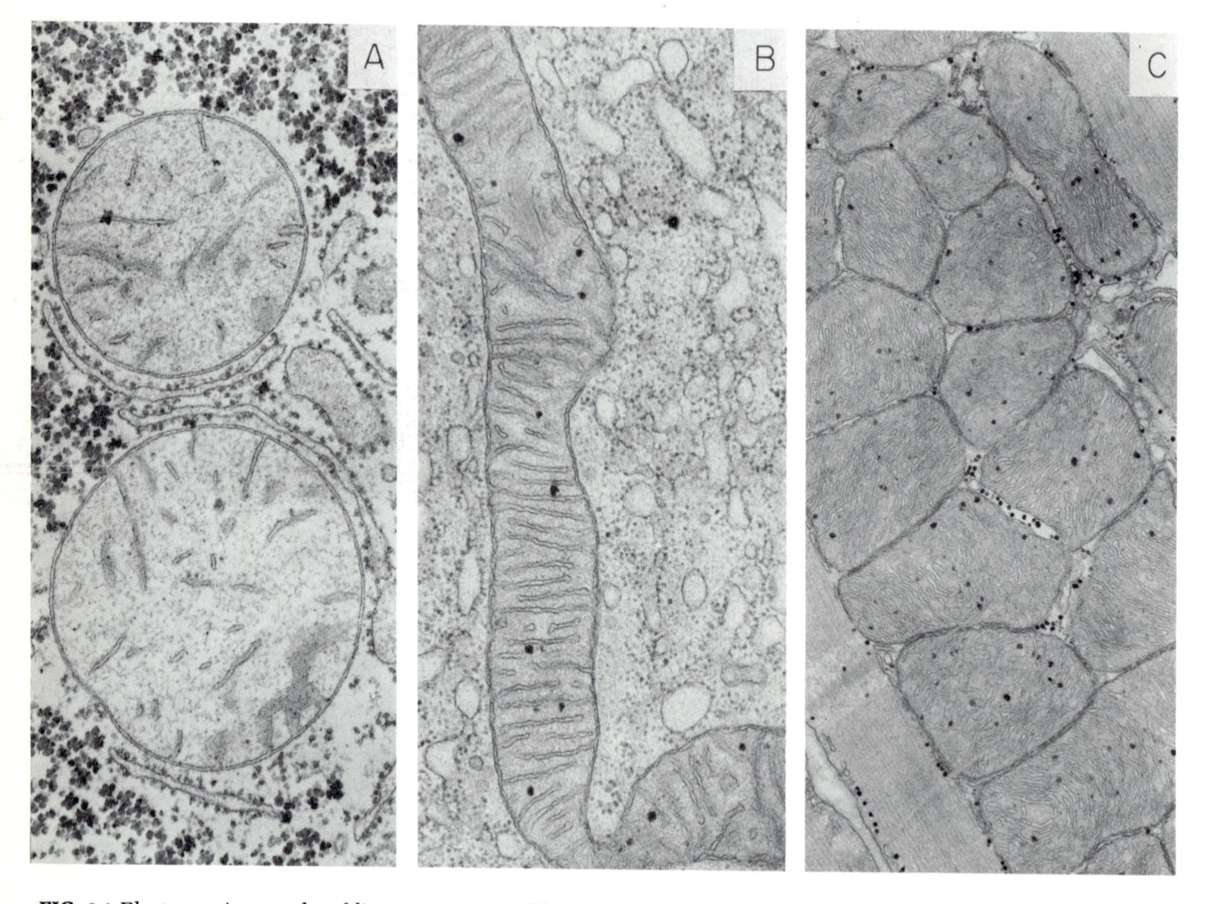

FIG. 2.1 Electron micrographs of liver, pancreas and heart tissues showing mitochondrial profiles with different ratios of inner to outer membranes. (A) Rat hepatocyte (×31,000), (B) guinea pig pancreatic acinar cell (×29,000), (C) rat myocardium (×21,000). (Courtesy of Dr. G. Palade.)

As a rule, the inner membrane always has a larger surface area than the outer membrane. The ratio of the two, however, may vary considerably depending on the tissue and type of cell. It is evident from the examples shown in Fig. 2.1 that some mitochondria have relatively more inner membrane than others. Muscle mitochondria have a larger number of more densely packed cristae than liver mitochondria. It will be seen later that the enzymes involved in oxidative phosphorylation are located in the inner membrane, and the self-evident generalization can be made that tissues such as muscle that demand the greatest output of oxidatively derived ATP will have mitochondria that are best adapted for this purpose. At the morphological level, this is expressed in a high concentration of cristae that house the enzymatic machinery for the production of ATP.

The outer and inner membranes form two internal compartments. The *intracristal* or *intermembrane* space is located between the outer and inner membranes, and the *intercristal* space or *matrix* is enclosed by the inner membrane (Fig. 2.2). In most mitochondria the intermembrane space is small compared to the matrix, but as will be seen, there are certain conditions that cause changes in the relative volumes of the two compartments.

B. Orthodox versus Condensed Conformations

The inner membrane can assume different spatial arrangements depending on the physiological state of the cell or the composition of the medium in which isolated mitochondria are suspended. In the orthodox conformation, the cristae are folded into sheets, and most of the internal space is occupied by the matrix. In the second configuration, referred to as *condensed*, the intermembrane space increases because of a separation of the cristae surfaces. The expansion of the intermembrane space occurs at the expense of a decrease in the volume of the matrix. In extremely condensed mitochondria, the inner membrane has the appearance of long tubules which wind through the interior to form a highly complex spaghetti-like network. (Fig. 2.3).

The orthodox and condensed conformations of rat liver mitochondria have been studied by Hackenbrock and have also been observed in other types of

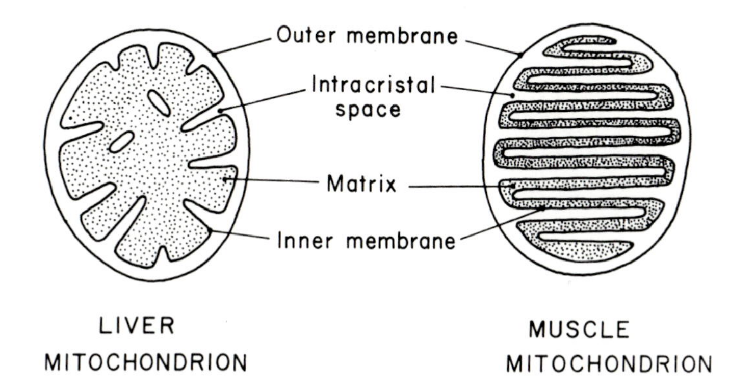

FIG. 2.2. Cross sections of liver and heart mitochondria showing the four compartments.

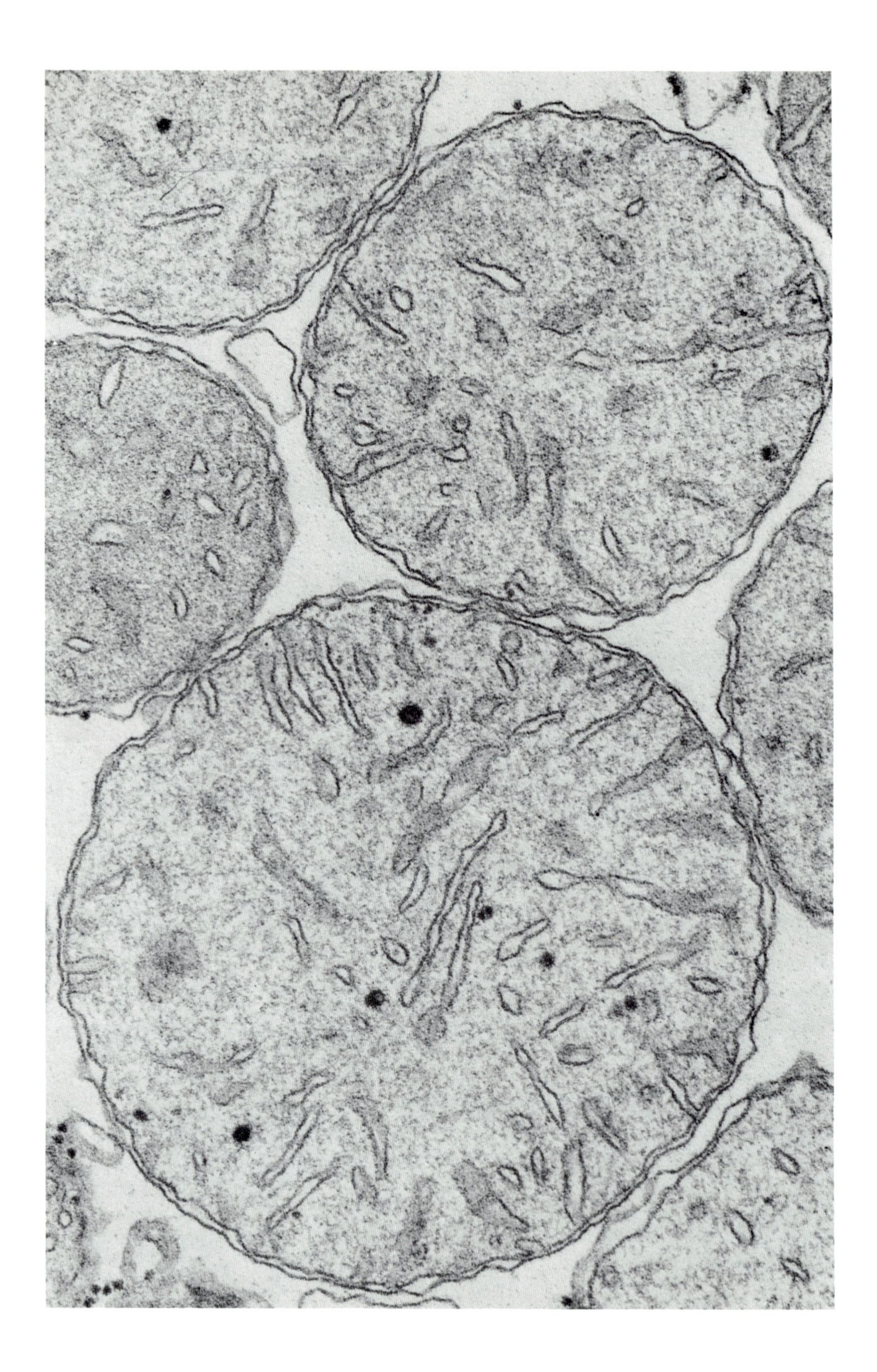

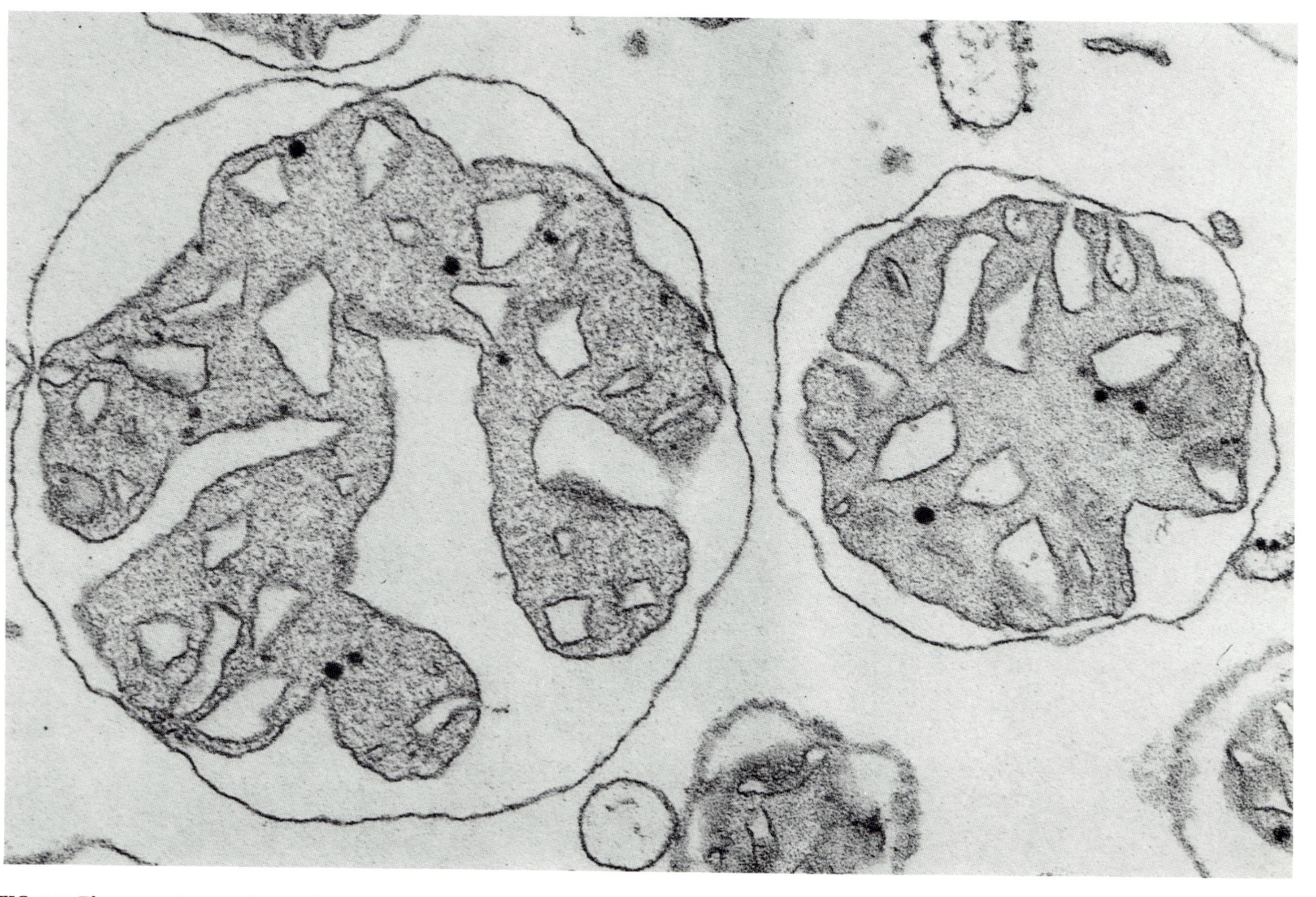

FIG. 2.3. Electron micrograph of isolated rat liver mitochondria fixed in the low-energy orthodox (upper) and high-energy condensed (lower) conformations. The magnification is $\times 110{,}000$ (courtesy of Dr. Charles Hackenbrock).

mitochondria. Hackenbrock found that during active respiration and oxidative phosphorylation most of the mitochondria are condensed. A reversible transition between the condensed and orthodox states can be induced by the addition or removal of ADP. These results suggested that the two conformations are related to the metabolic state of mitochondria or energetization of the inner membrane.

C. Mitochondria as Seen in Live Cells

Light microscopy, especially when used in conjunction with modern phase-contrast optics, provides an excellent means of examining unfixed cells. A number of cinematographic studies have been made of live cells under conditions of active metabolism. To those who are accustomed to seeing mitochondria and other cellular membranes in fixed tissue, these films are an eye-opening experience. They reveal that far from being the static structures observed in thin sections, mitochondria are extremely dynamic organelles capable of profound changes in size, form, and location. Mitochondria from different parts of the cell can stream through what appear to be cytoplasmic channels where they often meet and fuse into larger organelles. Such compound mitochondria can, at a later time, break up into a series of smaller particles which again become dispersed throughout the cell. Mitochondrial plasticity is also evidenced by recent studies of yeast in which three-dimensional reconstructions from serial sections have shown that in certain stages of growth, the cell contains a single mitochondrion presumed to arise from a coalescence of a large number of smaller mitochondria. In addition to fission and fusion, mitochondria are constantly contracting, expanding, and undergoing the same kind of shape changes seen in moving amoebas. The purpose and cause of this wild motion are not known at present.

A method for examining gross mitochondrial morphology in living cells, developed by Chen, takes advantage of the specific uptake by mitochondria of the fluorescent dye Rhodamine 123. Cells stained with this dye reveal the distribution of mitochondria against an empty background when viewed in a fluorescence microscope. Some examples of mitochondrial networks detected by this method in different cells are shown in Fig. 2.4. Chen and co-workers have found that conditions affecting the cytoskeletal system of the cell (colchicine) also cause changes in the shape and cellular distribution of mitochondria. This technique offers a new means for studying changes in mitochondrial morphology in response to various physiological and diseased states of cells.

D. Mitochondria with Unusual Morphologies

Under certain conditions of physiological stress, mitochondria become very large and are altered in their structure. This is observed when animals are placed on riboflavin- or copper-deficient diets. After prolonged deprivation of these essential dietary cofactors, mitochondria of rat liver have been reported to have ten times the volume seen in animals maintained on normal diets. Such mitochondria contain relatively little inner membrane and have the general appearance of being swollen.

Alterations in the gross morphology of mitochondria have also been correlated with aging and senescence. In *Drosophila*, mitochondria enlarge with age and contain glycogen granules. Similar observations have been reported for blowfly muscle mitochondria whose cristae are rearranged into lamellar structures (Fig. 2.5). The lamellar density increases as the muscle gets older,

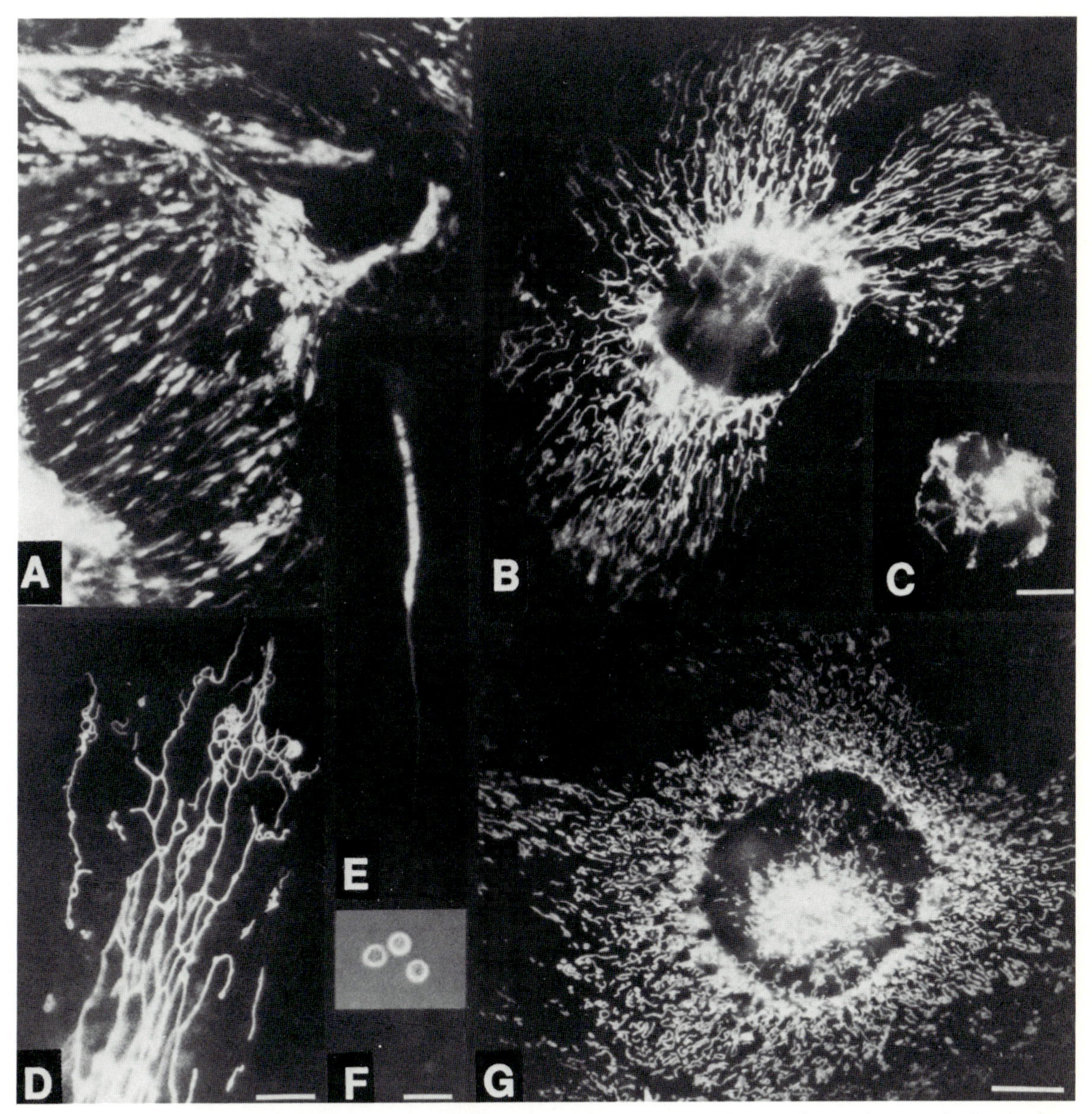

FIG. 2.4. Mitochondria of living cells stained with Rhodamine 123. (A) Rat heart, (B) marsupial kidney, (C) mouse B lymphocytes, (D) mouse 3T6 cells, (E) mouse sperm, (F) human erythrocytes in phase contrast (upper) and stained with Rhodamine (lower) (the erythrocytes lack mitochondria and therefore do not stain), (G) rat embryo fibroblast. The bars represent 15 μm in A, B, E, and G; 10 μm in C and F; 8 μm in D. (Courtesy of Dr. Lan Bo Chen.)

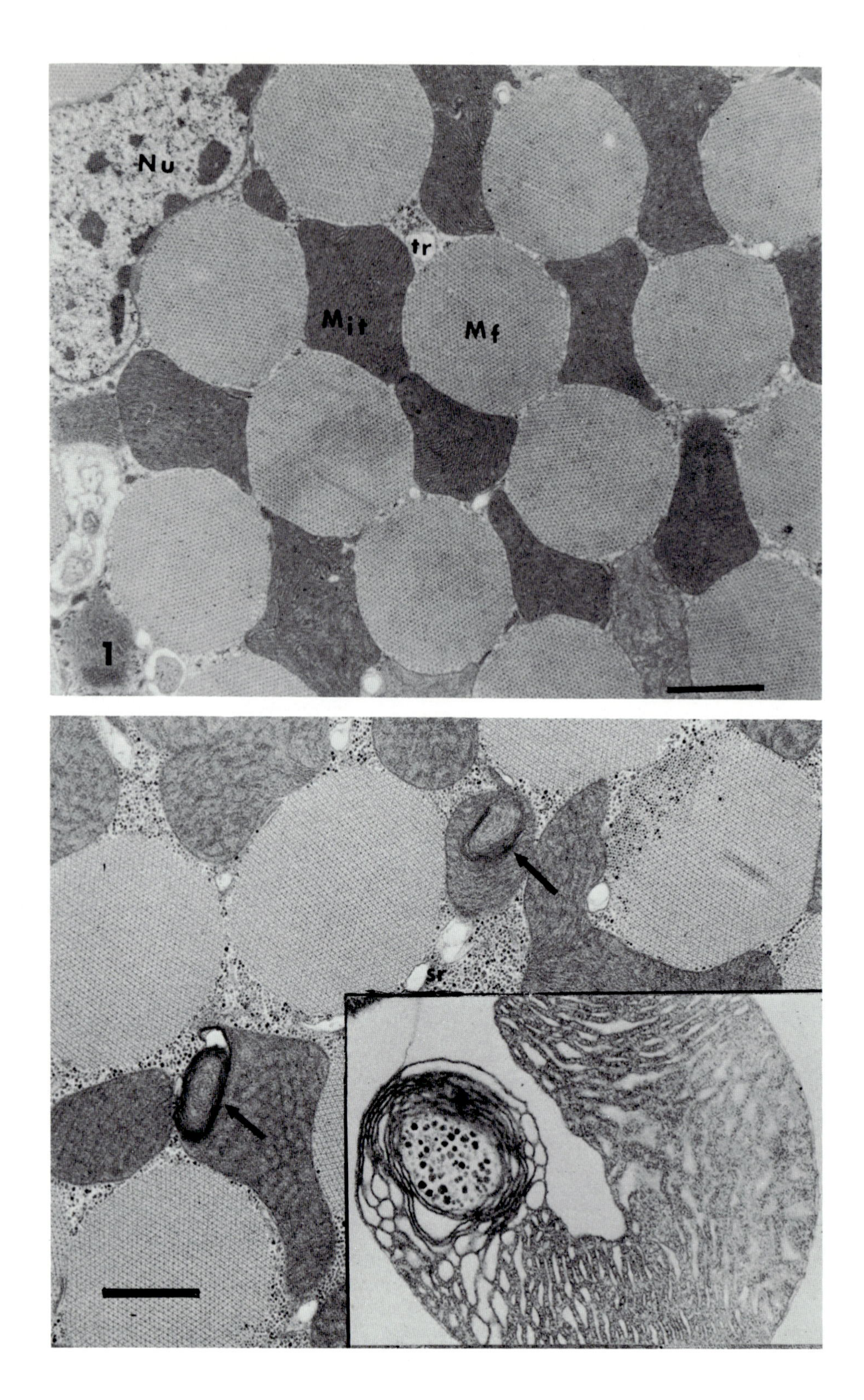
Nu
tr
Mit
Mf
1
sr

and there is an accompanying decrease in the respiratory capacity of the mitochondria. These changes are suggestive of a degenerative process.

Another interesting transformation in mitochondrial structure occurs during the differentiation of spermatocytes into mature spermatozoa. The process of spermatogenesis involves a growth of a spermatogonia cell into a primary spermatocyte followed by two meiotic divisions which produce four haploid spermatids. The spermatids then undergo a dramatic series of changes to become functional spermatozoa or sperm. The nucleus of the spermatid shrinks and moves towards one pole of the cell which eventually becomes the sperm head. Some of the cytoplasm of the spermatid forms a long tail which serves as a motility organelle. The mitochondria of the spermatid congregate in an intermediary zone between the head and the flagellum where they fuse into a structure known as the chondriosome. The changes in mitochondrial structure occurring during spermatogenesis in rat have been studied by the French cytologist André.

Spermatogonia have a relatively small number of mitochondria with normal morphology. During the beginning of the meiotic division, the mitochondria increase in number and aggregate into clusters of five to ten particles. There is also some change in the internal structure—the cristae begin to separate, and there is a condensation of the matrix. At the end of meiosis, the mitochondria are aligned along the internal periphery of the plasma membrane and have a very condensed matrix. During the maturation of the spermatid, the mitochondria move toward the tail and align themselves along the axis of the incipient flagellum. The matrix remains highly condensed, and there is a reduction in the size of the mitochondrial particles. In the last stage of spermatogenesis, the mitochondria fuse into large helix-shaped chondriosomes that are wound around the flagellum (Fig. 2.6). The chondriosomes occupy almost the entire space of the middle zone of the sperm and are very close to the flagellum. This probably insures a rapid diffusion of oxygen into the organelle and of ATP to the flagellum.

II. Subfractionation of Mitochondria

In the previous section it was pointed out that the arrangement of the two mitochondrial membranes creates two internal spaces. There are, therefore, a total of four compartments where an enzyme can be located in the organelle—in one of the two membranes or in one of the two internal spaces. It has been a matter of some interest to separate the inner and outer membranes and the components of the intra- and intercristal spaces and to study their enzymatic

←

FIG. 2.5. Electron micrographs of flight muscle mitochondria in young and aging blowflies. Upper: Transverse section of the dorsal longitudinal flight muscle from a 7-day adult fly. The mitochondria (Mt) exhibit a normal dense packing of cristae. Other structures are myofibrils (Mf), nucleus (Nu), and tracheoles (tr). Lower: Transverse section of flight muscle from a 34-day-old aging fly. Some of the mitochondria are seen to contain whorled structures (arrows). A higher magnification of an isolated mitochondrion from an aged fly is shown in the inset. The bars represent 1 μm. (Courtesy of Dr. Bertram Sacktor.)

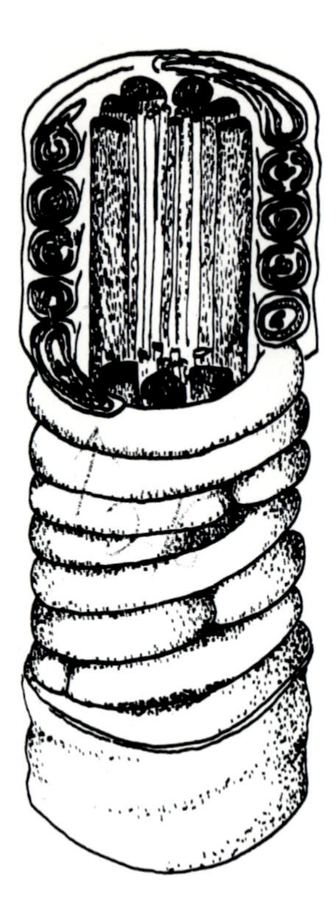

FIG. 2.6. Three-dimensional reconstruction of a midsection of a rat sperm. The chondriosomes are wound around the centrally located flagellum. (From Andre, 1962.)

composition. This has been done successfully with rat liver and other types of mitochondria.

Several methods have been described by which the two membranes as well as the soluble components can be separated. The procedures most commonly used involve an initial rupture of the outer membrane by mild detergent treatment or by osmotically induced swelling. The outer membrane, having a higher content of lipids and therefore a higher buoyant density, can be separated from the denser inner membrane on sucrose gradients.

A. Osmotic Method

This method employs dilute phosphate buffer which causes mitochondria to undergo large-amplitude swelling. In the swollen state, the outer membrane ruptures while the inner membrane remains osmotically intact. The medium is then made hypertonic with sucrose in order to contract the inner membrane plus its matrix components (mitoplast). The outer membrane at this point is almost completely pealed off and can be detached by mild sonic irradiation. The broken pieces of outer membrane reseal to form vesicles which can be isolated by centrifugation in a sucrose gradient. This method is illustrated in part A of Fig. 2.7.

B. Digitonin Method

Digitonin, a plant steroid with surface-active properties, has been used to selectively disaggregate the outer membrane. At concentrations of 1 mg per milligram of mitochondrial protein, the outer membrane of rat liver mitochondria is broken into smaller fragments which reseal into vesicles much in the same way as in the osmotic method. At this concentration of digitonin, the inner membrane remains intact and conserves its matrix. The low-digitonin treatment therefore results in a separation of the outer membrane from the inner membrane plus matrix and at the same time releases the constituents that are present in the intermembrane space. When the digitonin-treated mixture is centrifuged at low gravitational fields, the mitoplasts can be separated from the small outer membrane vesicles. The supernatant from the low-speed centrifugation contains the outer membrane and the components of the intracristal space. This fraction is centrifuged a second time at a higher speed to yield a pellet consisting of the outer membrane vesicles and a supernatant with the intracristal enzymes.

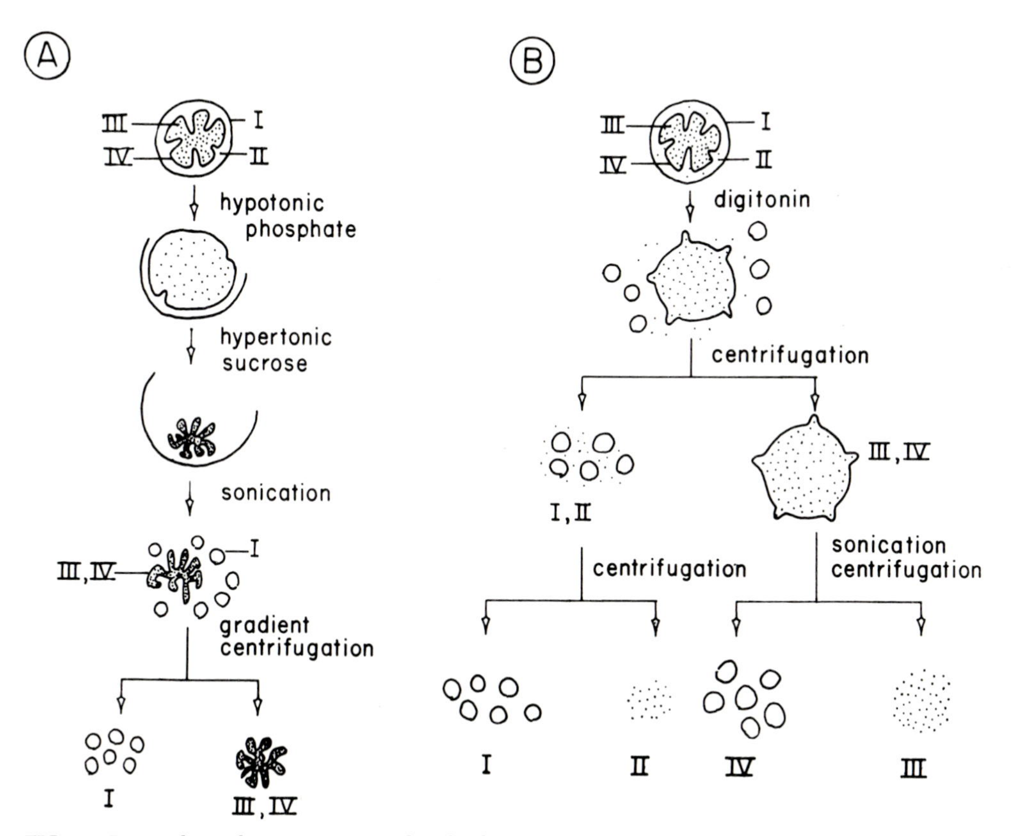

FIG. 2.7. Protocols used to prepare mitochondrial inner and outer membranes. A: Osmotic method. B: Digitonin method. The following notations are used to designate the mitochondrial compartments: I, outer membrane; II, intermembrane components; III, matrix components; IV, inner membrane. (Adapted from Ernster and Kuylenstierna, 1970; Schnaitman and Greenwalt, 1968.)

The mitoplasts can be further resolved into the inner membrane and matrix proteins by sonic irradiation, osmotic shock, or treatment with mild detergents. These procedures rupture the inner membrane, thereby releasing the soluble proteins of the matrix into the suspending medium. Analogous to the outer membrane, the inner membrane fragments reseal to form vesicles which can be separated from the soluble matrix proteins by sedimentation at high centrifugal speeds (part B of Fig. 2.7).

C. Fractionation of Mitochondria into Soluble and Membrane Components

Although the above procedures are useful for separating components localized in the different mitochondrial compartments, they have the disadvantages of being time consuming and also limited in terms of the amount of material that can be processed. For these reasons, most studies still employ preparations of submitochondrial particles obtained by direct sonic disruption of mitochondria. For such preparations, mitochondria are exposed to ultrasonic sound for a sufficient duration to fragment the two membranes and to release both intra- and intercristal proteins from their respective compartments into the suspending medium. The membrane vesicles (in this case, they represent a mixture of fragments from both membranes) are separated from the soluble components by differential centrifugation and can be used directly without further fractionation. These preparations, variously designated as electron transport particles (ETP) or submitochondrial particles (SMP), consist predominantly of fragments of the inner membrane, particularly when made from mitochondria that are relatively rich in inner membrane (Fig. 2.8).

III. Distribution of Mitochondrial Enzymes

Once procedures had been devised for separating the mitochondrial membranes and the components of the two soluble compartments, it became possible to determine the distribution of the various enzymes within the organelle. In such experiments, outer and inner membranes and the inter- and intracristal components are isolated, and their enzyme constituents are assayed. The dis-

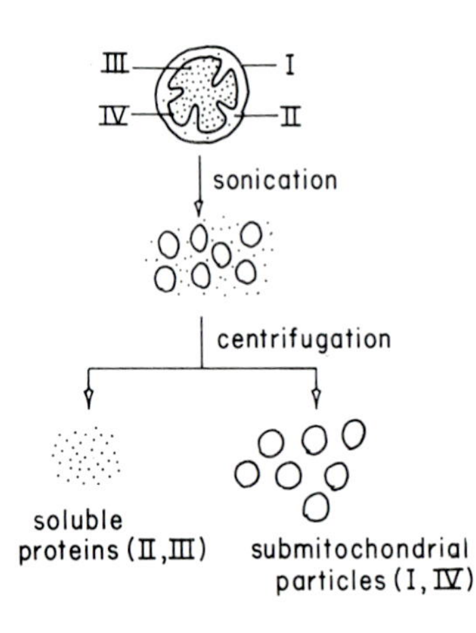

FIG. 2.8. Preparation of submitochondrial particles from sonically disrupted mitochondria.

tribution of some representative mitochondrial activities in the different subfractions obtained from rat liver mitochondria is shown in Table 2.1. In some instances, the assignments are not completely without ambiguity because of some degree of cross contamination of the two membranes. Nonetheless, most of the mitochondrial enzymes have now been localized. This information is summarized in Table 2.2.

A. Outer Membrane

The outer membrane contains a substantial number of enzymes. They do not constitute any integrated metabolic pathway, and consequently, the compositional data do not give us any clear idea about the function of this membrane. The two most frequently used marker enzymes for the outer membrane are monoamine oxidase and the rotenone-insensitive NADH–cytochrome *c* reductase. The latter activity is catalyzed by a flavoenzyme and cytochrome b_5 complex and is distinct from the NADH–cytochrome *c* reductase of the respiratory chain. A flavoprotein–cytochrome b_5 complex is also present in the endoplasmic reticulum. Although the two enzymes have similar properties, their identity has not been demonstrated.

The mitochondrial outer membrane also contains a number of enzymes concerned with phospholipid biosynthesis (glycerophosphate acyl transferases, fatty acid elongation system). Although mitochondria are capable of *de novo* synthesis of cardiolipin, the other major phospholipids are synthesized on endoplasmic reticulum. The outer membrane, however, does contain the enzymes for synthesizing phosphatidic acid from glycerol and fatty acids. It is possible the surface localization of these enzymes permits an easy access of this intermediate to enzymes of phospholipid metabolism in the endoplasmic reticulum.

B. Inner Membrane

Oxidative phosphorylation is the most important function of mitochondria. This process is catalyzed by four respiratory complexes and by the ATP syn-

TABLE 2.1. Recovery of Activities after Digitonin Treatment[a]

Enzyme	Percent of total activity		
	Intermembrane	Outer membrane	Mitoplasts
Adenylate kinase	120	3.1	0
Monoamine oxidase	16	63.6	19.3
Cytochrome oxidase	0.8	8.0	87.2
Malate dehydrogenase	8.0	1.1	96

[a]Rat liver mitochondria were treated with digitonin and centrifuged at 8000 × g. The pellet from this centrifugation corresponds to the mitoplast fraction. The supernatant was centrifuged at 144,000 × g. The pellet from the second high speed centrifugation corresponds to the outer membrane, and the supernatant to the intermembrane proteins. The activities are calculated as percent of the starting untreated mitochondria. (From Schnaitman and Greenawalt, 1968.)

thetase or oligomycin-sensitive ATPase. These five enzymes are located in the inner membrane and account for most of its membrane mass. In addition to the respiratory complexes and the ATPase, the inner membrane houses the pyridine nucleotide transhydrogenase, β-hydroxybutyrate dehydrogenase, and the enzyme systems responsible for the transport of ions, substrates, and nucleotides.

C. Intermembrane Space

Very few activities have been found in the compartment between the inner and outer membranes. The most important is adenylate kinase which probably functions in maintaining the proper balance of adenine nucleotides in the organelle.

D. Matrix

Most of the soluble enzymes of the mitochondrion are found in the space internal to the inner membrane. The list of activities is very long, but of partic-

TABLE 2.2. Distribution of Enzymes in Mitochondrial Compartments[a]

Outer membrane	Inner membrane	Intermembrane space	Matrix
NADH–cytochrome b_5 reductase	NADH–coenzyme Q reductase	Adenylate kinase (myokinase)	Pyruvate dehydrogenase complex
Cytochrome b_5	Succinate–coenzyme Q reductase	Nucleoside diphosphokinase	α-Ketoglutarate dehydrogenase complex
Monoamine oxidase	Coenzyme QH_2–cytochrome c reductase	Nucleoside monophosphokinase	Citrate synthase
Kynurine hydroxylase	Cytochrome oxidase		Aconitase
Glycerolphosphate acyl transferase	Oligomycin-sensitive ATPase		Malate dehydrogenase
Lysophosphatidyl acyl transferase	β-Hydroxybutyrate dehydrogenase		Isocitrate dehydrogenase (NAD)
Phosphatidate phosphatase	Pyridine nucleotide transhydrogenase		Isocitrate dehydrogenase (NADP)
Phospholipase A	Carnitine palmityl transferase		Fumarase
Nucleoside diphosphokinase	Ferrochetalase		Glutamate dehydrogenase
Fatty acid elongation	Adenine nucleotide carrier		Pyruvate carboxylase
			Aspartate aminotransferase
			Ornithine carbamyl transferase
			Carbamyl phosphate synthetase
			Fatty acyl-CoA synthetase
			Fatty acyl-CoA dehydrogenase
			Enoyl hydrase
			β-Hydroxyacyl-CoA dehydrogenase
			β-Ketoacyl-CoA thiolase
			Amino acid activating enzymes
			RNA polymerase
			DNA polymerase

[a]Based on studies of rat liver mitochondria. (From Ernster and Kuylenstierna, 1970.)

ular significance are the tricarboxylic acid cycle and fatty acid oxidation systems of enzymes. Among other important enzymes present in the matrix are glutamate dehydrogenase, the pyruvate dehydrogenase complex, carbamyl phosphate synthetase, and ornithine transcarbamylase. Finally, the matrix contains DNA, transfer RNAs, various aminoacyl transferases, DNA and RNA polymerases, and other components of the mitochondrial transcriptional and translational machineries.

IV. Composition of the Inner and Outer Membranes

Mitochondria are composed of proteins, lipids, nucleic acids, and small molecules such as substrates and cofactors. The major structural elements of the organelle are the proteins and lipids, the latter being for the most part phospholipids. The relative amounts of protein and lipid vary depending on how rich the mitochondria are in inner membrane. Since the phospholipids are associated exclusively with the membranes and are not found in the matrix, it stands to reason that mitochondria with a high concentration of cristae will also have a high content of phospholipids. Preparations of mitochondrial membranes free of the soluble constituents, however, have approximately the same amount of phospholipid, independent of the source of mitochondria.

A. Lipid Constituents

The major phospholipids of mitochondria are phosphatidyl choline, phosphatidyl ethanolamine, and cardiolipin (Fig. 2.9). Detailed lipid analyses have been done on different mitochondria; some of this information is summarized in Table 2.3. The lipid content of the fractionated inner and outer membranes has also been studied. Several major differences are evident from the data shown in Table 2.4. The first difference to be noted is the higher concentration

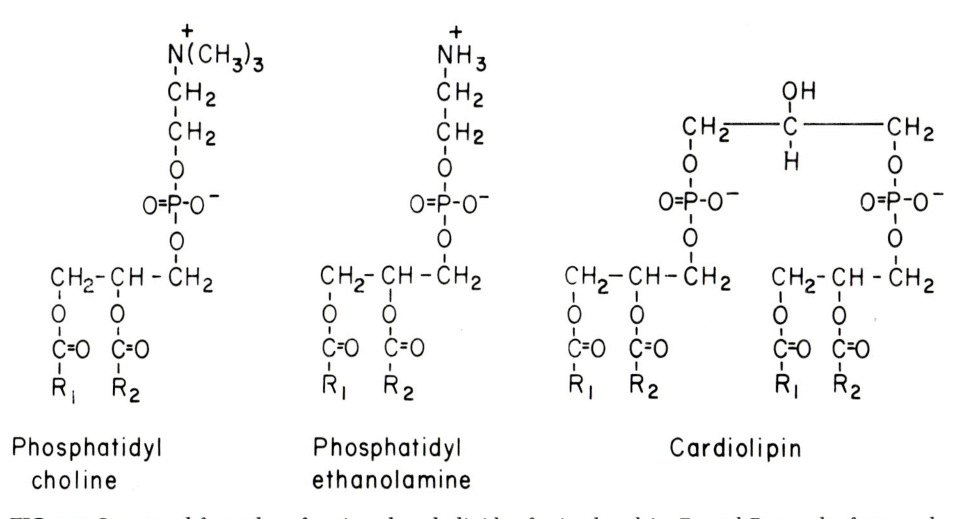

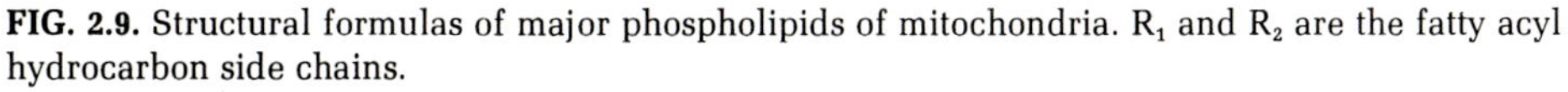

FIG. 2.9. Structural formulas of major phospholipids of mitochondria. R_1 and R_2 are the fatty acyl hydrocarbon side chains.

TABLE 2.3. Lipid Compositions of Bovine Heart, Kidney, and Liver Mitochondria[a]

Component	Heart	Kidney	Liver
Percent phospholipid by weight	25.3	19.6	14.9
	Percent of total phospholipids		
Phosphatidyl choline	40.8	39.6	43.4
Phosphatidyl ethanolamine	37.4	38.1	34.5
Cardiolipin	19.1	19.2	17.2
Phosphatidyl inositol	~3	~3	~3
Phosphatidyl serine	trace	trace	trace

[a]Data taken from Fleischer *et al.* (1967).

TABLE 2.4. Lipid Composition of Mitochondrial Membranes[a]

Lipid	Mitochondria	Inner membrane	Outer membrane	Microsomes
Phospholipids (mg/mg protein)	0.16	0.30	0.88	0.39
Cholesterol (μg/mg protein)	2.28	5.26	30.1	30.2
	Percent of total phospholipids			
Phosphatidyl choline	40.0	44.5	55.2	62.8
Phosphatidyl ethanolamine	28.4	27.7	25.3	18.6
Cardiolipin	22.5	21.5	3.2	0.5
Phosphatidyl inositol	7.0	4.2	13.5	13.4

[a]Based on studies with guinea pig liver mitochondria. (From Parson *et al.*, 1967.)

of total lipids in the outer membrane. This fact accounts for its lower density. The second difference is the absence of cardiolipin (the small amount that is present probably represents contamination by inner membrane) in the outer membrane. A comparison of the outer membrane with microsomes* reveals that their lipid compositions are quite similar.

B. Protein Constituents

In rat liver mitochondria, the outer membrane contains 5% of the total protein; the inner membrane represents 21%. The remaining proteins are almost all located in the matrix compartment. The relative distribution of proteins is different in heart mitochondria where only 30% of the total proteins are located in the matrix and the rest are almost all in the inner membrane. The proteins of the matrix are all soluble and can be fractionated by standard methods of protein fractionation. The membrane-bound proteins can be divided into two

*The microsomal fraction probably consists predominantly of rough and smooth endoplasmic reticulum, although other cellular membranes are also present in this crude fraction.

TABLE 2.5. External and Integral Proteins of the Mitochondria Inner Membrane

External	Integral
Succinate dehydrogenase	Cytochrome *b*
NADH dehydrogenase	Subunits 1, 2, and 3 of cytochrome oxidase
F_1 ATPase	Subunits 5, 7, and 9 of oligomycin-sensitive ATPase
Cytochrome *c*	Iron protein of coenzyme QH_2–cytochrome *c* reductase
Cytochrome c_1	
Subunits 4, 5, 6, and 7 of cytochrome oxidase	

groups: the external (extrinsic) and integral (intrinsic) proteins. External proteins are characterized by a rather loose association with the membrane and a solubility in water once they are separated from the membrane milieu. Integral proteins, on the other hand, are highly insoluble in water, and to separate them from other proteins and lipid components usually requires that the entire membrane be dissolved. Some examples of external and integral proteins of the mitochondrial inner membrane are listed in Table 2.5.

V. Current Views on the Structure of Biological Membranes

Concepts about the organization of proteins and lipids in biological membranes have undergone significant changes in recent years. One of the earliest models of membrane structure was proposed by Davson and Danielli in 1935. They envisioned that the phospholipids are arranged in a bimolecular leaflet with the polar ends facing the exterior and the fatty acyl chains forming an interior hydrophobic core (Fig. 2.10). Davson and Danielli further postulated that the proteins were bound to the two faces of the bilayer through electrostatic interactions between charged amino acid residues and the polar groups on the phospholipids. Although this model was modified in some of its details in subsequent years, the basic notions were retained until quite recently.

New biochemical and physical data, however, were difficult to reconcile with the Davson–Danielli type of model, and new proposals were made. The "fluid mosaic" model introduced by Singer and Nicolson in 1972 conforms with

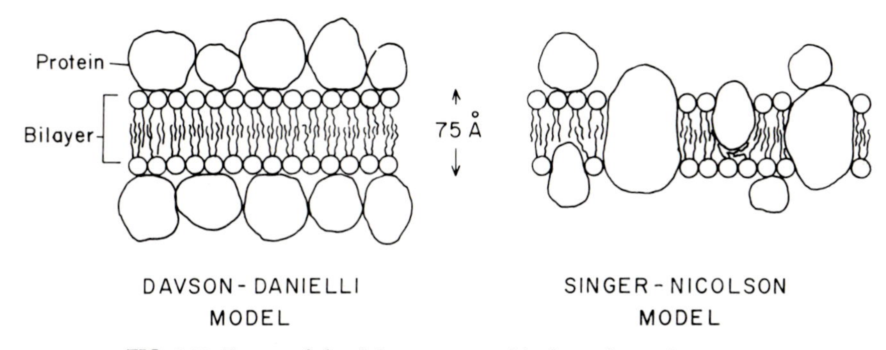

FIG. 2.10. Two models of the structure of biological membranes.

most of the evidence derived from studies of different membranes and has influenced much of the current thinking in this area. Some of the salient architectural features of this model will be reviewed briefly.

Singer and Nicolson have kept the bimolecular arrangement of the lipids as in the previous models. Instead of having the proteins bound to the two membrane surfaces, however, they postulated that globular proteins and enzymes are interspersed inside the lipid bilayer itself. This is particularly true of the integral class of proteins which are thought to interact with the fatty acids in the core of the bilayer by virtue of their own hydrophobic properties. Some proteins are more loosely associated with the membrane through electrostatic interactions with phospholipids and other proteins. This would be true of the external class of proteins which have a more polar character. The mosaic aspect of the model is based on the presumption that the proteins can be either entirely embedded in the bilayer, some traversing its entire width, or only partially embedded and, at the extreme, only superficially located on the membrane surfaces (Fig. 2.10).

A variety of independent lines of evidence support the basic structural features embodied in the "fluid mosaic" model. Only several of these will be mentioned here. Comparisons of the X-ray diffraction patterns of phospholipid bilayers and natural membranes indicate that the two are similar in their density profiles but that membranes have a considerably higher central density. This confirms the existence of a phospholipid bilayer but argues against an exclusive surface localization of the protein constituents.

A more dramatic demonstration that proteins are deeply embedded in the phospholipid bilayer comes from electron microscopic examinations of "freeze-fracture" membranes. The "freeze-fracture" technique has proven extremely useful for studying ultrastructural details of membranes. The principle of this method is illustrated in Fig. 2.11. Samples of isolated membranes or whole tissue are frozen at liquid nitrogen temperature and fractured with a sharp knife. When the knife edge comes in contact with the surface of a membrane, instead of cutting through it, a fracture occurs along a plane that corresponds to the hydrophobic core of the membrane. Thus, the two internal sur-

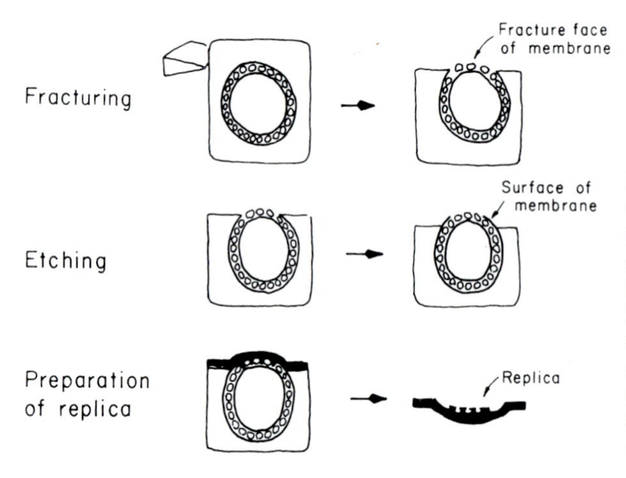

FIG. 2.11. Schematic representation of the freeze-fracture technique. A vesicular membrane showing the intramembrane particles is surrounded by ice. The fracture leads to an exposure of the convex fracture face. The etching removes some of the surrounding ice, exposing part of the convex face A of the membrane.

faces of the membrane become exposed: one is a convex fracture face (face A), and the second is a concave fracture face (face B). The fractured material is then subjected to a sublimation (etching) which causes some of the water around the fracture faces to be removed. This reveals the intact outer surface of the membrane. In the last step, the etched material is coated with carbon or a heavy metal to produce a replica which is then examined in the electron microscope. The special advantages of freeze-etching are that it avoids the use of fixatives or stains that may modify cellular structures, thus exposing to view the interior as well as exterior surfaces of membranes in the same specimen. An example of a mitochondrion visualized *in situ* by this technique is shown in Fig. 2.12. This photograph shows in a striking way that all the fracture faces are studded with particles of varying dimensions. These particles have been interpreted to be the globular proteins and enzymes that are embedded in the

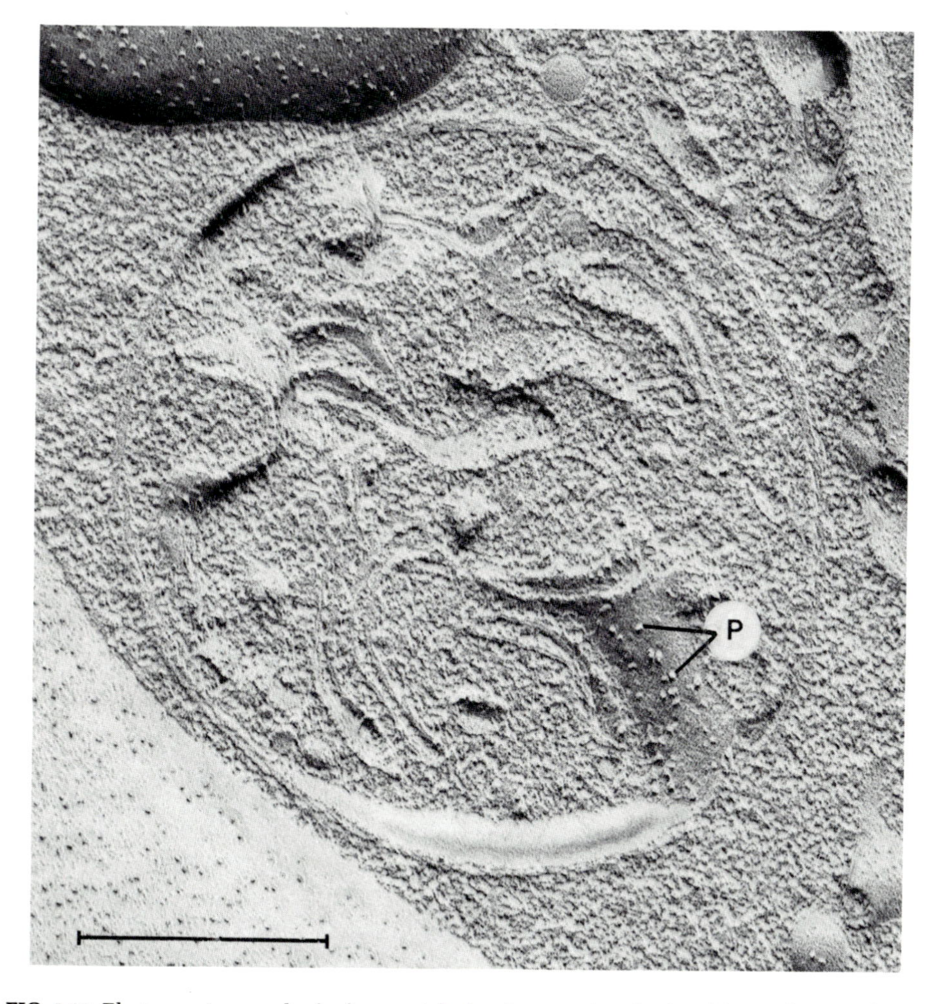

FIG. 2.12. Electron micrograph of a freeze-etched onion root tip mitochondrion. A fracture face of the inner membrane is seen to contain particles (P). The bar represents 0.3 μm. (Courtesy of Dr. Daniel Branton.)

bilayer. Surrounding the particles is a very smooth surface presumed to be the hydrophobic interior of the lipid bilayer.

An important aspect of the Singer and Nicolson model is the fluidity of its structural components. This notion is based on physical studies, especially those employing spin resonance probes to measure the rotational and translational freedoms of membrane lipids. This work, pioneered by McConnell and his colleagues, has demonstrated that the lipids are extremely mobile and capable not only of rotation around their axis but, more importantly, of diffusion at rapid rates through the plane of the membrane. The bimolecular leaflet is therefore a true sea of phospholipids in which individual molecules can move over large distances. Lateral diffusion has also been shown to be a property of the membrane proteins. We see, therefore, that biological membranes are dynamic assemblies not only at the gross morphological level but at the molecular level as well. The two phenomena are probably causally related.

VI. Fine Structure of Mitochondrial Membranes

The recognition that there is a close relationship between structure and function has underscored the need to understand the ultrastructure of membranes. In the case of mitochondria, most of the emphasis has been placed on the inner membrane since a great deal was already known about its functional properties. Both biochemical and electron microscopic studies encompassing a variety of new techniques have been brought to bear on the following questions:

1. What is the relationship of structural entities seen in the electron microscope to enzymes that are present in the membrane?
2. Are there any ultrastructural changes that can be detected under different metabolic states of the organelle?
3. How are the respiratory complexes and the ATPase oriented with respect to each other and to the two faces of the membrane?

The fine structure of the mitochondrial membranes will be reviewed here. The interpretation of the ultrastructure and the identity of the morphological units seen in the electron microscope will be left to subsequent chapters in which the various enzyme systems of the membrane are discussed.

A. Ultrastructure in Thin Sections

Under standard conditions of fixation, thin sections of both the outer and inner membranes are indistinguishable from other biological membranes. Each membrane consists of two dense lines with a spacing of 70–80 Å (see Fig. 1.4). This structure is also seen in phospholipid micelles and is believed to reflect the bimolecular arrangement of phospholipids. When the fixation time with osmium tetroxide is shortened, 90-Å particles are seen to be attached to the inner membrane surface facing the matrix. These particles are not seen on the outer membrane.

B. Ultrastructure in Negative Staining

In 1962, Fernandez-Moran used the negative staining technique to study mitochondrial ultrastructure. The principle of this technique is illustrated in Fig. 2.13. A dilute sample of the material to be examined is placed directly on the microscope grid and covered with a drop of a solution of the negative stain. The most commonly used stains are phosphotungstate and uranyl acetate. After drying, the grid can be immediately viewed in the microscope. The negative stain serves two purposes. It surrounds but does not penetrate the material because of its high molecular weight. A contrast is therefore created between the free areas on the grid that are occupied by the stain alone and the sample. Because the stain has a greater electron opacity than proteins and lipids, the latter appear lighter than the background in the final image. In addition to providing contrast, the stain protects the biological material against the destructive action of the electron beam and therefore eliminates the need to fix the sample. The advantages of this technique are that it is simple, quick, and achieves a high degree of resolution. Biological structures as small as 30 Å can be routinely observed.

Fernandez-Moran found that when mitochondria are negatively stained with phosphotungstate, the inner membrane is lined with closely spaced particles 90 Å in diameter. The particles face the matrix side and are attached to the inner membrane by means of thin stalks 50 Å in length and 20–25 Å in width (Fig. 2.14). The negative stain also reveals a fine structure in the membrane itself which appears to consist of a linear array of globular particles that have a cross-membrane dimension of 50–60 Å.

The projecting particles seen on the inner membrane were first called "elementary particles." This term, however, was later dropped, and they are now usually referred to as inner membrane spheres or headpieces. Following their discovery, the particles were thought to contain the entire assembly of the respiratory enzymes. This turned out to be incorrect, and it was subsequently shown by Racker and his colleagues that the particles correspond to the F_1 ATPase. The evidence for this will be discussed in Chapters 7 and 8.

Side view of
grid with specimen

Side view of
specimen with
negative stain

Top view of
specimen.
Final image.

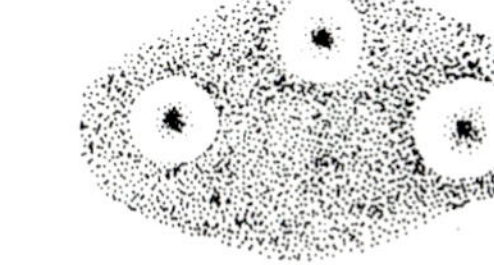

FIG. 2.13. Schematic representation of the negative staining technique.

In negatively stained preparations, the outer membrane has a smooth appearance, and there is no evidence of attached particles. Some recent electron microscopic studies of the rat liver outer membrane indicate the presence of pores. Whether these function as channels for the penetration of solutes is not clear at present.

C. Ultrastructure in Freeze-Etching

The morphology of mitochondrial membranes as revealed by the freeze-etching technique is consistent with the general features of the "fluid mosaic" model of membrane structure. The outer and inner membranes fracture along their planes, and the exposed faces are seen to contain globular particles. The

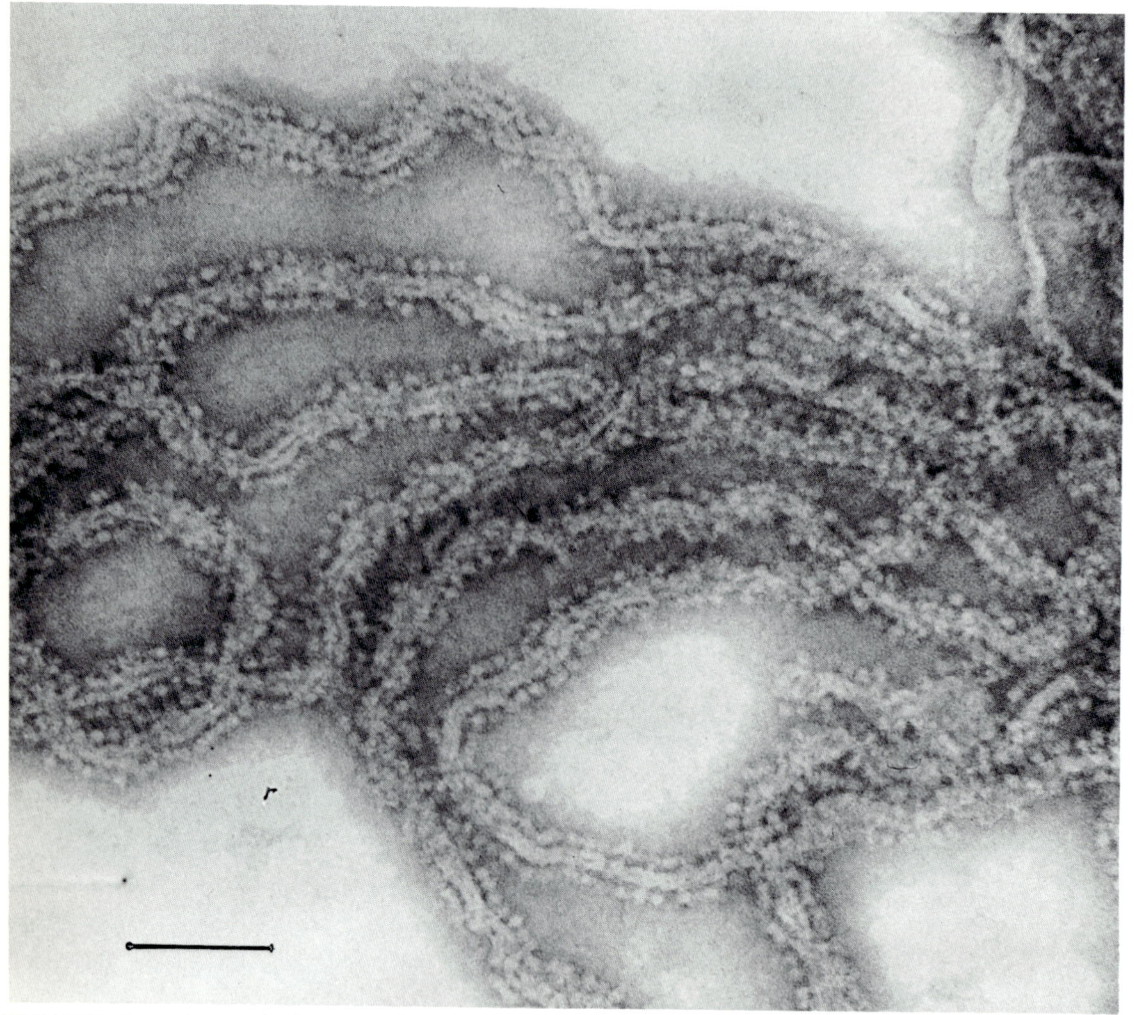

FIG. 2.14. Electron micrograph of a negatively stained mitochondrion from bovine heart. The ribbons represent two closely apposed cristae lined with particles 90 Å in diameter. The bar is 0.1μm. (Courtesy of Dr. Enrique Valdivia.)

particles on the fracture faces have size distributions that range from 50–150 Å in diameter. The most frequently observed particles have a mean diameter of 80 Å. Since the measurements are made on shadowed specimens and include the thickness of the metal coating, the actual size of the particles is smaller. If the metal contributes 20 Å, most of the globular proteins in the membrane would be about 60 Å in diameter.

Quantitative studies indicate that the number of intramembrane particles is different in the two fracture faces of each membrane. The outer membrane has been reported to have four times as many particles in the convex face, whereas the inner membrane appears to have more particles in the concave face. The segregation of intramembrane particles between the two fracture faces is thought to be influenced largely by their position relative to the longitudinal axis of the membrane. The observed differences in the distribution are therefore probably indicative of an asymmetric disposition of the protein over the width of the membrane.

The etched surfaces of the outer and inner membranes have a smooth appearance, and it is not clear at present why the 90-Å inner membrane spheres are not evident on the etched surface of the inner membrane.

Selected Readings

Andre, J. (1962) Contribution a la connaissance du chondriome. Etude de ses modifications ultrastructurales pendant la spermatogenese, *J. Cell Biol.* **6** (Suppl. 3):7.

Branton, D. (1966) Fracture faces of frozen membranes, *Proc. Natl. Acad. Sci. U.S.A.* **55**:1048.

Davson, H., and Danielli, J. F. (1943) *The Permeability of Natural Membranes,* Cambridge University Press, Cambridge.

Ernster, L., and Kuylenstierna, B. (1970) Outer membrane of mitochondria, in *Membranes of Mitochondria and Chloroplasts* (E. Racker, ed.), Van Nostrand Reinhold, New York, pp. 172–212.

Fernandez-Moran, H., Oda, T., Blair, P. V., and Green, D. E. (1964) A macromolecular repeating unit of mitochondrial structure and function. Correlated electron microscopic studies of isolated mitochondria and submitochondrial particles of beef heart muscle, *J. Cell Biol.* **22**:63.

Frye, C. D., and Edidin, M. (1970) The rapid intermixing of cell surface antigens after formation of mouse–human heterokaryons, *J. Cell Biol.* **7**:319.

Fleischer, S., Rouser, G., Fleischer, B., Casu, A., and Kritchevsky, G. (1967) Lipid composition of mitochondria from bovine heart, liver and kidney, *J. Lipid Res.* **8**:170.

Hackenbrock, C. R. (1968) Ultrastructural basis for metabolically linked mechanical activity of mitochondria. II. Electron-transport linked ultrastructural transformations in mitochondria, *J. Cell Biol.* **37**:345.

Johnson, L. V., Walsh, M. L., and Chen, L. B. (1980) Localization of mitochondria in living cells with rhodamine 123, *Proc. Natl. Acad. Sci. U.S.A.* **77**:990.

Lehninger, A. L. (1964) *The Mitochondrion,* W. A. Benjamin, New York.

Melnick, D. L., and Packer, L. (1971) Freeze-fracture faces of inner and outer membranes of mitochondria, *Biochim. Biophys. Acta* **253**:503.

Moor, H. (1966) Use of freeze-etching in the study of biological ultrastructure, *Int. Rev. Exp. Pathol.* **5**:179.

Munn, E. A. (1974) *The Structure of Mitochondria,* Academic Press, New York.

Palade, G. E. (1953) An electron microscope study of the mitochondrial structure, *J. Histochem. Cytochem.* **1**:188.

Parsons, D. F., Williams, G. R., Thompson, W., Wilson, D., and Chance, B. (1967) Improvements in the procedure for purification of outer and inner membrane. Comparison of the outer membrane with smooth endoplasmic reticulum, in *Mitochondrial Structure and Compartmentation* (E. Quagliariello, S. Papa, E. C. Slater and J. M. Tager, eds.), Adriatica Press, Bari, pp. 29–73.

Racker, E., Tyler, D. D., Estabrook, R. W., Conover, T. E., Parsons, D. F., and Chance, B. (1965) Correlation between electron transport activity, ATPase and morphology of submitochondrial particles, in *Oxidases and Related Redox Systems* (T. E. King, H. S. Mason and M. Morrison, eds.), John Wiley & Sons, New York, pp. 1077–1094.

Sacktor, B., and Shimada, Y. (1972) Degenerative changes in the mitochondria of flight muscle from aging blowflies, *J. Cell Biol.* **52**:465.

Schnaitman, C., and Greenawalt, J. W. (1968) Enzymatic properties of the inner and outer membranes of rat liver mitochondria, *J. Cell Biol.* **38**:158.

Singer, S. J., and Nicolson, G. L. (1972) The fluid mosaic model of the structure of cell membranes, *Science* **175**:720.

Stoeckenius, W. (1970) Electron microscopy of mitochondria and model membranes, in *Membranes of Mitochondria and Chloroplasts* (E. Racker, ed.), Van Nostrand Reinhold, New York, pp. 53–90.

3

Oxidative Pathways of Mitochondria

ATP is the immediate source of energy used by cells to do chemical, mechanical, and osmotic work. Obligate and facultative anaerobic organisms can survive on the ATP synthesized during the glycolytic breakdown of sugars to pyruvic acid.

$$C_6H_{12}O_6 + 2ADP + 2NAD^+ + 2P_i \rightarrow$$

$$2CH_3{-}\overset{\overset{\displaystyle O}{\|}}{C}{-}COOH + 2ATP + 2NADH + 2H^+ + 2H_2O \quad [3.1]$$

The yield of ATP in glycolysis is 2 moles per mole of sugar metabolized. The reduced pyridine nucleotide formed during glycolysis must be reoxidized; this usually occurs in one of several ways. Pyruvate can be decarboxylated to acetaldehyde which is then reduced to ethanol by the action of alcohol dehydrogenase. The reduction of acetaldehyde requires NADH and provides a mechanism for the regeneration of NAD^+. Alternatively, pyruvic acid can be reduced to lactic acid in an NADH-dependent reaction catalyzed by lactate dehydrogenase.

In aerobic microorganisms and in eucaryotic cells, pyruvate is further oxidized to CO_2 and water. The aerobic utilization of pyruvate in eucaryotes takes place in mitochondria and is very important because it provides an additional source of ATP far in excess of that obtained from glycolysis. The amount of ATP formed during the complete oxidation of glucose has been estimated to be 38 moles per mole of sugar, a yield 19 times greater than that of glycolysis.

The participation of mitochondria in aerobic metabolism starts with the oxidative decarboxylation of pyruvate to acetyl-CoA by a soluble multiprotein enzyme, the pyruvate dehydrogenase complex. The further oxidation of acetyl-CoA is mediated by two related pathways. The first, known as the tricarboxylic acid (TCA) cycle, oxidizes acetyl-CoA to CO_2 and, in the process, reduces NAD^+ to NADH. The second pathway involves a series of oxidation–reduction reactions whose net result is the reoxidation of NADH to NAD^+ and the reduction of oxygen to water. This is catalyzed by the respiratory chain, and it is here that the energy of oxidation of substrates is finally trapped and used for ATP synthesis.

The central intermediate in the oxidative metabolism of mitochondria is acetyl-CoA which can be derived not only from sugars but also from fatty acids and amino acids. Fatty acids are degraded to acetyl-CoA by the β-oxidation pathway whose enzymes are located in the mitochondrial matrix. Fatty acids are a particularly rich source of acetyl-CoA during starvation when there is a deficiency of carbohydrates, and stored fats serve as the principal source of energy. Even though amino acid degradation occurs mostly in the cytoplasm, mitochondria also play an important role in this process. There are mitochon-

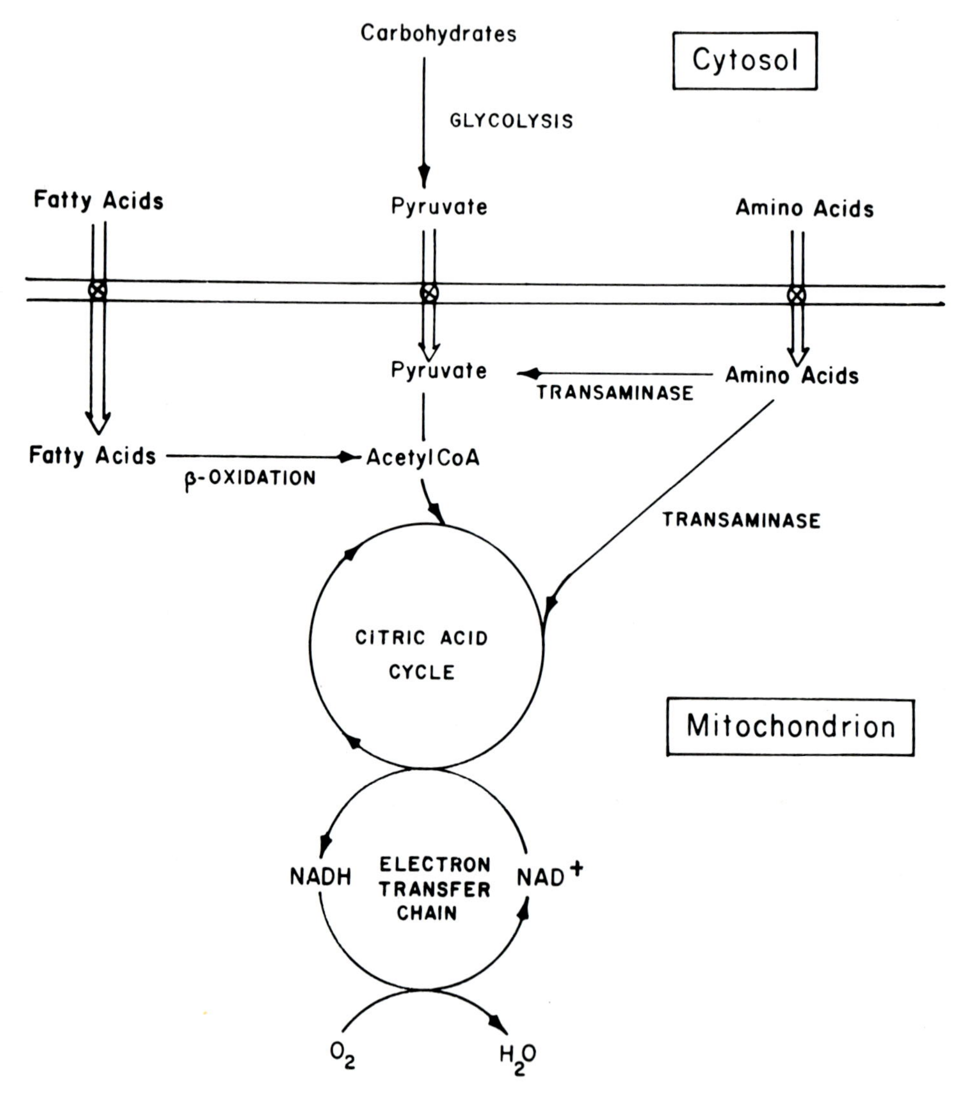

FIG. 3.1. Relationship among the major oxidative pathways of the mitochondrion.

drially located transaminases that convert certain amino acids to the corresponding keto acids which are either intermediates of the TCA cycle or can enter the cycle through pyruvic acid. The relationship among the different pathways is outlined in the scheme of Fig. 3.1.

In this chapter, we shall first consider those reactions that lead to the production of acetyl-CoA. This will be followed by a discussion of the TCA cycle reactions concerned with the metabolism of acetyl-CoA. The oxidation events embodied in these mitochondrial pathways culminate in the generation of NADH. The way in which NADH is oxidized by the respiratory chain will be dealt with in the next chapter.

I. Conversion of Pyruvic Acid to Acetyl-CoA

Pyruvic acid is oxidatively decarboxylated to acetyl-CoA according to the following reaction:

$$CH_3-\overset{\overset{\large O}{\|}}{C}-COOH + CoA(SH) + NAD^+ \rightarrow$$

$$CH_3-\overset{\overset{\large O}{\|}}{C}\sim SCoA + CO_2 + NADH + H^+ \quad [3.2]$$

This reaction is catalyzed by a high-molecular-weight multienzyme complex, the pyruvate dehydrogenase complex, located in the mitochondrial matrix. The catalytic mechanism and the physical properties of the complex have been extensively studied by Reed and co-workers. It is composed of three enzymes which are physically associated with each other in a very specific way. The individual enzymes making up the complex are (1) pyruvate dehydrogenase, (2) dihydrolipoyl transacetylase, and (3) dihydrolipoyl dehydrogenase. The three enzymes carry out a coupled sequence of reactions in which none of the intermediates are released from the complex. The fact that the products formed at each step remain protein bound probably facilitates their utilization by the next enzyme because of the internal organization of the complex and the spatial proximity of the catalytic sites. The sequence of reactions catalyzed by the complex is shown in Fig. 3.2.

A. Structure of the Pyruvate Dehydrogenase Complex

Much of our current knowledge of the structure of this multienzyme complex has come from studies of the *E. coli* complex. The complex of *E. coli* has a molecular weight of 4×10^6 and consists of 24 subunits of dihydrolipoyl transacetylase, each having a molecular weight of 40,000; it also contains 24 subunits of pyruvate dehydrogenase and 24 subunits of dihydrolipoyl dehydrogenase with molecular weights of 90,000 and 55,000, respectively. The complex of mammalian mitochondria is larger (8.5×10^6 daltons) and has a somewhat

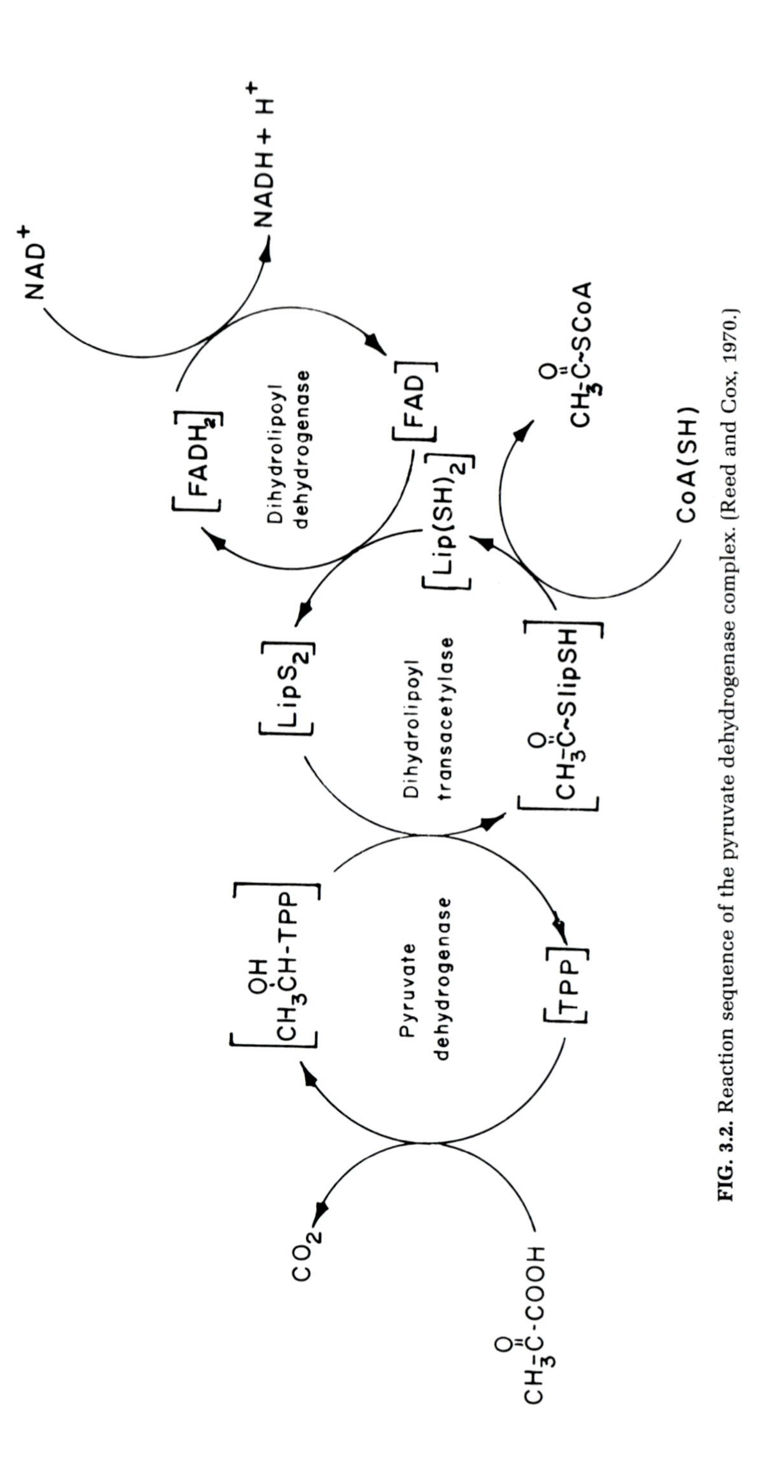

FIG. 3.2. Reaction sequence of the pyruvate dehydrogenase complex. (Reed and Cox, 1970.)

different stoichiometry of the three enzymes. It is estimated to have 60 subunits of transacetylase (50,000 daltons), 20 subunits of pyruvate dehydrogenase (154,000 daltons), and five subunits of dihydrolipoyl dehydrogenase (110,000 daltons).

From reconstitution and electron microscopic evidence, Reed has deduced that the transacetylase subunits form a core structure around which are arranged the subunits of the pyruvate and dihydrolipoyl dehydrogenases. Each subunit of the dihydrolipoyl transacetylase contains one molecule of lipoic acid covalently bound to the enzyme through an amide bond with the ϵ-amino group of a lysine residue. Reed has postulated that the arm (nine carbons plus one nitrogen) connecting the ring structure of lipoic acid to the protein acts as a pendulum that swings between the active centers of the pyruvate and dihydrolipoyl dehydrogenases. This mechanism is illustrated in Fig. 3.3.

B. Regulation of the Mammalian Pyruvate Dehydrogenase Complex

In addition to the catalytic subunits, the mammalian complex has been shown to have two other enzymes with regulatory functions. One of the regu-

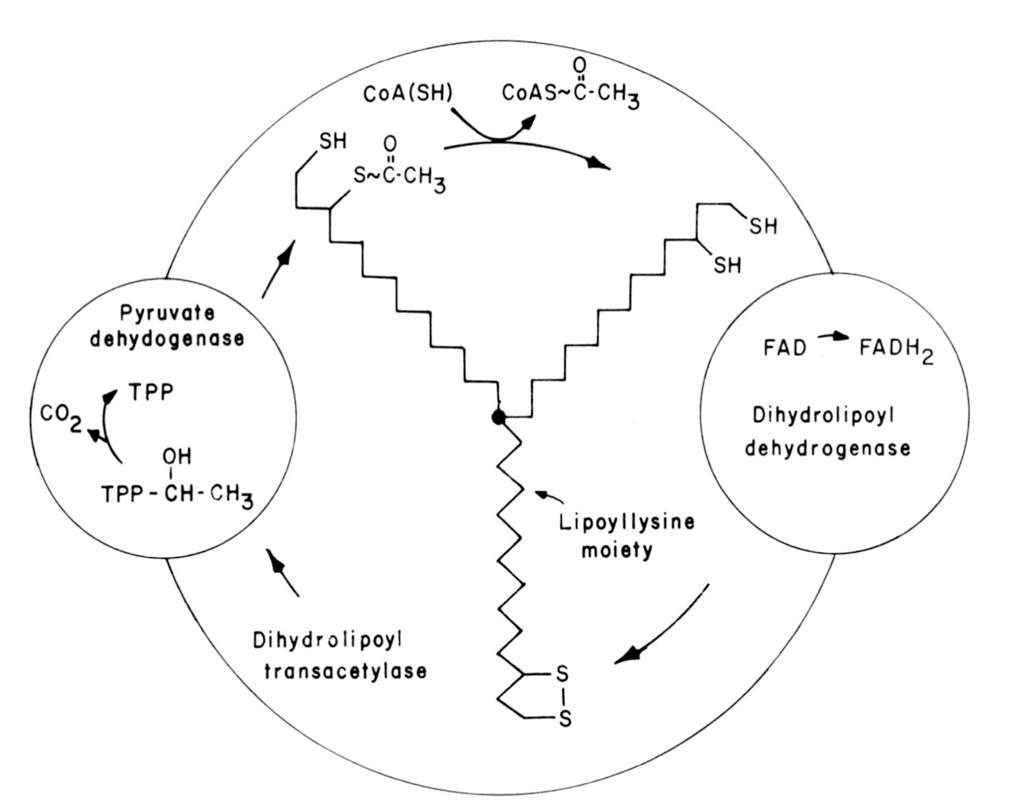

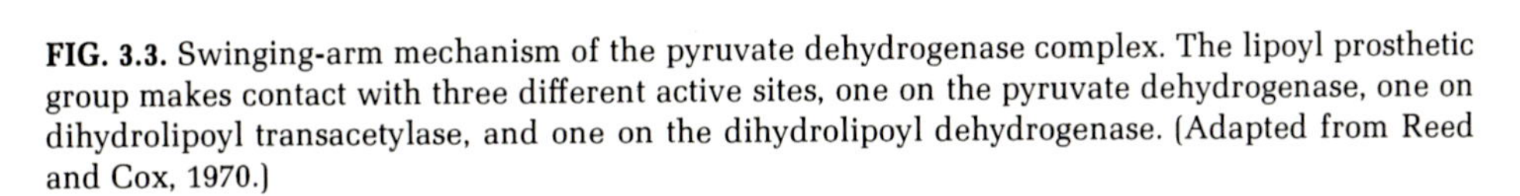
FIG. 3.3. Swinging-arm mechanism of the pyruvate dehydrogenase complex. The lipoyl prosthetic group makes contact with three different active sites, one on the pyruvate dehydrogenase, one on dihydrolipoyl transacetylase, and one on the dihydrolipoyl dehydrogenase. (Adapted from Reed and Cox, 1970.)

latory enzymes is pyruvate dehydrogenase kinase. In the presence of ATP and low concentrations of magnesium, the kinase phosphorylates pyruvate dehydrogenase and thereby makes it enzymatically inactive. The second regulatory enzyme, pyruvate dehydrogenase phosphatase, converts the enzyme to its catalytically active form by removing the phosphate from the enzyme. The phosphatase requires calcium for optimal activity.

A consequence of the requirements for the phosphorylation and dephosphorylation reactions is that when the intramitochondrial pool of ATP is high, i.e., under conditions of active oxidative phosphorylation, the activity of the pyruvate dehydrogenase complex is suppressed, and conversely, when the pool of ATP is low, the oxidative utilization of pyruvate is enhanced by the activation of the complex. This type of regulatory mechanism appears to be absent in *E. coli.*

II. β-Oxidation of Fatty Acids

Fatty acids constitute an important metabolic fuel in microorganisms and in higher plants and animals. The oxidative metabolism of fatty acids occurs in mitochondria where they are first converted through the β-oxidation pathway to acetyl-CoA and then further oxidized to CO_2 by the TCA cycle. The β-oxidation system consists of four separate enzymes functioning in a repetitive cycle or spiral, each turn resulting in a shortening of the fatty acyl-CoA chain by two carbon units. After one cycle, the products formed are acetyl-CoA and a new acyl-CoA two carbons shorter than the starting acyl-CoA (Fig. 3.4). The

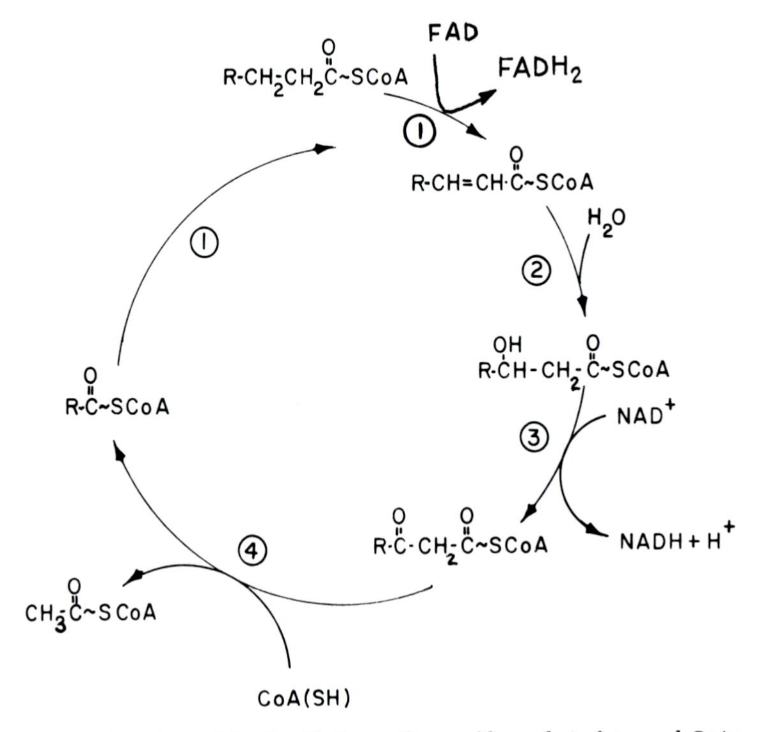

FIG. 3.4. Reactions of the β-oxidation pathway of long-chain fatty acyl-CoAs.

acyl-CoA produced at each turn remains bound to the enzyme and enters the cycle again. This process continues until the fatty acid molecule is completely degraded to acetyl-CoA (Fig. 3.4).

The enzymes of β-oxidation are (1) fatty acyl-CoA dehydrogenase which introduces an unsaturated double bond at the α,β position, (2) enoyl hydratase which forms the β-hydroxyacyl-CoA, (3) β-hydroxyacyl-CoA dehydrogenase which oxidizes the hydroxyl compound to the corresponding keto compound, and (4) β-ketoacyl-CoA thiolase which cleaves (the thiolytic reaction uses the SH group of coenzyme A) the keto ester to acetyl-CoA and the new acyl-CoA.

A. Entry of Fatty Acids into the Mitochondrion

Free fatty acids as such cannot penetrate the mitochondrial inner membrane to their site of oxidation. They are first esterified to coenzyme A, a reaction involving the formation of a high-energy thioester bond. The esterification reaction occurs at the expense of the hydrolysis of ATP to AMP and pyrophosphate.

$$RCH_2\overset{\overset{\displaystyle O}{\|}}{C}OH + CoA(SH) + ATP \rightarrow RCH_2\overset{\overset{\displaystyle O}{\|}}{C}{\sim}SCoA + AMP + PP_i \qquad [3.3]$$

The formation of acyl-CoA is catalyzed by enzymes located both in mitochondria and the cytoplasm and generically known as acyl thiokinases.

The transport of the acyl-CoA into the matrix compartment occurs by an interesting mechanism involving the transfer of the acyl group from the thioester to a new compound, carnitine. The acyl-carnitine then crosses the inner membrane and is once again converted to acyl-CoA, the substrate for the first enzyme of β-oxidation. This transport system, known as the carnitine shuttle, is show in more detail in Fig. 3.5. The conversion of long-chain acyl-CoAs to acyl-carnitine is achieved by an enzyme, long-chain acyl-CoA carnitine acyl-CoA transferase, located in the inner membrane.

$$R{-}\overset{\overset{\displaystyle O}{\|}}{C}{\sim}SCoA + (CH_3)_3N^+{-}CH_2{-}\underset{\underset{\displaystyle OH}{|}}{CH}{-}CH_2{-}CO_2H \rightarrow$$

$$(CH_3)_3N^+{-}CH_2{-}\underset{}{\overset{\overset{\overset{\displaystyle R}{|}}{\overset{\displaystyle C=O}{|}}}{\overset{\displaystyle O}{|}}}{CH}{-}CH_2CO_2H + CoA(SH) \qquad [3.4]$$

The inner membrane is permeable to acyl-carnitine and allows the esterified fatty acyl group to enter the matrix compartment. There, the fatty acyl group is transferred to intramitochondrial CoA by the action of a second long-chain fatty acyl-carnitine acyl-CoA transferase. Since the inner membrane is impermeable to free carnitine, a mechanism exists for replenishing the external pool of carnitine. This is thought to occur by the transfer of acetyl-carnitine from the matrix to the outside of the mitochondrion. Acetyl-CoA formed during β-oxidation is converted to acetyl-carnitine, the reaction being catalyzed by a short-chain carnitine acyl-transferase. The acetyl-carnitine can leave the matrix, and once it is outside, free carnitine is regenerated by a transfer of the acetyl group to CoA. The overall process, therefore, requires four separate carnitine acyl-transferases distributed between the two sides of the membrane and achieves a balance of carnitine inside and outside the inner membrane (Fig. 3.5).

B. Enzymes of β-Oxidation

The four enzymes of β-oxidation are soluble proteins that are located in the matrix compartment. There is no evidence at present that the β-oxidation enzymes are associated in a physical complex.

1. Fatty Acyl-CoA Dehydrogenase

Three different acyl-CoA dehydrogenases have been isolated from mitochondria. Each is a flavoprotein with an approximate molecular weight of 200,000. The three enzymes show different specificities with respect to the length of the fatty-acyl chains. Dehydrogenase G is active only with short-chain acyl-CoAs (C_4–C_6), dehydrogenase Y1 with medium-chain-length acyl-CoAs (C_6–C_{12}), and dehydrogenase Y2 with medium- to long-chain acyl-CoAs (C_8–C_{18}). The enoyl-CoA product formed in the reaction has a *trans* double bond in the α,β position of the fatty acid molecule. The hydrogen acceptor is the FAD prothetic group of the dehydrogenase. In the *in vitro* assay of the enzyme, FAD

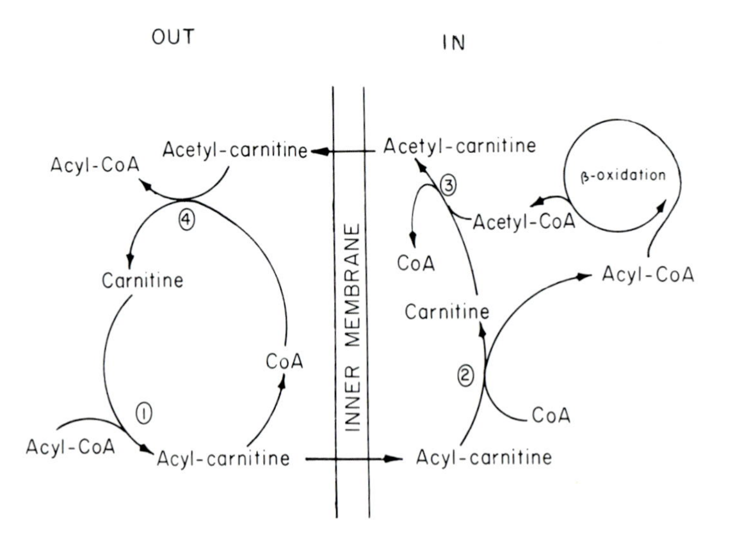

FIG. 3.5. Carnitine shuttle for the transfer of fatty acyl-CoAs into the mitochondrion. The four enzymes participating in the cycle are: 1 and 2, long-chain carnitine acyl-transferases; 3 and 4, short-chain carnitine acyl-transferases. (Adapted from Bressler, 1970.)

is reoxidized by an artificial acceptor such as phenazine methosulfate. In mitochondria, however, the reduced flavin is reoxidized by another flavoprotein enzyme called electron transfer flavoprotein (ETF) which is linked to the terminal electron transfer chain of mitochondria. The natural acceptor, therefore, is molecular oxygen. This aspect of β-oxidation will be discussed in more detail in the next chapter when we consider the mitochondrial electron transfer chain.

2. Enoyl-CoA Hydratase

The hydratase is a soluble protein with a molecular weight of 100,000. The enzyme shows a broad specificity towards varying chain length fatty acyl-CoAs. The product of the reaction is the *trans* isomer of β-hydroxyacyl-CoA.

3. β-Hydroxyacyl-CoA Dehydrogenase

As in the case of the hydratase, β-hydroxyacyl-CoA dehydrogenase is equally active with short- and long-chain acyl-CoAs. The enzyme catalyzes the oxidation of the hydroxyacyl-CoA to the corresponding β-ketoacyl-CoA. The acceptor for the reducing equivalents is NAD^+. This step of β-oxidation, therefore, produces NADH that can be oxidized by the electron transfer chain.

4. β-Ketoacyl-CoA Thiolase

The last step in β-oxidation is a thiolytic cleavage of the ketoacyl-CoA to acetyl-CoA and a two-carbon-shorter acyl-CoA. The β-ketoacyl-CoA thiolase catalyzing this reaction is not very specific for chain length but has been found to be slightly less active with longer-chain fatty acids.

C. Ancillary Enzymes of β-Oxidation

Saturated fatty acids with an even number of carbon atoms can be completely converted to acetyl-CoA by the β-oxidation pathway. Many naturally occuring fatty acids, however, have an odd number of carbons or have *cis* unsaturated double bonds. There are special enzymes that, in concert with the β-oxidation system, allow such fatty acids to be totally degraded to acetyl-CoA.

1. Enoyl-CoA Isomerase

This enzyme catalyzes an isomerization of a double bond from a Δ^3 *cis* to a Δ^2 *trans* configuration.

$$\underset{\Delta^3\ cis}{R{-}CH_2{-}CH{=}CH{-}CH_2{-}\underset{\underset{O}{\|}}{C}{\sim}SCoA} \rightarrow \underset{\Delta^2\ trans}{R{-}CH_2{-}CH_2{-}CH{=}CH{-}\overset{\overset{O}{\|}}{C}{\sim}SCoA} \quad [3.5]$$

The Δ^2 *trans* isomer, being the natural substrate for the enoyl-CoA hydratase (see above), can then be further degraded by the β-oxidation pathway.

2. Propionyl-CoA Carboxylase

Fatty acyl-CoAs with an odd number of carbons are oxidized by the β-oxidation pathway to propionyl-CoA. This compound is carboxylated to D-methylmalonyl-CoA, a reaction catalyzed by propionyl-CoA carboxylase.

$$CH_3CH_2\overset{\overset{\displaystyle O}{\|}}{C}{\sim}SCoA + CO_2 + ATP \rightarrow$$

$$H{-}\underset{\underset{\displaystyle COOH}{|}}{\overset{\overset{\displaystyle CH_3}{|}}{C}}{-}CH_2{-}\overset{\overset{\displaystyle O}{\|}}{C}{\sim}SCoA + ADP + P_i \qquad [3.6]$$

A racemase converts the D to the L isomer. Finally, L-methylmalonyl-CoA is isomerized to succinyl-CoA which, after deacylation, can enter the TCA cycle.

$$CH_3{-}\underset{\underset{\displaystyle H}{|}}{\overset{\overset{\displaystyle COOH}{|}}{C}}{-}CH_2{-}\overset{\overset{\displaystyle O}{\|}}{C}{\sim}SCoA \rightarrow HOOC{-}CH_2{-}CH_2{-}\overset{\overset{\displaystyle O}{\|}}{C}{\sim}SCoA \qquad [3.7]$$

III. Utilization of Ketone Bodies

The synthesis and utilization of ketone bodies are important mitochondrial pathways that are closely related to β-oxidation. Ketone bodies are a mixture of acetoacetate, β-hydroxybutyrate, and acetone. Most animals, when deprived of a source of carbohydrates, depend on stored fats for their source of energy. Under these circumstances, liver mitochondria make ketone bodies that are transported through the circulatory system to other tissues where they can be further oxidized by mitochondria. Most of the acetoacetate is made in liver through the condensation of two molecules of acetyl-CoA formed during β-oxidation of fatty acids. This reaction is catalyzed by an acetyltransferase.

$$2\ CH_3{-}\overset{\overset{\displaystyle O}{\|}}{C}{\sim}SCoA \rightarrow CH_3{-}\overset{\overset{\displaystyle O}{\|}}{C}{-}CH_2{-}\overset{\overset{\displaystyle O}{\|}}{C}{\sim}SCoA + CoA(SH) \qquad [3.8]$$

The acetoacetyl-CoA cannot be directly deacylated to the free acid, since the necessary enzyme, acetoacetyl-CoA deacylase, does not appear to exist. Instead, acetoacetyl-CoA reacts with another molecule of acetyl-CoA to form β-hydroxymethyl glutaryl-CoA. β-Hydroxymethyl glutaryl-CoA is enzymatically converted to acetoacetate and acetyl-CoA. The second most important ketone body, β-hydroxybutyrate, arises from the reduction of acetoacetate which is catalyzed by β-hydroxybutyrate dehydrogenase. The sequence of reactions is:

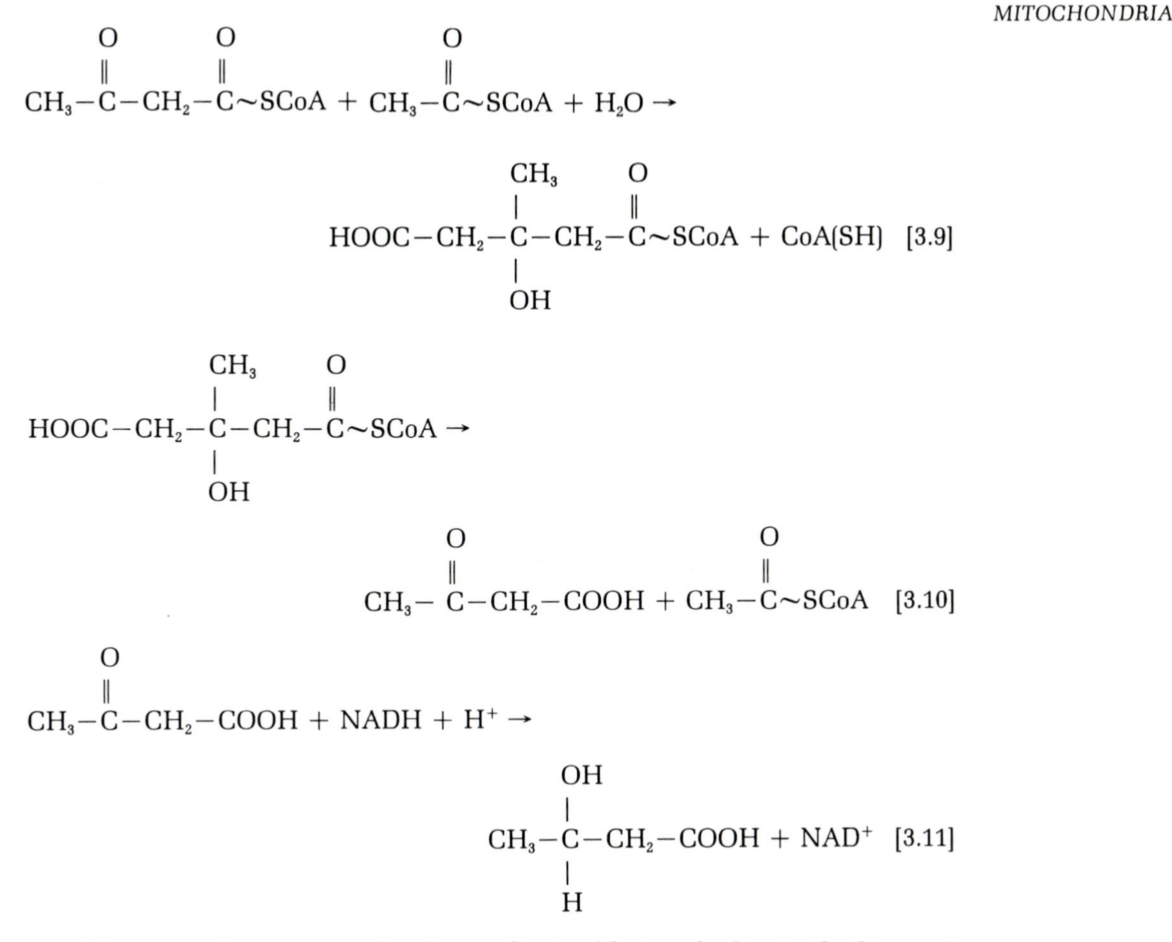

$$CH_3-\overset{\overset{O}{\|}}{C}-CH_2-\overset{\overset{O}{\|}}{C}\sim SCoA + CH_3-\overset{\overset{O}{\|}}{C}\sim SCoA + H_2O \rightarrow HOOC-CH_2-\underset{\underset{OH}{|}}{\overset{\overset{CH_3}{|}}{C}}-CH_2-\overset{\overset{O}{\|}}{C}\sim SCoA + CoA(SH) \quad [3.9]$$

$$HOOC-CH_2-\underset{\underset{OH}{|}}{\overset{\overset{CH_3}{|}}{C}}-CH_2-\overset{\overset{O}{\|}}{C}\sim SCoA \rightarrow CH_3-\overset{\overset{O}{\|}}{C}-CH_2-COOH + CH_3-\overset{\overset{O}{\|}}{C}\sim SCoA \quad [3.10]$$

$$CH_3-\overset{\overset{O}{\|}}{C}-CH_2-COOH + NADH + H^+ \rightarrow CH_3-\underset{\underset{H}{|}}{\overset{\overset{OH}{|}}{C}}-CH_2-COOH + NAD^+ \quad [3.11]$$

These reactions account for the synthesis of ketone bodies in the livers of animals that depend on their stored fats for their energy needs. Under such conditions, brain tissue can survive on β-hydroxybutyrate which is oxidized to acetoacetate with a concomitant production of 1 mole equivalent of NADH.

$$CH_3-\underset{\underset{H}{|}}{\overset{\overset{OH}{|}}{C}}-CH_2-COOH + NAD^+ \rightarrow CH_3-\overset{\overset{O}{\|}}{C}-CH_2-COOH + NADH + H^+ \quad [3.12]$$

The acetoacetate formed in this reaction is converted to acetoacetyl-CoA by accepting the CoA moiety from succinyl-CoA. Acetoacetyl-CoA in turn is thiolytically cleaved to give two molecules of acetyl-CoA which can be further metabolized by the TCA cycle. The enzyme that catalyzes the transfer of the CoA moiety from succinyl-CoA to acetoacetate is absent in liver. This may explain why liver is the main ketone-body-producing tissue.

IV. Metabolism of Amino Acids

Most of the energy of living systems is obtained from the oxidation of carbohydrates and fats. Amino acids, however, can serve as an additional important food. The major catabolic pathways for the breakdown of amino acids are found in the cytoplasm and will not be delineated here. Mitochondria, however, share in the process, and the extent to which they do will be outlined.

A. Amino Acids as a Source of Acetyl-CoA

The catabolism of a large number of amino acids leads either directly or indirectly to the formation of acetyl-CoA. There exist catabolic pathways for the conversion of some amino acids, referred to as ketogenic, to acetoacetyl-CoA; among these are leucine, phenylalanine, tryptophan, lysine, and tyrosine. As seen already, there are several ways for converting acetoacetyl-CoA to acetyl-CoA. Another large group of amino acids including alanine, glycine, cysteine, serine, and threonine are degraded to pyruvate which can be oxidatively decarboxylated to acetyl-CoA by the mitochondrial pyruvate dehydrogenase complex.

B. Amino Acids as a Source of Tricarboxylic Acid Cycle Intermediates

Some amino acids can be transaminated either by cytoplasmic or mitochondrially located transaminases (amino transferases) to keto acid intermediates of the TCA cycle. Glutamate and aspartate are directly transaminated to α-ketoglutarate and oxalacetate, respectively.

There are also pathways in the cytoplasm for converting arginine and histidine to glutamic acid and thence by transamination to α-ketoglutarate. Other TCA cycle intermediates such as succinyl-CoA can be generated from isoleucine, valine, and methionine or from phenylalanine and tyrosine.

C. The Glutamate Cycle

Mitochondria contain nonspecific transaminases capable of transferring the amino group of most amino acids to α-ketoglutarate. The products of the reaction are a keto acid (the deaminated amino acid) and glutamate. This is an important reaction because of the presence in mitochondria of glutamate dehy-

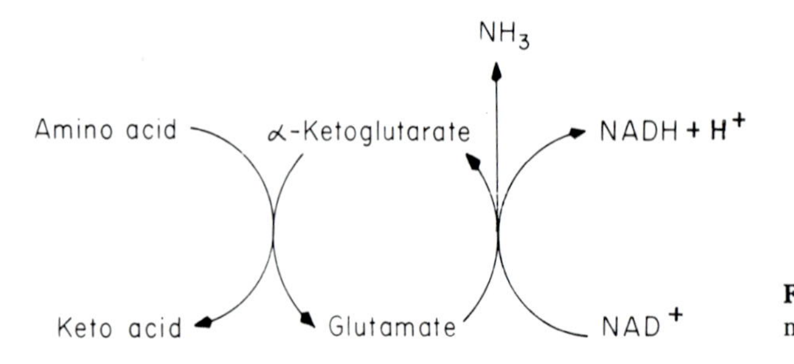

FIG. 3.6. Glutamate cycle of the mitochondrion.

drogenase which catalyzes an oxidative deamination of glutamate to α-ketoglutarate. The net result of this cyclic process is the oxidation of the amino acid to the corresponding keto acid, the reduction of NAD^+ to NADH, and the release of ammonia (Fig. 3.6). The glutamate cycle, therefore, is a means of deaminating amino acids to keto acid intermediates of the TCA cycle. The ammonia produced from the oxidation of glutamate is extremely toxic and is eliminated in the form of urea. Two important enzymes of the urea cycle are located in the matrix of liver mitochondria. Carbamylphosphate synthetase catalyzes the reaction

$$NH_3 + HCO_3^- + 2ATP + H_2O \rightarrow NH_2-\overset{\overset{\displaystyle O}{\|}}{C}-PO_3^- + 2ADP + P_i \qquad [3.13]$$

in which ammonia is condensed with HCO_3^- to form carbamylphosphate. The synthetase has an absolute requirement for acetylglutamate which is thought to act as an allosteric activator. Carbamylphosphate synthetase has a molecular weight of 160,000 and is the largest protein present in the matrix. It is an inducible enzyme whose synthesis is greatly enhanced in animals fed on a high-protein diet. In rats, it can make up as much as 15% of the total protein of liver mitochondria. It is suspected although not proven that the synthesis of carbamylphosphate is the regulatory step in the urea cycle.

The condensation of carbamylphosphate with ornithine, the next reaction of the urea cycle, is also brought about in the mitochondrial matrix. This reaction is catalyzed by ornithine transcarbamylase. The product formed, citrulline, is transported to the cytoplasm where the cycle is completed.

V. Tricarboxylic Acid Cycle

Most of the discussion so far has centered around the production of acetyl-CoA, either from carbohydrates, fatty acids, or from amino acids. The further oxidation of acetyl-CoA to CO_2 and water is achieved by a cyclic process involving eight separate catalytic steps. This pathway, known alternatively as the citric acid or tricarboxylic acid cycle, was shown by Kennedy and Lehninger to be present in mitochondria.

The scheme of reactions encompassed by the TCA cycle is shown in Fig. 3.7. The first reaction catalyzed by citrate synthase results in the condensation of acetyl-CoA with oxalacetate to form citric acid. The citrate is then converted to a mixture of *cis*-aconitate and isocitrate. This step is carried out by a single enzyme called aconitase. In the next reaction, isocitrate dehydrogenase oxidizes isocitrate to oxalsuccinate. This is the first step at which NAD^+ is reduced to NADH. The oxalsuccinate remains enzyme bound to isocitrate dehydrogenase and is further decarboxylated to α-ketoglutarate. The fourth reaction involves a conversion of α-ketoglutarate to succinyl-CoA with the production of the second equivalent of NADH. The oxidation of α-ketoglutarate to succinyl-CoA is catalyzed by the α-ketoglutarate dehydrogenase complex. This is a multienzyme complex whose structure and mechanism of action are very similar to that of the pyruvate dehydrogenase complex. In the fifth step, succinyl-CoA is deacylated by succinate thiokinase to the free acid. The energy of the thioester bond is conserved through the esterification of GDP and P_i to form GTP. The terminal phosphate of GTP can be transferred to ADP by a nucleoside diphosphokinase. The formation of GTP by this mechanism is known as "substrate-level phosphorylation." The next reaction involves the oxidation of succinate to fumarate. This is catalyzed by succinate–coenzyme Q reductase, the only enzyme of the TCA cycle that is not soluble and is found in the inner

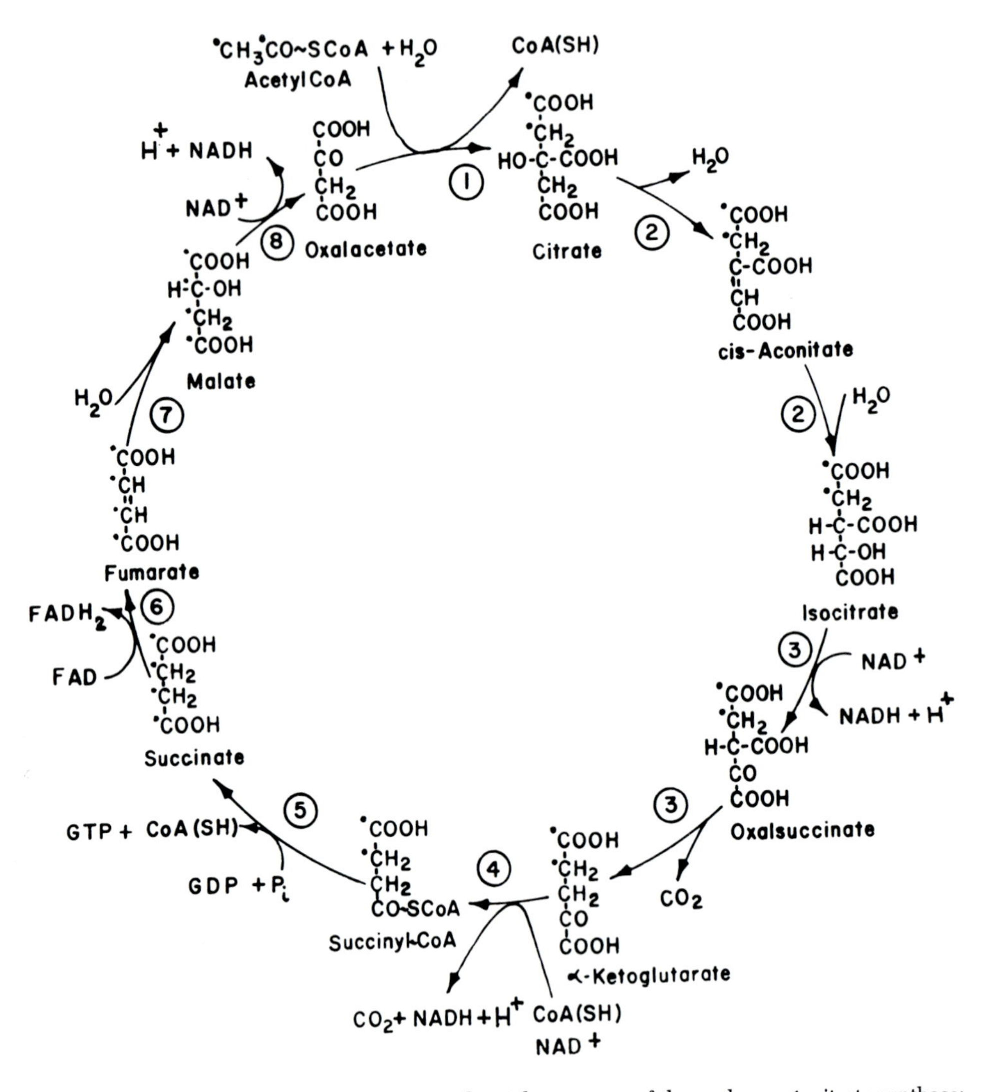

FIG. 3.7. Tricarboxylic acid cycle reactions. The eight enzymes of the cycle are: 1, citrate synthase; 2, aconitase; 3, isocitrate dehydrogenase; 4, α-ketoglutarate dehydrogenase complex; 5, succinate thiokinase; 6, succinate–coenzyme Q reductase; 7, fumarase; 8, malate dehydrogenase. The marked carbons indicate the distribution of the two carbons of acetate after the first turn of the cycle.

membrane. In the seventh reaction, fumarate is hydrated to malate by fumarase. The eighth and last step is the oxidation of malate to oxalacetate, thus completing the cycle. This oxidation reaction is catalyzed by malate dehydrogenase and yields the third equivalent of NADH.

The eight reactions of the cycle achieve the oxidation of one molecule of acetyl-CoA and the release of eight hydrogens.

$$CH_3-\overset{\overset{\displaystyle O}{\|}}{C}\sim SCoA + GDP + 3NAD^+ + FAD + P_i + 2H_2O \rightarrow$$

$$2CO_2 + GTP + 3NADH + FADH_2 + 2H^+ + CoA(SH) \quad [3.14]$$

Of the eight reducing equivalents, six are used to reduce NAD^+, and two are used to reduce the flavin adenine nucleotide prosthetic group of succinate dehydrogenase. The fate of these hydrogens will be traced in the next chapter when we consider the oxidation-reduction reactions of the electron transfer chain.

A. Enzymes of the Tricarboxylic Acid Cycle

As has already been pointed out, all of the enzymes of the TCA cycle with the exception of succinate-coenzyme Q reductase are soluble proteins that are found in the matrix fraction of mitochondria. Some of the key features of these enzymes follow.

1. Citrate Synthase (Condensing Enzyme)

Citrate synthase is a soluble protein with a molecular weight of 100,000. The condensation of acetyl-CoA with oxalacetate has a ΔG^0 of approximately -8 kcal because of the exergonic hydrolysis of a thioester bond. The equilibrium of the reaction is therefore predominantly in favor of the formation of citrate. The condensation is thought to involve the removal of a hydrogen from the methyl group of acetyl-CoA followed by an attack of the resultant carbanion on the carboxyl group of oxalacetate. The exact mechanism, however, is not known at present.

2. Aconitase

This is a soluble enzyme that catalyzes the dehydration of both citrate and isocitrate. As a result, an equilibrium mixture of citrate, *cis*-aconitate, and isocitrate is established. Under neutral conditions of pH, the equilibrium mixture consists of 91% citrate, 7% isocitrate, and 3% *cis*-aconitate.

3. Isocitrate Dehydrogenase

Mitochondria contain two types of isocitrate dehydrogenases—one uses NAD^+ as the hydrogen acceptor, and the other $NADP^+$. There is also an extramitochondrial NADP-dependent isocitrate dehydrogenase differing from the mitochondrial enzyme in its catalytic and physical properties. Even though there appear to be three different enzymes, only the NAD-dependent dehydrogenase is present exclusively in mitochondria.

It has been of some interest to determine which of the two mitochondrial dehydrogenases (if not both) participate in the TCA cycle. Although the activity of the NADP-dependent enzyme is somewhat higher in mitochondria of most tissues, it is now fairly certain that the principal enzyme responsible for the oxidation of isocitrate is the NAD-dependent dehydrogenase. This is based on the observation that the activity of the NAD- but not the NADP-linked isocitrate dehydrogenase is markedly stimulated by ADP in mammalian mitochondria. When assayed at low levels of isocitrate in the presence of ADP, the NAD-dependent enzyme oxidizes isocitrate with high rates. ADP has been found to act as a modulator of the enzyme by inducing a reversible aggregation of the monomer to an active dimer form. It is interesting that this phenomenon is prevented by NADH.

Isocitrate dehydrogenase catalyzes two separate reactions. The substrate is first oxidized to oxalsuccinate in the presence of NAD^+. The oxalsuccinate is then decarboxylated to α-ketoglutarate in a reaction requiring Mg^{2+}. In the absence of the divalent metal, there is an accumulation of oxalsuccinate, and very little α-ketoglutarate is formed.

4. α-Ketoglutarate Dehydrogenase Complex

The α-ketoglutarate dehydrogenase complex of mammalian mitochondria has a molecular weight of 2.7×10^6. It is composed of three separate enzymes: (1) α-ketoglutarate dehydrogenase, (2) dihydrolipoyl transsuccinylase, and (3) dihydrolipoyl dehydrogenase. The complex has been resolved into the component enzymes, and these in turn have been used to reconstitute the native complex. From such studies, it has been learned that the quaternary structure of the complex is quite similar to that of the pyruvate dehydrogenase complex. There are four transsuccinylase subunits making up the core, and the α-ketoglutarate and dihydrolipoyl dehydrogenase subunits form the outer structure of the complex.

The oxidation of α-ketoglutarate to succinyl-CoA occurs by the same reaction mechanism described for the oxidation of pyruvate to acetyl-CoA (see Section I). α-Ketoglutarate is first decarboxylated by the α-ketoglutarate dehydrogenase with the formation of succinyl semialdehyde which is bound to the thiamine pyrophosphate prosthetic group of the enzyme. The succinyl semialdehyde is then transferred to lipoic acid by the transsuccinylase. This transfer results in the oxidation of the semialdehyde to the acid at the expense of the reduction of lipoic acid to dihydrolipoic acid. The succinyldihydrolipoic acid ester reacts with free coenzyme A to form succinyl-CoA. Dihydrolipoic acid is reoxidized to lipoic acid by dihydrolipoyl dehydrogenase, an FAD-containing enzyme that uses NAD^+ as the hydrogen acceptor. Unlike the pyruvate dehydrogenase complex, there does not appear to be a phosphorylation and dephosphorylation type of regulatory mechanism operating in the α-ketoglutarate dehydrogenase complex.

5. Succinate Thiokinase

Succinate thiokinase catalyzes a conversion of succinyl-CoA to free succinic acid and coenzyme A. The standard free energy of hydrolysis of the thioester bond is sufficiently high (−7.5 kcal) to allow for the esterification of GDP and P_i. Studies on the mechanism of catalysis of succinyl thiokinase have shown that there is a protein-bound phosphorylated intermediate formed during the reaction. The following sequence of reactions has been proposed on the basis of partial and exchange reactions observed with the enzyme:

$$\mathrm{HOOC{-}CH_2{-}CH_2{-}\overset{\overset{\displaystyle O}{\|}}{C}{\sim}SCoA + Enz + P_i \rightleftharpoons Enz \cdots HOOC{-}CH_2CH_2{-}\overset{\overset{\displaystyle O}{\|}}{C}{\sim}OP + CoA(SH)} \quad [3.15]$$

$$\mathrm{Enz \cdots HOOC{-}CH_2CH_2{-}\overset{\overset{\displaystyle O}{\|}}{C}{\sim}OP \overset{Mg^{2+}}{\rightleftharpoons} Enz{\sim}P + HOOC{-}CH_2{-}CH_2{-}COOH} \quad [3.16]$$

$$\mathrm{Enz{\sim}P + GDP \rightleftharpoons Enz + GTP} \quad [3.17]$$

The mechanism involves the formation of a succinyl phosphate ester at the active site of the enzyme. The high-energy phosphate is transferred to a histidine residue on the enzyme to form histidine phosphate. In the last step, the high-energy phosphoryl group is transferred from histidine to the external acceptor, GDP, yielding 1 mole of GTP per mole of succinyl-CoA hydrolyzed.

6. Succinate-Coenzyme Q Reductase

This is an insoluble enzyme tightly bound to the mitochondrial inner membrane. Its properties will be discussed in the next chapter when we consider the respiratory chain.

7. Fumarase

Fumarase is a 200,000-dalton enzyme composed of four identical subunits. It catalyzes the addition of water to the double bond of fumarate in a *trans* configuration. The equilibrium constant has been found to be approximately 4 under physiological conditions of pH. Although ATP decreases the affinity of the enzyme for fumarate, the physiological significance of this observation is uncertain.

8. Malate Dehydrogenase

Two malate dehydrogenases have been purified from mammalian tissue: one is intramitochondrial, and the other cytoplasmic in origin. The two enzymes have different physical and catalytic properties, and only the mitochondrial dehydrogenase is believed to function in the TCA cycle. One of the differences lies in the inhibitory effect of oxalacetate on the mitochondrial but not the cytoplasmic malate dehydrogenase. The equilibrium constant for malate oxidation is in favor of the formation of malate rather than oxalacetate. Since oxalacetate is effectively removed during the normal functioning of the cycle, malate can be oxidized at high rates.

B. Regulation of the Tricarboxylic Acid Cycle

A number of factors have been proposed to regulate the rate at which the TCA cycle operates. Among the most important are (1) the availability of NAD^+ for the dehydrogenation steps of the cycle, (2) the rate at which acetyl-CoA is synthesized and intermediates of the cycle are drained for biosynthetic reactions, and (3) the intramitochondrial concentrations of ADP and ATP.

The NADH formed during the oxidation of acetyl-CoA is channeled into the terminal respiratory pathway where it is oxidized to NAD^+, a process coupled to the synthesis of ATP. The electron transfer reactions are very tightly coupled to the phosphorylation mechanism. The notion of tight coupling will be discussed in more detail later—suffice it to say that when the concentration of ADP relative to ATP is low (i.e., when there is a deficiency of the phosphate acceptor ADP), the rate of oxidation of NADH is significantly depressed. In well-coupled mitochondria, the ratio of the oxidation rates in the presence (state 3 respiration) and absence (state 4 respiration) of ADP is referred to as the respiratory control (*R.C.*). Intact mitochondria can have *R.C.* values as high as 10. *In vivo*, this difference is probably even larger. We can now see how the intramitochondrial level of ATP will determine the rate at which several interrelated pathways function. The more rapidly ATP is used for biosynthetic and other work activities of the cell, the higher the rate of oxidation of NADH because of the availability of ADP for oxidative phosphorylation (this also applies for the oxidation of succinate). High rates of NADH oxidation in turn promote increased flux through the TCA cycle.

The intramitochondrial levels of ADP and ATP will also influence the TCA cycle activity at the levels of the pyruvic dehydrogenase complex and the NAD-dependent isocitrate dehydrogenase. High concentrations of ATP will tend to favor phosphorylation and inactivation of the pyruvate dehydrogenase complex through the kinase-mediated phosphorylation of the dehydrogenase. The supply of acetyl-CoA will therefore become limiting in the citrate synthase reaction. A similar effect is seen with the NAD-dependent isocitrate dehydrogenase which is more active at high ADP-to-ATP ratios. Both enzymes, therefore, act to limit pyruvate utilization when the mitochondrial concentration of ATP is high. The NAD-dependent isocitrate dehydrogenase is also activated by citrate and isocitrate. Since the aconitase equilibrium is such that there is a 15-fold excess of citrate, the *in vivo* activation is probably attributable primarily to citrate. This is a positive control in which the substrate for a reaction promotes its own utilization.

The rate-limiting step in the TCA cycle is the condensation of acetyl-CoA and oxalacetate. This conclusion is based on the early observation that in respiring tissue, the rate of oxygen consumption is enhanced by the addition of each of the intermediates of the cycle. This means that under steady-state conditions, the TCA cycle activity is not limited by the activity of any of the enzymes that intervene between citrate and oxalacetate. The rate of citrate synthesis is probably determined by the concentration of the reactants, oxalacetate and acetyl-CoA, as well as modulators of the citrate synthase. The synthesis of acetyl-CoA by the pyruvate dehydrogenase complex was already indicated to be regulated by ATP. The recent finding that citrate synthase is also inhibited by ATP shows again how the final product of a long sequence of reactions regulates one of the earlier steps in the metabolic pathway. Some of the control points discussed in this section are summarized in Fig. 3.8.

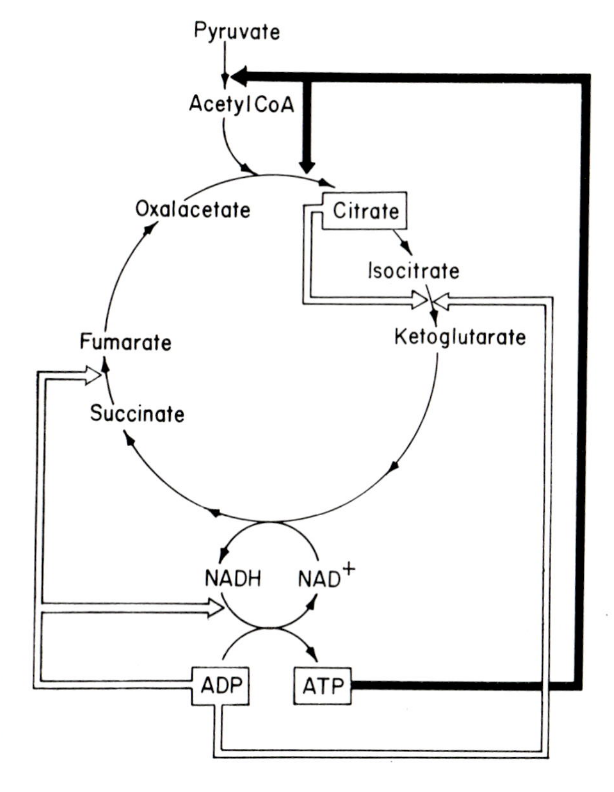

FIG. 3.8. Regulation of the TCA cycle. Activation by ADP and citrate is shown by the clear arrows. Inhibition of key reactions by ATP is indicated by the dark arrows.

VI. Yield of ATP from the Aerobic Oxidation of Glucose

Although the mechanism of oxidative phosphorylation will be discussed in Chapter 6, it may be appropriate at this point to consider the efficiency of energy conservation during the aerobic utilization of glucose.

A. Free Energy Change of ATP Hydrolysis

The three phosphoric acid groups of adenosine-5′-triphosphate are linked to each other through pyrosphophate bonds (Fig. 3.9). The hydrolysis of the terminal (γ) phosphate group is accompanied by a large negative change in standard free energy (G^0). The common value of the ΔG^0 for the hydrolytic reaction is -7.3 kcal/mole, indicating that it is an exergonic process in which the products ADP and P_i are considerably more stable than ATP. Several explanations have been advanced to account for the large negative change in the free energy of hydrolysis of ATP. At neutral pH, three of the phosphate oxygens are completely ionized, thus conferring an instability to the molecule because of charge-repulsion effects. The instability is partially relieved when the γ phosphate group is hydrolyzed. Second, there are fewer resonance forms of phosphate when it is present in a pyrophosphate linkage. The hydrolytic products therefore have a greater resonance of stabilization and are at a lower energy level than ATP.

B. ATP Yield from Glucose

The ATP synthesized during the aerobic oxidation of glucose arises from substrate-level and oxidative phosphorylation. Substrate-level phosphorylation occurs both in the glycolytic and TCA pathways. In glycolysis, there are two ATPs produced for each molecule of glucose converted to pyruvate. Similarly, in the TCA cycle, GTP is formed by a substrate-level phosphorylation mechanism at the succinyl thiokinase step. In the present context, GTP may be considered equivalent to ATP, since the two nucleoside triphosphates are readily interconvertible through the nucleoside diphosphate kinase reaction. Since glucose gives rise to two succinyl-CoAs, there are two ATPs formed from substrate-level phosphorylation in the TCA cycle.

Most of the energy released from the oxidation of glucose is salvaged by

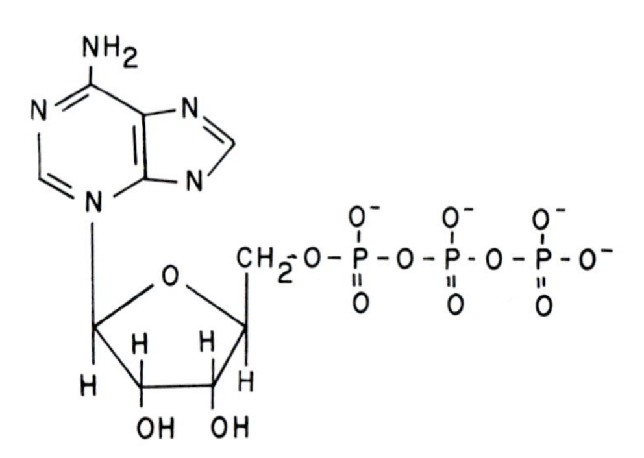

FIG. 3.9. Chemical structure of ATP.

TABLE 3.1. ATP Yield during the Aerobic Oxidation of Glucose

	Moles/mole of glucose	
	NADH	ATP
Glycolysis (glucose → 2 pyruvates)	2 × 3 →	2 6
Pyruvate dehydrogenase complex (2 pyruvates → 2 acetyl CoAs)	2 × 3 →	0 6
TCA cycle (2 acetyl CoAs → $4CO_2$)	6 × 3 →	2 (GTP) 18
	2 $FADH_2$ × 2 →	4
Total ATP		38

oxidative phosphorylation. The NADH formed in glycolysis, in the pyruvate dehydrogenation reaction, and in the dehydrogenation steps of the TCA cycle is reoxidized by oxygen, a process that is coupled to the synthesis of 3 mole equivalents of ATP. The second source of oxidatively derived ATP comes from the oxidation of $FADH_2$ that is formed at the succinate dehydrogenase step. Unlike the oxidation of NADH, that of $FADH_2$ is coupled to the synthesis of only two mole equivalents of ATP.

Glucose is potentially capable of yielding 38 ATPs (Table 3.1). Four ATPs are derived from substrate-level phosphorylation, and the rest from oxidative phosphorylation. In actuality, however, the yield of ATP may be less when the oxidation of extramitochondrial NADH involves the glycerol phosphate shuttle (see Chapter 9, Section VIIIA) in which one ATP is lost for each NADH oxidized. Tissues such as muscle that are thought to utilize this shuttle suffer a deficit of two ATPs, the net yield being reduced to 36.

We can now calculate the efficiency of energy conservation for the overall process. Assuming a ΔG^0 of 7.3 kcal/mole for the synthesis of ATP from ADP and P_i, the formation of 38 ATPs corresponds to a total gain of 277 kcal in standard free energy. Since the combustion of glucose to CO_2 and water has a ΔG^0 of -686 kcal/mole, the efficiency with which energy is conserved is equal to 277/686 or 40%. Under most physiological conditions, however, the intracellular concentration of ATP relative to ADP is high. Consequently, the free energy of formation of ATP is greater than 7.3 kcal/mole, and the efficiency of the process is more than 40%.

Selected Readings

Bressler, R. (1970) Physiological-chemical aspects of fatty acid oxidation, in *Lipid Metabolism* (S. J. Wakil, ed.), Academic Press, New York, pp. 49–77.

Green, D. E., and Allmann, D. W. (1968) Fatty acid oxidation, in *Metabolic Pathways* (D. M. Greenberg, ed.), Vol. II, Academic Press, New York, pp. 1–67.

Krebs, H. A. (1948–1949) The tricarboxylic acid cycle, *Harvey Lect.* **44**:165.

Lehninger, A. L. (1964) *The Mitochondrion*, W. A. Benjamin Co., New York.

Lehninger, A. L. (1976) *Biochemistry*, Worth Publishers, New York.

Lowenstein, J. M. (1967) The tricarboxylic acid cycle, in *Metabolic Pathways* (D. M. Greenberg, ed.), Vol. I, Academic Press, New York, pp. 146–270.

Lusty, C. L. (1978) Carbamoyl phosphate synthetase of rat liver mitochondria. Purification, properties and polypeptide molecular weight, *Eur. J. Biochem.* **85**:373.

Mahler, H. R., and Cordes, E. H. (1966) *Biological Chemistry*, Harper & Row, New York.

Meister, A. (1965) *Biochemistry of the Amino Acids*, Vols. 1 and 2, Academic Press, New York.

Ratner, S. (1973) Enzymes of arginine and urea synthesis, *Adv. Enzymol.* **39**:1.

Reed, L. J., and Cox, D. J. (1970) Multienzyme complexes, in *The Enzymes* (P. D. Boyer, ed.), Vol. 1, Academic Press, New York, pp. 213–266.

Wakil, S. J. (1970) Fatty acid metabolism, in *Lipid Metabolism* (S. J. Wakil, ed.), Academic Press, New York, pp. 1–48.

4

The Electron Transfer Chain

In Chapter 3 we traced how mitochondria participate in the aerobic breakdown of pyruvate, fatty acyl-CoAs, and amino acids. The NADH formed during β-oxidation and the tricarboxylic acid (TCA) cycle reactions must be reoxidized to NAD^+ in order for the two pathways to continue operating. This is accomplished by a sequence of oxidation–reduction events in which the final hydrogen acceptor is molecular oxygen, and the product is water. This pathway, known as the respiratory or electron transfer chain, is also involved in the oxidation of the TCA cycle intermediate, succinate.

In addition to forming NAD^+, the respiratory chain performs a second very important function. When NADH and succinate are oxidized, the energy that is released, rather than being lost as heat, is conserved in the form of the terminal pyrophosphate bond of ATP. The process by which the synthesis of ATP is coupled to the oxidation of NADH and succinate is known as oxidative phosphorylation.

Much of modern research on mitochondria has been aimed at understanding the enzymes that carry out electron transfer and the coupled synthesis of ATP. It is now generally accepted that the fundamental oxidative and phosphorylative events are catalyzed by five enzyme complexes. Four of these are respiratory complexes, and the fifth is an ATP synthetase or ATPase complex. In this chapter we shall consider the function, composition, and structure of the respiratory complexes—a discussion of the ATP synthetase is deferred to Chapter 7.

The four respiratory complexes are located in the mitochondrial inner membrane and are its most important functional and structural constituents. Together they comprise the entire enzymatic machinery that is responsible for the oxidation of NADH and succinate. There are two other components that function in electron transfer by acting as shuttles between the complexes. These electron-shuttling molecules, known as "mobile" carriers, are cytochrome *c*, a low-molecular-weight hemoprotein, and coenzyme Q, a lipoidal quinone. In the oxidized state, cytochrome *c* and coenzyme Q function as electron acceptors; in the reduced state, as electron donors. These molecules may be viewed as being both substrates and products of the respiratory complexes.

Each respiratory complex represents a defined segment of the overall electron transfer chain. The first enzyme, known as NADH-coenzyme Q reductase or complex I, oxidizes NADH and reduces coenzyme Q. The second enzyme, succinate-coenzyme Q reductase also known as complex II, carries out a similar reaction except that the hydrogen donor is succinate produced in the TCA cycle. The products of the reaction catalyzed by complex II are fumarate and coenzyme QH_2. Coenzyme QH_2 is reoxidized by complex III, coenzyme QH_2-cytochrome *c* reductase. The acceptor in this reaction is cytochrome *c*. The fourth and last enzyme in the sequence is cytochrome oxidase or complex IV which transfers the reducing equivalents from ferro-cytochrome *c* to oxygen. This is the step at which water is formed. Later in this chapter, it will be evident that each of the four respiratory complexes is in itself a mini-electron-transfer chain, containing different proteins that function as electron carriers.

As can be seen from the scheme of Fig. 4.1, the four respiratory complexes form a branched pathway which converges at the level of coenzyme Q, the common acceptor that links the two dehydrogenase complexes with the cytochrome *c* reductase. From a historical standpoint and also for the purpose of a formal treatment of this topic, it is advantageous to first describe the known carriers of the chain before relating them to the respiratory complexes.

I. Carriers of the Electron Transfer Chain

The earlier work on the terminal respiratory pathway of mitochondria focused on the identification of the various carriers of the chain and on the order in which they undergo oxidation-reduction changes before the reducing equivalents are finally combined with oxygen. Keilin succeeded in providing the first skeleton of the chain on the basis of his simple and elegant spectroscopic studies. Later efforts, especially in the laboratories of Chance, Green, and Beinert, were instrumental in the discovery of new carriers that added to a much more complete picture of the chain.

There are three types of carrier proteins that are known at present—flavoenzymes, cytochromes, and metalloproteins. These differ in their prosthetic groups, being flavin nucleotides in flavoenzymes, heme in cytochromes, and iron-sulfur centers or copper in the metalloproteins. The distinction, however, is not entirely clear-cut, since some of the flavoproteins also contain iron-sulfur centers. In addition to the protein carriers, the chain contains two low-molecular-weight carriers, NAD^+ and coenzyme Q, which do not appear to be associated with any proteins.

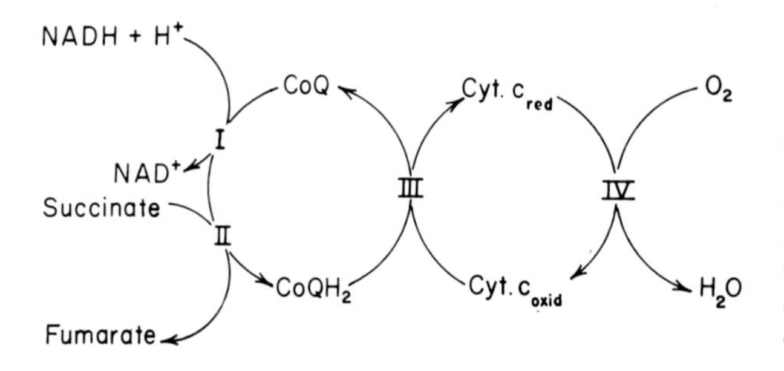

FIG. 4.1. Reactions of the electron transfer chain. The four respiratory complexes are: I, NADH-coenzyme Q reductase; II, succinate-coenzyme Q reductase; III, coenzyme QH_2-cytochrome *c* reductase; IV, cytochrome oxidase.

A. Flavoenzymes

Flavoenzymes are prevalent carriers that participate in many oxidation–reduction reactions involving the transfer of hydrogen. All known flavoenzymes have either flavin mononucleotide (FMN) or flavin adenine dinucleotide (FAD) as prosthetic groups (Fig. 4.2).

The reduction of the flavin prosthetic group often proceeds in two steps. The isoalloxazine ring is first reduced to a semiquinone intermediate which is very unstable and becomes further reduced to the dihydroquinone (Fig. 4.3). Each oxidation state of the flavin moiety has a characteristic spectrum. Oxidized FMN and FAD absorb visible light with maxima at 370 and 450 nm. The semiquinone has an additional absorption band at 590 nm. The fully reduced flavin, on the other hand, absorbs only at 370 nm. These chromophoric properties of the flavin account for the color changes of flavoproteins which are yellow in the oxidized form and are bleached on reduction.

The electron transfer chain contains two flavoproteins—NADH dehydrogenase which accepts hydrogen from NADH and has FMN as its prosthetic group and succinate dehydrogenase, an FAD enzyme, which accepts hydrogen from succinate. Both NADH and succinate dehydrogenases have been extensively studied, and numerous procedures have been devised for their isolation. Since the flavoenzymes are no longer associated with their natural acceptors, they are usually assayed by measuring the rate with which NADH or succinate reduces various artificial acceptors such as ferricyanide, phenazine methosulfate, dichloroindophenol, and other dyes. Depending on the method of purification, NADH and succinate dehydrogenases exhibit different acceptor specificities. Often the best artificial acceptors for the isolated dehydrogenase react very poorly when used to assay the enzyme in its natural environment in the membrane. This probably means that the catalytic properties of the enzyme are

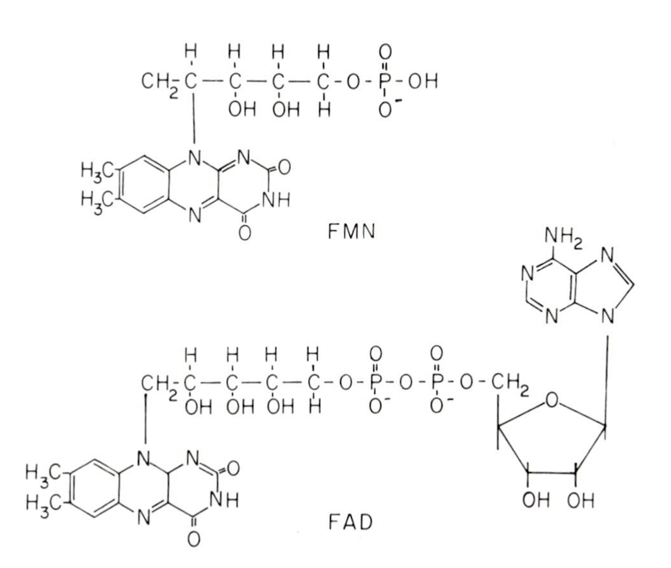

FIG. 4.2. Structural formulas of flavin mononucleotide and flavin adenine dinucleotide.

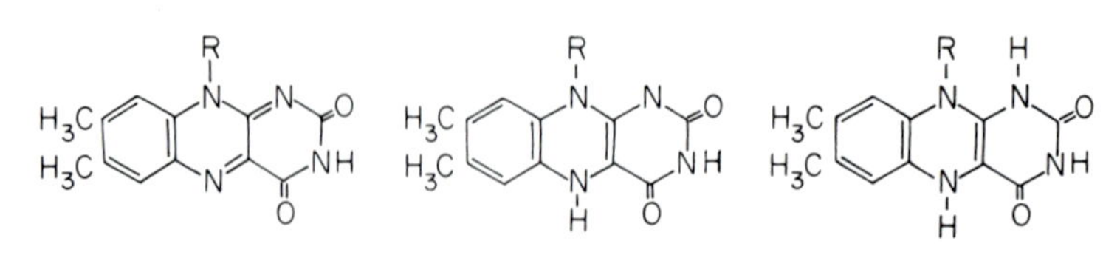

FIG. 4.3. Oxidation–reduction states of the isoalloxazine ring of flavin. Left, oxidized; middle, semiquinone; right, reduced.

modified when it is separated from the other protein or proteins with which it is associated in the membrane, and as a result, electron-donating groups that are normally masked become exposed and accessible to the artificial acceptors. For a long time this was a cause of much confusion concerning the nature and identity of the "true" respiratory chain dehydrogenase.

1. NADH Dehydrogenase

This flavoenzyme has been isolated from mitochondria and from NADH–coenzyme Q reductase with chaotropic agents,* by treatment with acidic ethanol at elevated temperatures, and by phospholipase digestion. These procedures cause the enzyme to be released as a soluble protein which can be further purified by various fractionation procedures. The purest preparations of NADH dehydrogenase contain 14 nmoles of FMN per milligram of protein and four times as much nonheme iron and acid-labile sulfur. The purified protein is water soluble and has an estimated molecular weight of 70,000. The spectrum of the dehydrogenase is shown in Fig. 4.4. The absorption in the visible range results from both the FMN and the nonheme iron component. By treating the enzyme with mersalyl which destroys the iron–sulfur centers, NADH dehydrogenase acquires a spectrum that is characteristic of flavoproteins.

NADH dehydrogenase catalyzes the reduction of ferricyanide, menadione, cytochrome *c*, and coenzyme Q. The reduction of cytochrome *c* and coenzyme Q by the isolated dehydrogenase, however, is thought to proceed by a catalytic mechanism that is different from that of mitochondria or of the isolated respiratory complexes. Some of the salient properties of the enzyme are summarized in Table 4.1.

2. Succinate Dehydrogenase

This component of the electron transfer chain has been studied in many laboratories. The earliest preparations of the enzyme were obtained by extraction of mitochondrial particles with 20% ethanol at pH 9.0 followed by purification on calcium phosphate gels. When obtained by this procedure, succinate

*Chaotropic agents are chemicals that disorganize (from the word *chaos*) membrane-associated proteins. They are powerful solubilizing agents for the external class of proteins. The best chaotropes are salts of the Hofmeister series (NaBr, NaI, $NaClO_4$, NaSCN) and reagents such as urea and guanidine hydrochloride.

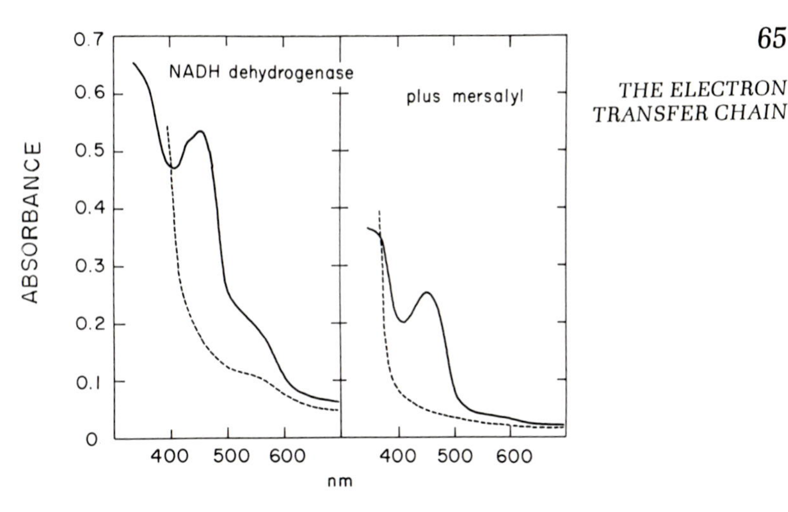

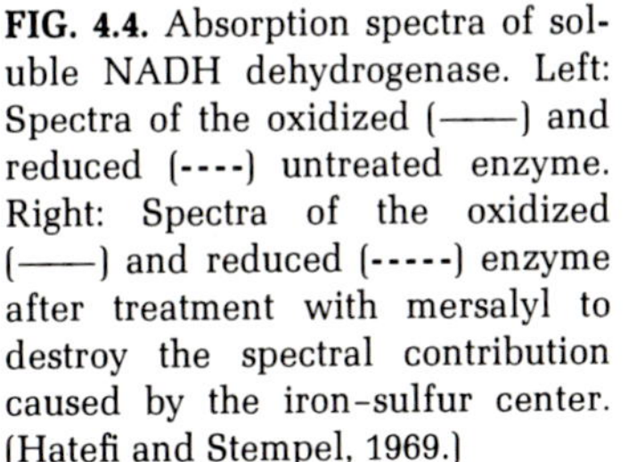

FIG. 4.4. Absorption spectra of soluble NADH dehydrogenase. Left: Spectra of the oxidized (——) and reduced (----) untreated enzyme. Right: Spectra of the oxidized (——) and reduced (-----) enzyme after treatment with mersalyl to destroy the spectral contribution caused by the iron–sulfur center. (Hatefi and Stempel, 1969.)

dehydrogenase contains FAD, nonheme iron, and acid-labile sulfur in a ratio of 1:4:4. Keilin and King purified succinate dehydrogenase by a similar procedure but included succinate in the extraction step and kept the enzyme under nitrogen. This type of preparation had an FAD-to-nonheme iron ratio of 1:8. King showed that the enzyme with the higher iron content was capable of restoring the ability of alkali-extracted mitochondrial membranes to oxidize succinate by oxygen and referred to the preparation as being "reconstitutively" active.

Hatefi and co-workers have recently isolated a pure "reconstitutively" active succinate dehydrogenase from complex II by extraction with $NaClO_4$ and precipitation with ammonium sulfate. This enzyme contains 10 nmoles of FAD, 80 nmoles of iron, and 80 nmoles of labile sulfur per milligram of protein and has a molecular weight of 100,000. It consists of two nonidentical subunits of 70,000 and 27,000 daltons. The subunits have been separated, and their flavin and iron contents studied. The 70,000-dalton protein contains flavin and iron in a ratio of 1:4. The 27,000-dalton protein is free of flavin and contains 110

TABLE 4.1. Properties of Soluble NADH Dehydrogenase[a]

Enzymatic activity	Specific activity (μmoles/min per mg)
NADH → $K_3Fe(CN)_6$	215
NADH → menadione	170
NADH → 2,6-dichloroindophenol	100
NADH → coenzyme Q_1	160
NADH → cytochrome c	43
Molecular weight	70,000
FMN content	13.5–14.5 nmoles/mg protein
Nonheme iron content	60–65 ng-atoms/mg protein
Acid-labile sulfur content	56–60 nmoles/mg protein

[a]From Hatefi and Stempel (1969).

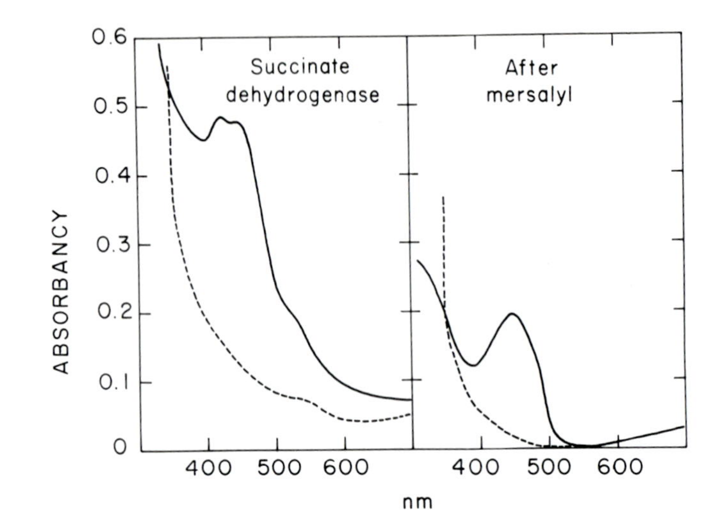

FIG. 4.5. Absorption spectrum of soluble succinate dehydrogenase. Left: Spectra of the oxidized (——) and reduced (----) untreated enzyme. Right: Spectra of the oxidized (——) and reduced (----) enzyme after mersalyl treatment. (Davis and Hatefi, 1971.)

nmoles of nonheme iron and an equal amount of labile sulfur per milligram of protein. The 70,000-dalton flavoprotein is inactive in reconstituting succinate oxidase in alkali-extracted particles. At present it appears, therefore, that the minimal requirements for an active succinate dehydrogenase are the 70,000-dalton flavoprotein and the 27,000-dalton iron protein. The spectrum of succinate dehydrogenase is shown in Fig. 4.5. The absorbance in the 400- to 500-nm range of the spectrum is caused by the FAD and the iron–sulfur centers.

The FAD of succinate dehydrogenase is covalently bound to the 70,000-dalton polypeptide. The chemical nature of the binding has been studied by Ehrenberg, Singer, and Hemmerich. A pentapeptide with the sequence ser-his-thr-val-ala and containing FAD was obtained after proteolytic digestion of succinate dehydrogenase. The FAD was established to be covalently linked to the histidyl residue as shown in Fig. 4.6.

Purified succinate dehydrogenase catalyzes the oxidation of succinate to fumarate with either ferricyanide or phenazine methosulfate as acceptors. Unlike complex II, the soluble dehydrogenase does not reduce coenzyme Q,

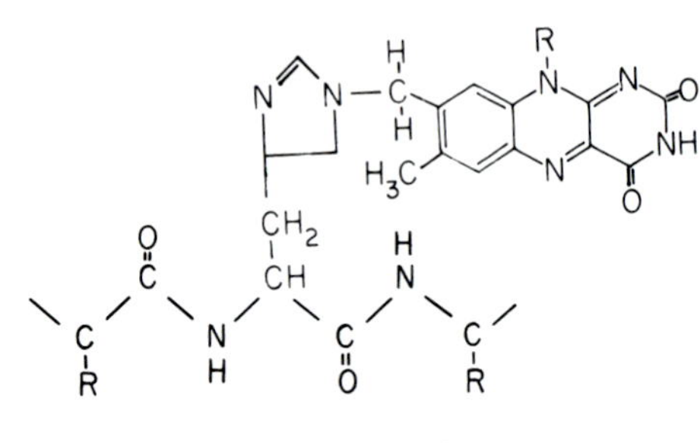

FIG. 4.6. Covalent linkage of the FAD group of succinate dehydrogenase to a histidyl residue. (Salach *et al.*, 1972.)

TABLE 4.2. Properties of Soluble Succinate Dehydrogenase[a]

Enzymatic activity	Specific activity (μmoles/ min per mg)
Succinate → $K_3Fe(CN)_6$	13.5
Succinate → phenazine methosulfate	67–78
Molecular weight	97,000
FAD content	10.3 nmoles/mg protein
Nonheme iron content	70–80 ng-atoms/mg protein
Acid-labile sulfur content	70–80 nmoles/mg protein

[a]From Davis and Hatefi (1971).

nor is it inhibited by 2-theonoyl trifluoroacetone, a specific inhibitor of succinate–coenzyme Q reductase. Both soluble and membrane-bound succinate dehydrogenases are activated by a number of compounds (e.g., succinate, ATP, coenzyme Q). The activation has been correlated with a release of oxalacetate which is bound to a sulfhydryl group on the enzyme.

Some properties of succinate dehydrogenase are listed in Table 4.2.

B. Cytochromes

Cytochromes are hemoproteins that undergo one-electron oxidation–reduction changes in the iron of their heme prosthetic groups. Two different kinds of heme are found in the cytochromes of the mitochondrial respiratory chain. Cytochromes of the *b* and *c* type contain ferroprotoporphyrin, also known as protoheme. In cytochromes of the *a* type, the prosthetic function is a heme *a* which differs from protoheme in two of the substituent groups on the porphyrin ring (Fig. 4.7). The central iron of heme has a coordination number of six and is chelated to the four nitrogens of the porphyrin ring. In *b*- and *c*-type cytochromes, the fifth and sixth ligands are nitrogen or sulfur of amino

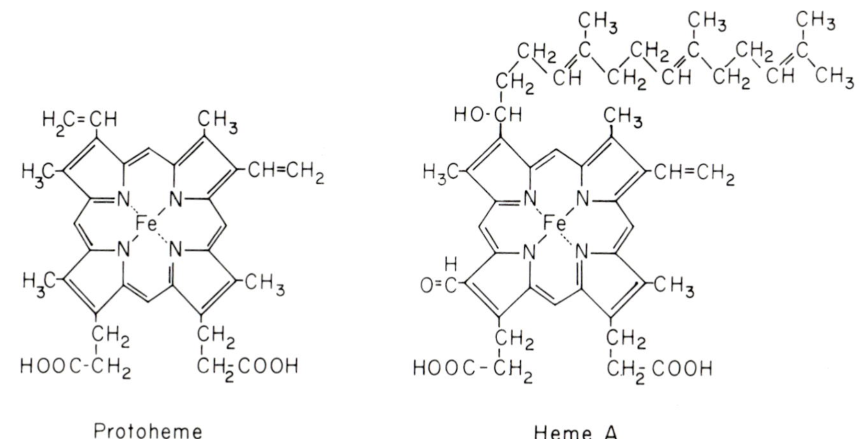

FIG. 4.7. Structural formulas of protoheme and of heme *a*.

acid groups of the protein. In cytochrome a_3, however, the sixth ligand is molecular oxygen.

Cytochromes are beautiful colored proteins that absorb light in the visible range of the spectrum. In the reduced form, the cytochrome absorption spectrum exhibits three bands called α, β, and γ or Soret. The exact wavelengths at which these bands have their maxima are characteristic properties of each cytochrome. By definition, the α band absorbs light at the highest, the γ band at the lowest, and the β band at an intermediate wavelength.

1. Cytochromes *a* and a_3

Cytochromes *a* and a_3 are catalytic constituents of the terminal respiratory complex, cytochrome oxidase. Neither of the two cytochromes has been purified, and it is still not known whether they are physically distinct proteins or whether there is a single hemoprotein whose heme *a* prosthetic group, depending on the environment or its ligands, acquires different spectral properties that correspond to cytochromes *a* and a_3. It is known, however, that both cytochromes have the same prosthetic group, heme *a* (Fig. 4.7). The heme is not covalently bound and can be quantitatively extracted with organic solvents under acidic or basic conditions. The ligands of the fifth and sixth coordinates of the heme iron of cytochrome *a* are probably basic amino acids such as lysine or histidine. In the case of cytochrome a_3, the fifth coordinate is again the nitrogen of histidine or lysine, and the sixth is molecular oxygen that becomes catalytically reduced to water in the last step of electron transport.

The visible absorption bands of cytochromes *a* and a_3 occur at similar wavelengths, and, consequently, they are not resolved in spectra of either oxidized or reduced mitochondria or of cytochrome oxidase. The composite spectrum of the two cytochromes in their naturally occurring ratio is exemplified by the spectrum of purified cytochrome oxidase (Fig. 4.8). The absorption maxima of the α, β, and γ bands of the reduced cytochromes are at 605, 517, and 445 nm, respectively.

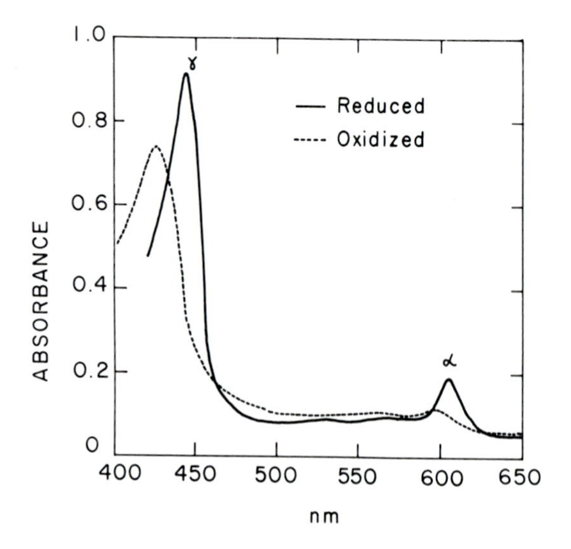

FIG. 4.8. Oxidized and reduced spectra of cytochrome oxidase purified from beef heart mitochondria.

It is of interest to trace how the two cytochromes were discovered. Keilin observed that aerobic cells contain several *a*-type cytochromes only one of which reacts with carbon monoxide (Fig. 1.2). The carbon monoxide-binding component was named cytochrome a_3, whereas the unreactive component was designated as cytochrome *a*. The cytochrome a_3–CO adduct had absorption properties that were similar to the photochemical action spectrum reported by Warburg for the Atmungsferment–CO complex. Since it was known that carbon monoxide inhibits respiration by competing for oxygen, cytochrome a_3 was postulated to be the terminal carrier which reacts with oxygen. The distinction between cytochromes *a* and a_3 is, therefore, a spectral one, based on the fact that only part of the absorption at 605 nm and 445 nm is lost when mitochondria or cytochrome oxidase are poisoned with carbon monoxide.

Since cytochromes *a* and a_3 have not been physically separated, their spectra have only been measured by indirect means. This can be done by differential spectrophotometry of cytochrome oxidase treated under conditions that cause the two cytochromes to assume different oxidation–reduction states. The various forms of cytochrome oxidase and the conditions under which these are obtained are listed below.

1. In the presence of oxygen and absence of substrate, the enzyme is fully oxidized, and both cytochromes are in the ferric state ($a^{3+}a_3^{3+}$).
2. When the enzyme is reduced with sodium dithionite, the *a* and a_3 components are fully reduced to the ferrous state ($a^{2+}a_3^{2+}$).
3. Treatment of the reduced cytochrome oxidase with carbon monoxide causes the cytochrome a_3–CO adduct to be formed ($a^{2+}a_3^{2+}$–CO).
4. Under anaerobic conditions, carbon monoxide induces a reduction of cytochrome a_3 which then forms the adduct. In this preparation, cytochrome *a* remains in the ferric state ($a^{3+}a_3^{2+}$–CO).
5. Cyanide is an effective ligand of the heme *a* of cytochrome a_3 but not that of cytochrome *a*. When oxidized enzyme is treated with cyanide, a cytochrome a_3–CN complex is formed ($a^{3+}a_3^{3+}$–CN).
6. The cytochrome a_3 component of the cyanide complex can be reduced enzymatically without affecting the oxidation state of cytochrome *a* ($a^{3+}a_3^{2+}$–CN).

We can now proceed to see how these preparations are used to obtain the spectra of cytochromes *a* and a_3. In the top part of Fig. 4.9 is shown the difference spectrum of reduced minus oxidized cytochrome oxidase. This spectrum represents the difference in absorbance of both cytochromes *a* and a_3. The second part of Fig. 4.9 shows a spectrum of the carbon monoxide adduct of cytochrome oxidase. This direct spectrum differs from that of the control enzyme (Fig. 4.8) in some bleaching of absorbance in the 605- and 445-nm bands and the appearance of new bands at 590 and 430 nm, corresponding to the α and γ bands of cytochrome a_3–CO. The absorption properties of cytochrome a_3–CO are more evident in the difference spectrum (4 − 1). The spectrum of cytochrome *a* alone can be measured by recording the difference in absorption between preparations 3 and 4 or preparations 6 and 5. This spectrum is shown in the fourth part of Fig. 4.9. Cytochrome *a* is seen to have an α band with a relatively high extinction. The spectrum of cytochrome a_3 shown in the bottom part of Fig. 4.9 is obtained by subtracting the absorption caused by cytochrome *a* from the spectrum of cytochrome oxidase. In contrast to cytochrome *a*, cytochrome a_3 has a weakly absorbing α band. The spectral properties of the two cytochromes are summarized in Table 4.3. Based on their relative extinction coefficients and assuming a 1:1 stoichiometry, it is possible to calculate the relative contribution of each component to the absorption spectrum of cytochrome oxidase. At 605 nm, cytochrome *a* contributes 80% and cytochrome a_3 only 20% of the total absorbance. Both components contribute equally to the γ band at 445 nm.

2. Cytochrome *b*

Cytochromes of the *b* type are characterized by the positions of their α bands which are usually in the 555- to 567-nm region of the visible spectrum. These carriers are found in the electron transfer chains of bacteria and higher organisms. The heme prosthetic group of the *b* cytochromes is protoheme that is noncovalently bound to the protein. The fifth and sixth coordinates of the iron are probably ligated to the nitrogen of histidine.

Mammalian mitochondria contain at least three spectrally resolvable species of cytochrome *b*. Chance first noted that the oxidation of the α band of

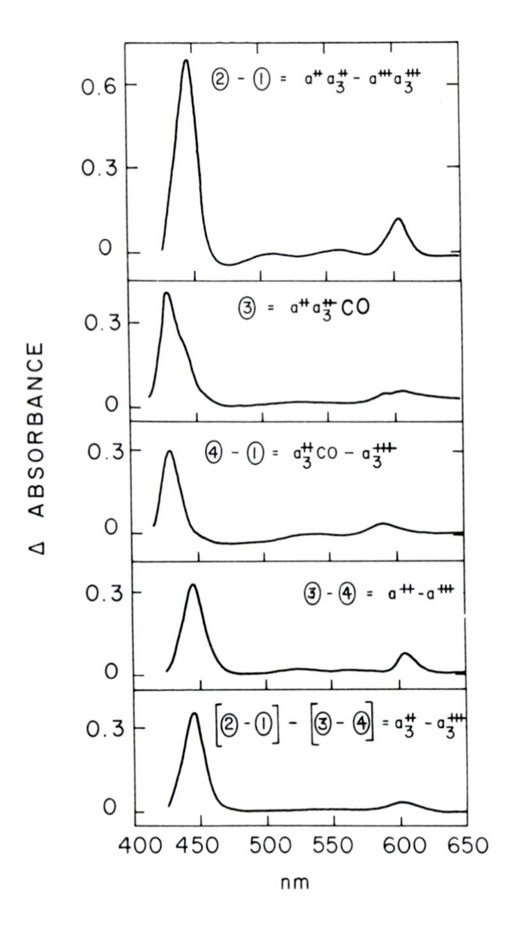

FIG. 4.9. Difference spectra of cytochrome oxidase. The treatments used to obtain the various preparations are described in the text.

TABLE 4.3. Spectral Properties of Cytochromes *a* and a_3[a]

Difference spectrum	Position of bands		Ratio of absorbance
	α	γ	γ/α
a^{2+} minus a^{3+}	605	445	3.9
a_3^{2+} minus a_3^{3+}	605	445	18.0
$a^{2+}a_3^{2+}$ minus $a^{3+}a_3^{3+}$	605	445	7.0
a_3^{2+}-CO minus a_3^{3+}	590	432	7.1

[a]From Tzagoloff and Wharton (1965).

cytochrome *b* in well-coupled pigeon heart mitochondria follows a biphasic time course. About 50% of the cytochrome *b* is rapidly oxidized, whereas the remaining cytochrome *b* is oxidized at a much slower rate. In uncoupled mitochondria, the oxidation of cytochrome *b* occurs at a constant rate. The faster reacting component, designated cytochrome b_K, was found to have an absorption maximum at approximately 560–562 nm, whereas the slower component, cytochrome b_T, had an α band that absorbed maximally at 566 nm. On the basis of these findings, Chance concluded that b_T functions in both electron transport and coupling and that b_K functions in transport only (see Chapter 6 for crossover points in coupled mitochondria). The existence of these two different cytochromes has been confirmed in many laboratories, and their oxidation–reduction properties are noted in Table 4.4. Recently, Hatefi and co-workers have established that cytochromes b_K and b_T are present in complex III of the respiratory chain.

In addition to b_K and b_T, mitochondria have another cytochrome *b* whose

TABLE 4.4. Properties of Mitochondrial *b*-Type Cytochromes

Name	Absorption maxima of reduced band: 23°C			77°K	Reducibility
	α	β	γ	α	
b, b_K, or b_{562}	562	532	430	560	NADH and succinate in coupled and uncoupled mitochondria
					NADH and succinate in presence of antimycin
b_T or b_{566}	566	538	432	562.5	NADH and succinate in coupled mitochondria
				(555)	NADH and succinate in presence of antimycin
b_{558}	558			557.5	Poorly reducible

α band absorbs maximally at 558 nm. A distinct peak corresponding to cytochrome b_{558} can be demonstrated in reduced versus oxidized spectra of mitochondria or submitochondrial particles by fourth-derivative analysis of the normal absorption spectrum. This component has also been found in complex II of the respiratory chain. The function of cytochrome b_{558} has not been elucidated. Initially, it was thought that the absorbance at 558 nm was caused by cytochrome b_K or by a degradation product of b_K. Recently, however, it has been shown that under certain conditions, cytochrome b_{558} undergoes oxidation–reduction changes independent of b_K and b_T. The possibility that the 558-nm species is a genuine carrier of the chain is therefore still a viable one.

Mitochondrial *b*-type cytochromes are very hydrophobic proteins which have proven very difficult to purify. Nonetheless, cytochrome *b* has been isolated in different stages of purity from beef heart, yeast, and *Neurospora crassa* mitochondria. Weiss and co-workers have purified cytochrome *b* from *N. crassa* and determined its molecular weight to be 30,000. This value was obtained both by gel filtration and gel electrophoresis of the SDS-depolymerized protein. The molecular weight value agrees with the minimum molecular weight based on the heme content of the protein (34 nmoles per milligram of protein). Recent data on the DNA sequences of the yeast and human cytochrome *b* genes (see Chapter 11, Section VIIC), however, indicate the molecular weight to be 44,000. The discrepancy in molecular weights is probably caused by underestimations of the true size because of anomalous binding of SDS to this hydrophobic protein. Cytochrome *b* purified from yeast mitochondria has a heme content of 50–60 nmoles per milligram of protein. This would suggest the presence of two heme prosthetic groups per polypeptide chain.

It is therefore tempting to speculate that the spectral species corresponding to cytochromes b_K and b_T is a single protein with two hemes, each in a different environment. The problem of the number of distinct chemical species of cytochrome *b* is still inconclusive. The visible spectrum of the purified cytochrome (Fig. 4.10) shows absorption maxima that have slightly different positions from

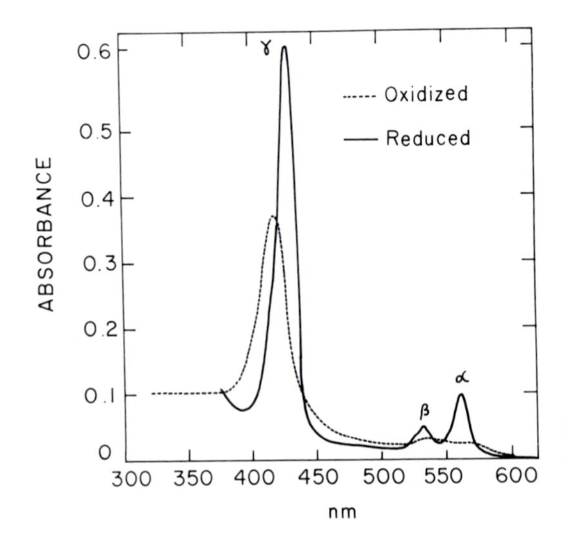

FIG. 4.10. Oxidized and reduced spectra of cytochrome *b* purified from *Neurospora crassa* mitochondria. (Weiss and Ziganke, 1974.)

those of b_K, b_T or b_{558}. This evidence as well as other differences such as the redox potential and the ability of the isolated hemoprotein to react with carbon monoxide (a property not shared by the native protein) suggests that the environment of the heme has been modified during isolation. A positive identification of the purified cytochrome *b* with the species detected in spectra of mitochondria is, therefore, still not a resolved issue.

3. Cytochromes *c* and c_1

Cytochromes of the *c* type are obligatory carriers of bacterial, plant, and animal electron transfer chains. In the reduced state, their α bands have absorption maxima at 550–557 nm. The prosthetic group of these cytochromes is a modified protoheme in which the vinyl side chain is covalently linked to the protein. There are two different *c*-type cytochromes in the respiratory pathway of mitochondria: cytochromes *c* and c_1.

Cytochrome *c*. This is one of the best-studied proteins. It is readily extractable from mitochondria with salt or under acid conditions. Because of its basic isoelectric point, it can be purified on acidic ion-exchange columns. Unlike the other carriers of the chain, cytochrome *c* is a stable, water-soluble protein that retains its native properties after isolation.

The spectrum of cytochrome *c* is shown in Fig. 4.11, and some of its properties are listed in Table 4.5. In addition to the α, β, and γ bands, which in reduced cytochrome *c* have maxima at 550, 521, and 415 nm, a fourth band is seen at 315 nm (δ band).

The amino acid sequences of some 70 different cytochromes *c* have been determined, and from the observed variations, a phylogenetic tree showing the evolution of the protein has been constructed. Mammalian cytochrome *c* (horse heart) consists of 104 amino acids and has a molecular weight of 12,400. The three-dimensional or tertiary structures of the cytochrome from different

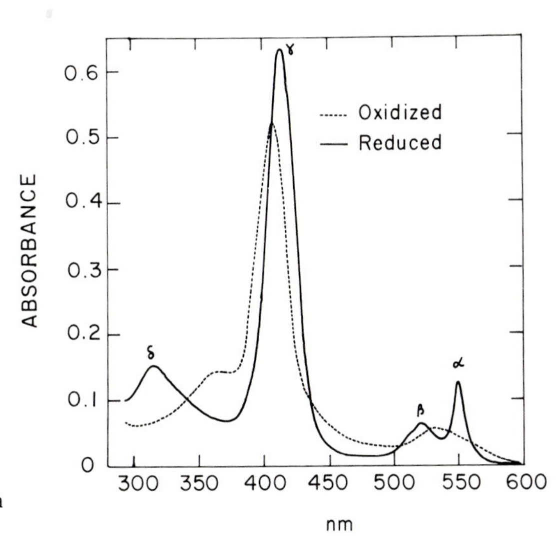

FIG. 4.11. Oxidized and reduced spectra of cytochrome *c*.

sources (horse, tuna, bonito) have been found to be quite similar (Fig. 4.12). Most of the acidic and basic polar amino acid groups occur on the exterior of the protein, which accounts for its solubility in water. The heme group is covalently tied to the protein by thioether bonds with two cysteine residues at positions 14 and 17. The heme is further stabilized by ligation of the fifth and sixth axial coordinates of the iron to histidine and methionine residues and through hydrogen bonding of one of the propionic acid side chains with tryptophan and tyrosine. This structure appears to guarantee a high degree of immobilization of the heme.

There are several aspects of the tertiary structure that have been interpreted to be important in the mechanism of electron transport. Although most of the heme is buried in a hydrophobic crevice, one edge of the porphyrin ring and one of the propionic side chains are exposed to the solvent (dark edge in Fig. 4.12). The tertiary structure also reveals a hydrophobic channel in the upper part of the molecule which provides an alternate access to the heme. It has been suggested that the exposed porphyrin edge could accept electrons directly or that amino acids along the channel may act as electron conductors which transfer the electrons to the centrally located heme.

Cytochrome *c* mediates the transfer of electrons between complexes III and IV. The interaction of cytochrome *c* with these enzymes seems to involve different parts of the molecule. For example, chemical modification of some amino acid residues (lysine) leads to loss of enzymatic reducibility of the cytochrome but does not affect its ability to react with cytochrome oxidase. Similarly, monospecific antibodies directed against different antigenic sites of the protein inhibit binding to complex III but not to cytochrome oxidase.

Cytochrome c_1. This carrier was discovered by Okunuki and co-workers. It is a catalytic subunit of complex III of the respiratory chain. Cytochrome c_1 has been purified from beef heart and from yeast mitochondria. The spectrum is similar to that of cytochrome *c* except that the absorption maxima of the α and γ bands are displaced by about 2 nm toward the red (Table 4.5). The heme prosthetic group of cytochrome c_1 is covalently bonded to two different cysteine

TABLE 4.5. Spectral Properties of Mitochondrial "c" Type Cytochromes

	Absorption maxima of reduced bands			
	23°C			77° K
	α	β	γ	α
Cytochrome *c*	550	520.5	416	449 546 538
Cytochrome c_1	553	520	418	552

residues of the apoprotein, and the iron is thought to be complexed to a methionine as is the case in cytochrome *c*.

The molecular weight of cytochrome c_1 has been estimated to be 30,000 under dissociating conditions. Although the purified protein is soluble in water, it has been found to aggregate into dimers and higher polymers in the absence of bile salts or detergents. Most of the catalytic properties of the native protein are preserved after isolation. Ferrocytochrome c_1 can reduce cytochrome *c* with a second-order rate constant of 3.3×10^6 $M^{-1}sec^{-1}$. The midpoint redox potential of the purified cytochrome has been calculated to be 225 mV which is in agreement with the potential measured in mitochondria. These and other properties of the isolated protein indicate that it is in a fairly native state.

C. Metalloproteins

Metalloproteins constitute an important class of carriers of the respiratory chain. Three different types of such carriers are known at present: (1) flavoenzymes with nonheme iron–sulfur centers, (2) proteins with only iron–sulfur centers, and (3) a copper protein. The study of iron–sulfur centers was

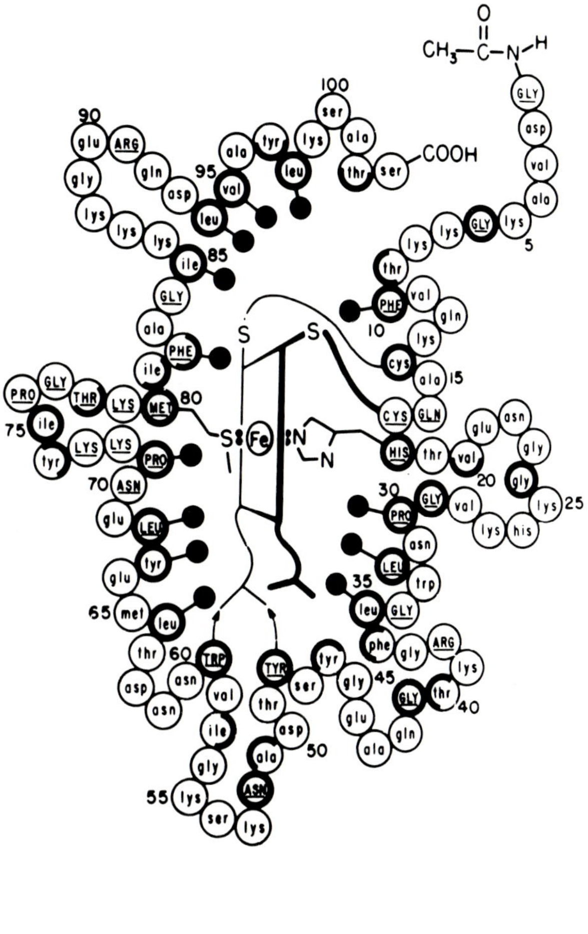

FIG. 4.12. Tertiary structure of cytochrome *c*. Heavily outlined circles represent buried amino acid residues. The black dots indicate side chains that face the buried heme moiety. (Dickerson and Timkovitch, 1975.)

pioneered in the laboratory of Beinert who established their role in electron transfer. Since the detection of this class of carriers depends on electron spin resonance spectroscopy (EPR), this technique will be briefly described.

Molecules such as free radicals or certain metal ions that contain an unpaired electron can absorb radiation in the microwave frequency through the interaction of the magnetic moment of the electron with the magnetic field. In the absence of a magnetic field, the unpaired electron has a double degenerate spin state with angular momentum quantum numbers of $m_s = \pm\frac{1}{2}$. In the presence of a magnetic field, the degeneracy is resolved such that the electron magnetic moment is aligned with the field ($m_s = -\frac{1}{2}$) in the low energy level and is opposed to the field in the high energy state ($m_s = +\frac{1}{2}$). The energy absorbed in the transition from the low to the high state is expressed by the formula, $E = \beta Hg$, where β is the Bohr magneton, H is the field strength, and g is a spectroscopic splitting factor. The g factor is a dimensionless constant which, for a free electron with no orbital angular momentum, has a value of 2.0023. In free radicals, the g value is usually close to that of a free electron.

When the unpaired electron is in an unsymmetrical field, the g value deviates from that of a free electron, and additional bands are seen. An electric field that has a two-dimensional symmetry results in two absorption bands, one caused by interaction with a magnetic field perpendicular to the plane of symmetry ($g_x = g_y = g_\perp$) and a second lesser band caused by the field that is parallel to the plane ($g_z = g_\parallel$). Three absorption bands (g_x, g_y, and g_z) are seen in orthorhombic fields. The EPR spectroscopy also detects interactions between unpaired electrons and nuclei that have a magnetic moment. Such interactions cause "hyperfine" splitting of the EPR signal. The expected number of absorption bands is equal to $2I + 1$, where I is the nuclear spin. For example, hydro-

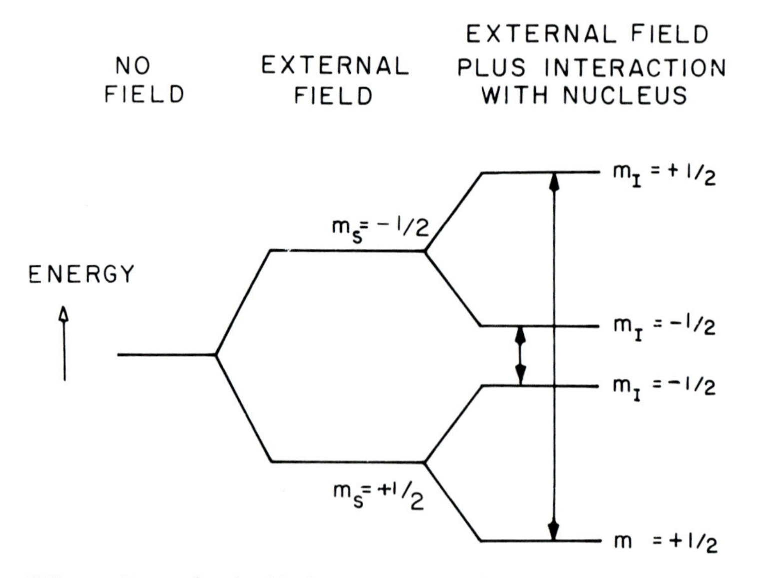

FIG. 4.13. Energy levels of hydrogen. Arrows indicate permitted transitions.

gen has a nuclear spin of ½ and consequently shows two EPR signals. The four energy states of the electron in the hydrogen atom and the two permissible transitions are shown in the energy level diagram of Fig. 4.13.

Electron spin resonance spectra are usually recorded as first derivatives of the absorption spectrum. This is convenient for purposes of calculating the g values at the absorption maxima. In the first-derivative curves, the absorption maximum is observed at the intersection of the signal with the abscissa. Shoulders in the absorption spectrum show downward deflections in the first-derivative curve but do not intersect the abscissa (see Fig. 4.14).

The different iron–sulfur (FeS) centers that are currently believed to function in electron transfer outnumber the cytochromes and other known carriers of the chain. Because only a few nonheme iron proteins have been isolated from mitochondria, our knowledge of these carriers is still fragmentary.

Iron–sulfur centers are associated with three segments of the respiratory chain: NADH–coenzyme Q reductase, succinate–coenzyme Q reductase, and coenzyme QH_2–cytochrome *c* reductase. In the reduced state, the iron exhibits a low-temperature EPR spectrum that is characterized by two minor signals (g_x and g_z) and a prominent signal (g_y) with maximal absorption at 1.85 to 1.95 gauss. The standard midpoint potentials of these carriers have been studied in several laboratories (see Section II) by EPR spectroscopy and found to range from −300 to +280 mV. Some properties of the iron–sulfur centers in the mitochondrial electron transfer chain are listed in Table 4.6. Based on the stoichiometry of iron to sulfur, the concentration of iron in the few proteins that have been purified, and the EPR and visible absorption spectra, the respiratory chain FeS centers have two or four irons per center and are probably structurally related to the centers of plant or bacterial ferredoxin (Fig. 4.15).

A brief review of the isolated metalloproteins follows.

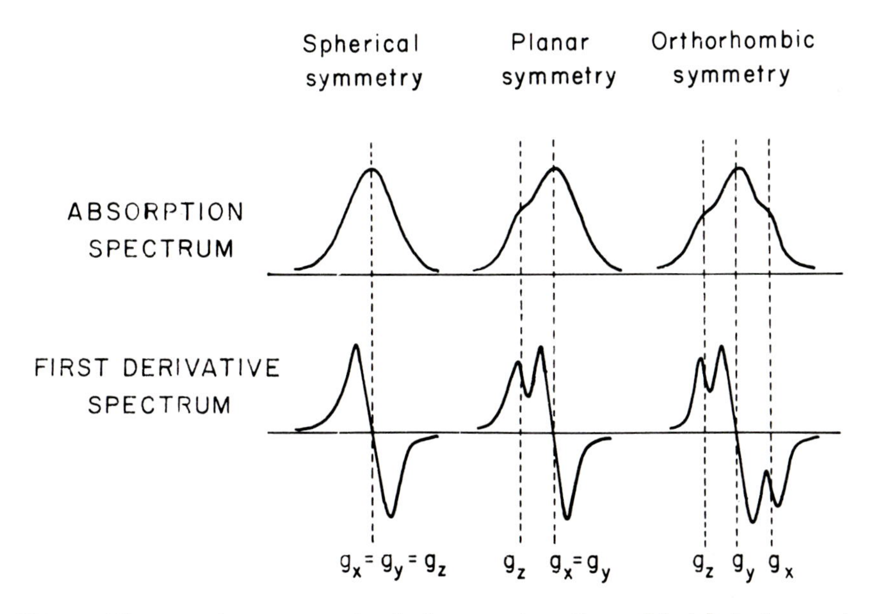

FIG. 4.14. Electron spin resonance signals shown as absorption and first derivative spectra.

TABLE 4.6. Properties of the Iron–Sulfur Centers[a]

Center	Complex	Field positions (g values)			Number of irons and sulfurs per center
		g_z	g_y	g_x	
FeS 1[b]	I	2.03	1.94	1.94	4:4
FeS 2	I	2.05	1.93	1.93	2:2
FeS 3	I	2.10	1.94	1.88	4:4
FeS 4	I	2.04	1.93	1.86	4:4
FeS_{s-1}	II	2.03	1.93	1.91	2:2
FeS_{s-2}	II	2.03	1.93	1.91	2:2
FeS_{s-3}	II	2.02	2.01	1.99	4:4
FeS_{III}	III	2.025	1.89	1.81	2:2

[a]Based on data of Orme-Johnson *et al.* (1974) and Ohnishi (1979).
[b]This iron–sulfur center has been resolved into two separate tetranuclear centers designated as N1*a* and N1*b* by Ohnishi (1979). Most of the EPR signal of the FeS 1 center is contributed by N1*a*. Center N1*b* is not reducible by NADH in complex I and its role in electron transfer is still ambiguous. Ohnishi has also reported the presence of still another center N5 in complex I. Its function as an electron carrier on the main pathway is also under dispute.

1. NADH Dehydrogenase

The soluble preparations of this enzyme contain nonheme iron and acid-labile sulfur. The NADH dehydrogenase isolated by the procedure of Hatefi and Stempel has one FeS center with four irons and four sulfurs. From EPR data, this center appears to correspond to FeS 1 (see Table 4.6).

2. Succinate Dehydrogenase

The FeS centers of succinate dehydrogenase have been studied by Ohnishi and King. According to these authors, the flavoenzyme contains two iron–sulfur

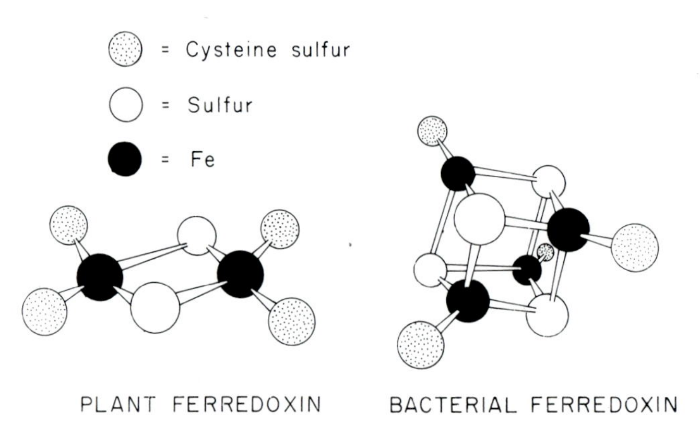

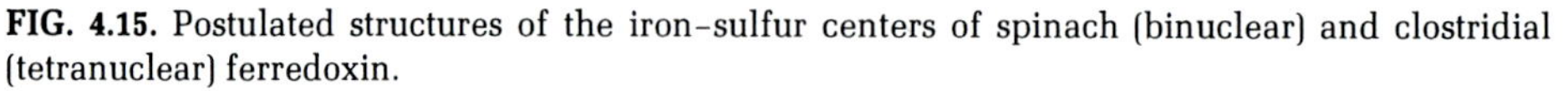

FIG. 4.15. Postulated structures of the iron–sulfur centers of spinach (binuclear) and clostridial (tetranuclear) ferredoxin.

centers, FeS_{s-1} and FeS_{s-2}. Each is of the plant ferredoxin type with two irons and two sulfurs per center.

3. Nonheme Iron Protein of Complex I

This soluble protein has been isolated from NADH-coenzyme Q reductase by dissociation of the complex with $NaClO_4$ and fractionation with ammonium sulfate. The best preparations of the protein are still not pure and have a nonheme iron and sulfur content of 30–40 nmoles per milligram protein. The visible absorption spectrum (Fig. 4.16) is typical of other nonheme iron proteins, and the EPR spectrum indicates that it contains center FeS 2.

4. Nonheme Iron Protein of Complex II

In addition to FeS_{s-1} and FeS_{s-2} which are associated with the 70,000-dalton flavoprotein, succinate-coenzyme Q reductase contains a third center, FeS_{s-3}, with a midpoint potential of 60 mV. Davis and Hatefi have purified a soluble protein with a molecular weight of 27,000 that contains only trace amounts of flavin but has 110 nmoles of nonheme iron and 110 nmoles of acid-labile sulfur per milligram of protein. The 27,000-dalton protein is tightly complexed with the flavoprotein and appears to be essential for succinate dehydrogenase activity. The EPR spectrum of the purified protein indicates that it contains center FeS_{s-3}. The center is probably of the bacterial ferredoxin type with four irons and four sulfurs per center.

5. Nonheme Iron Protein of Complex III

This protein was purified from coenzyme QH_2-cytochrome *c* reductase by Rieske and co-workers. The purification depended on a succinylation of the protein which otherwise is insoluble in water because of its hydrophobic properties. The molecular weight has been estimated to be 26,000, and the nonheme iron and sulfur contents are 2 moles each per mole of protein. The EPR signal

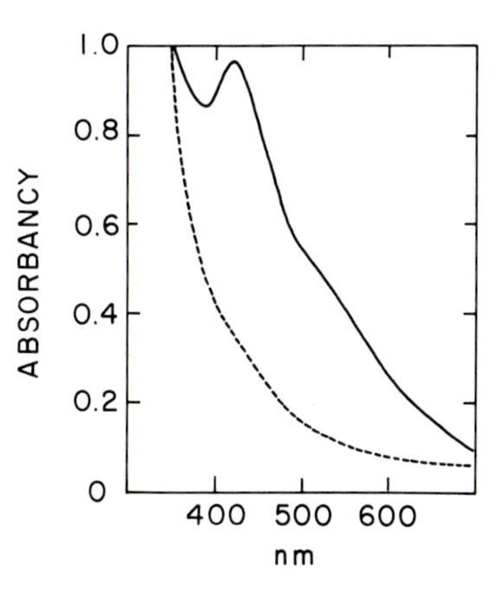

FIG. 4.16. Absorption spectrum of the nonheme iron protein purified from NADH-coenzyme Q reductase. Oxidized (——); reduced (----). (Hatefi, 1976.)

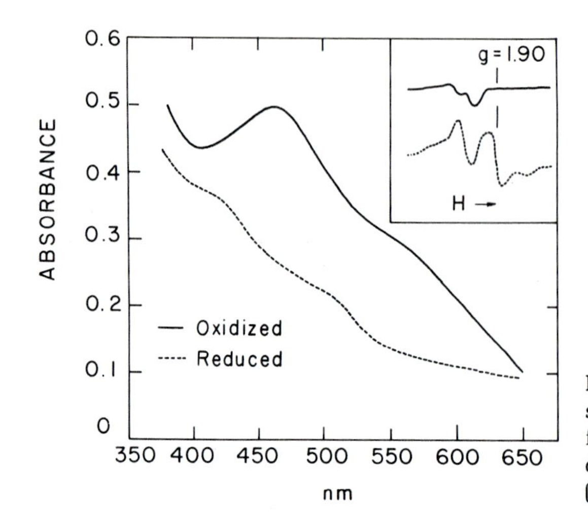

FIG. 4.17. Absorption spectrum and EPR signal of nonheme iron protein purified from coenzyme QH_2-cytochrome *c* reductase. Oxidized (——); reduced (----). (Rieske, 1976.)

of this FeS center is evident in whole mitochondria as well as isolated complex III, since its main signal occurs at g = 1.90, and is well separated from the other FeS centers which overlap at g = 1.94. In Fig. 4.17 are shown the EPR and visible spectra of the purified succinylated protein. Rieske and others have titrated the g = 1.94 signal and arrived at midpoint potentials of +280 mV in mitochondria and +230 mV in the isolated protein. This high electropositive value for the midpoint potential is unusual for iron–sulfur centers.

6. Copper Protein

Cytochrome oxidase contains copper in an amount roughly equivalent to the heme *a*. This copper undergoes oxidation–reduction changes with the same kinetics as the cytochrome components and is now accepted to have a function in the electron transfer mechanism of the enzyme. The copper of cytochrome oxidase has been correlated with a broad absorption band that is found in the infrared region of the visible spectrum. This band absorbs with a maximum at 830 nm in oxidized cytochrome oxidase and is bleached on reduction.

Beinert and co-workers have studied the EPR properties of the copper component in submitochondrial particles and in purified cytochrome oxidase. Both types of preparations show similar EPR spectra with signals at $g_\perp = 2.03$ and $g_\parallel = 2.27$ (Fig. 4.18). The EPR signals are lost when the copper is reduced to the cuprous state. An interesting aspect of the spectrum is that only 40–50% of the total copper can be accounted for by the EPR signal. Since all the copper is in the cupric state, this would suggest that the unpaired electron of one of the copper atoms is shared with some other group on the protein.

The position of the prominent EPR signal at g = 2.03 indicates that the unpaired electron is highly delocalized (like a free electron) and has been interpreted by Beinert as indicating a high extent of covalency in the binding of the copper to the protein. This is also supported by an absence of hyperfine structure in the $g_\parallel$ region of the spectrum (hyperfine structure is usually observed in inorganic copper complexes).

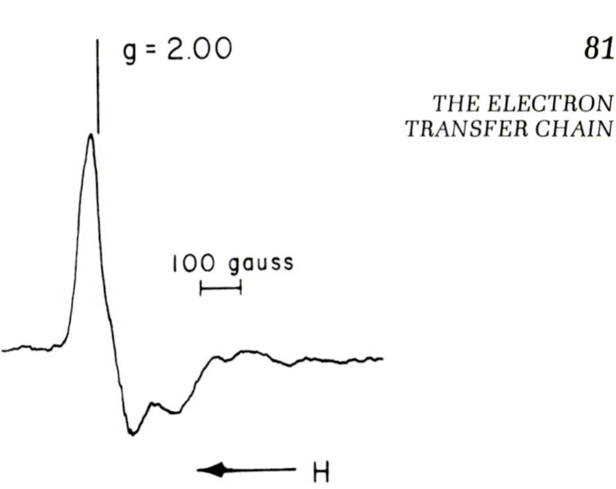

FIG. 4.18. Electron spin resonance signal of copper in beef-heart cytochrome oxidase. (Beinert and Palmer, 1965.)

MacLennan and Tzagoloff have purified a protein from cytochrome oxidase that has a high copper content (80 nmoles per milligram of protein) and does not contain heme *a*. The minimum molecuar weight of the protein based on the copper content is 12,500. Although this component could be a true copper protein of the enzyme, the evidence is still inconclusive, since the EPR spectrum of the copper in the isolated protein is modified and exhibits a hyperfine structure that is not seen in the signal of the native copper. More work is presently needed on this problem.

D. Low-Molecular-Weight Carriers

The respiratory chain contains two carriers that are not associated with proteins: nicotinamide adenine dinucleotide (NAD^+) and coenzyme Q. The NAD^+ links various primary dehydrogenases (Table 4.7) with complex I of the respiratory chain. The reduced nucleotide donates hydrogen to the FMN prosthetic group of the NADH dehydrogenase. Coenzyme Q is a lipid-soluble quinone that accepts hydrogen from different flavoenzymes (Table 4.7) and donates electrons to complex III. Both NAD^+ and coenzyme Q have been extensively characterized, and some of their properties will be reviewed.

TABLE 4.7. NAD- and Coenzyme Q-Linked Dehydrogenases of Mitochondria

Dehydrogenases that reduce NAD^+	Dehydrogenases that reduce coenzyme Q
α-Ketoglutarate dehydrogenase complex	NADH–coenzyme Q reductase
Pyruvate dehydrogenase complex	Succinate–coenzyme Q reductase
Malate dehydrogenase	α-Glycerophosphate dehydrogenase
Isocitrate dehydrogenase	Electron transfer flavoprotein (ETF)
Glutamate dehydrogenase	
β-Hydroxybutyrate dehydrogenase	
Hydroxyacyl-CoA dehydrogenase	

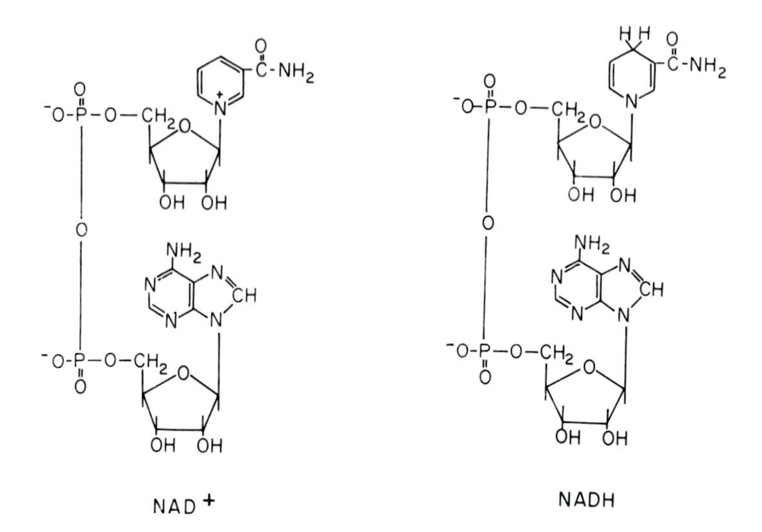

FIG. 4.19. Structural formulas of NAD^+ and NADH.

1. NAD^+

This is a water-soluble compound consisting of nicotinamide mononucleotide and adenylic acid covalently linked through a pyrophosphate bond (Fig. 4.19). NAD^+ has an absorption band at 260 nm. On reduction to NADH, the 260-nm absorption decreases, and a new band with a maximum at 340 nm appears (Fig. 4.20). The changes in absorbance at 340 nm provide a convenient

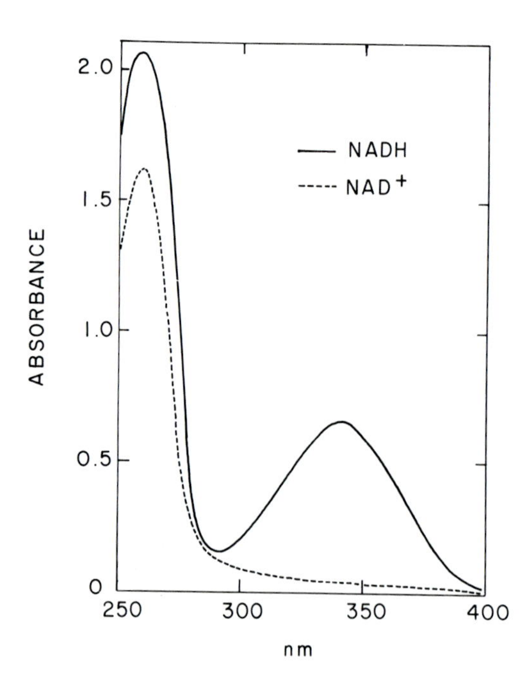

FIG. 4.20. Absorption spectra of NAD^+ and NADH.

means of following enzymatic reactions in which the nucleotide becomes either oxidized or reduced. Reduction of NAD^+ results in the addition of a hydrogen at position 4 of the pyridine ring in the nicotinamide moiety. The reduced carbon is asymmetric, and two isomers (α or β) can be formed. Almost all the known dehydrogenases carry out the reduction of NAD^+ in a stereospecific fashion and yield either the α or β stereoisomer of NADH.

Most mitochondria contain a tenfold excess of NAD over the other carriers of the chain, for example, cytochrome c (see Section VI). The NAD^+ is fixed in the internal matrix compartment and, under normal circumstances, does not leak out. A partial explanation for this lies in a lack of permeability of the nucleotide across the inner membrane. For this reason also, mitochondria do not oxidize externally added NADH. Mitochondria can be depleted of their endogenous NAD^+ after swelling or if they are incubated at elevated temperatures in the presence of phosphate. After these treatments, NAD-linked substrates such as malate, glutamate, and β-hydroxybutyrate are no longer oxidized. The internal pool of NAD^+ and the ability of mitochondria to oxidize these substrates are restored if the depleted particles are incubated with exogenous NAD^+ and ATP.

2. Coenzyme Q (Ubiquinone)

This compound was discovered independently in three different laboratories in three consecutive years (1955–1957). The isolation of the quinone from beef heart mitochondria by Crane and co-workers in 1957 had the greatest impact, since it was shown soon afterwards to be implicated in electron transfer. All coenzyme Qs have a common structure consisting of a 2,3-dimethoxy-5-methyl-1,4-benzoquinone with a polyisoprenoid side chain (Fig. 4.21). The naturally occuring coenzyme Qs differ in the length of the side chain which varies from six to ten isoprenoid units. These compounds are named according to the number of isoprenoid units; for example, coenzyme Q_{10} has a chain with ten isoprenoids. The alternate nomenclature adopted by Morton uses the name ubiquinone and designates the compound on the basis of the number of carbons in the side chain. Coenzyme Q_{10} by this convention is ubiquinone-50. Almost all mammalian mitochondria contain coenzyme Q_{10}; the exception to this is rat mitochondria which contain predominantly coenzyme Q_9.

Naturally occurring coenzyme Qs, because of the length of their isoprenoid side chains, are insoluble in water. There are synthetic analogues, however,

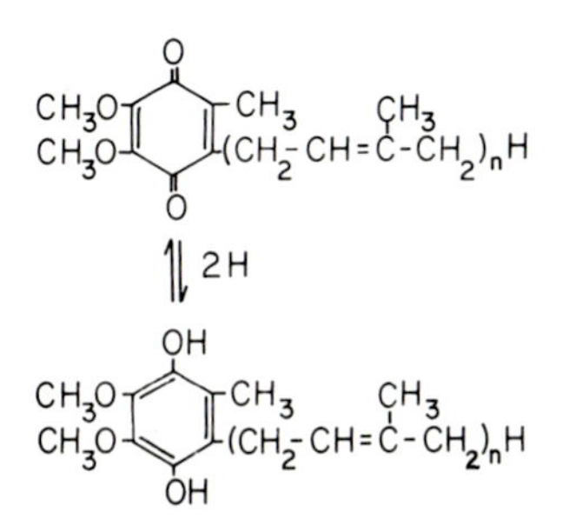

FIG. 4.21. Structural formulas of oxidized and reduced coenzyme Q.

with one or two isoprenoid units which are partially soluble in water and are quite useful for assaying various respiratory complexes for which coenzyme Q serves as an electron acceptor or donor.

Coenzyme Q has a yellow color in the oxidized form because of a visible absorption band at 405 nm. A second absorption band occurs in the ultraviolet region and has a maximum at 275 nm (Fig. 4.22). On reduction, the yellow color is lost, and the absorption spectrum shows a new band with a maximum at 290 nm.

There is now a substantial body of evidence indicating that coenzyme Q is an essential carrier of the electron transfer chain. The most convincing evidence comes from depletion and reconstitution experiments. Lester and Fleischer showed that extraction of coenzyme Q from mitochondria with isooctane leads to a loss of succinate oxidase activity in mitochondria. These authors further demonstrated that the activity is restored to near normal levels when coenzyme Q is added back to the extracted particles. A similar reconstitution was not possible for the NADH oxidase because of the denaturation of the NADH–coenzyme Q reductase complex by the organic solvent. In 1966, however, Szarkowska showed that it is possible to extract coenzyme Q from lyophilized mitochondria with pentane without denaturation of the NADH–coenzyme Q reductase. The ability of the particles to oxidize succinate and NADH was lost after the extraction, and both activities were restored with exogenously added coenzyme Q. This simple and conclusive experiment is shown in Table 4.8.

II. Redox Potentials of the Electron Transfer Chain Carriers

A traditional approach in studying the sequence of the electron transfer chain carriers has been to measure their midpoint or standard oxidation–reduction potentials which allows them to be placed on an electropotential scale. Information about the midpoint potentials of the carriers is also useful in establishing the spans that are capable of yielding sufficient energy for ATP synthesis. In recent years, there has been a resurgence of interest in such studies, particularly since many new carriers have been discovered (iron–sulfur centers, new species of cytochrome *b*) and improved methods have been devel-

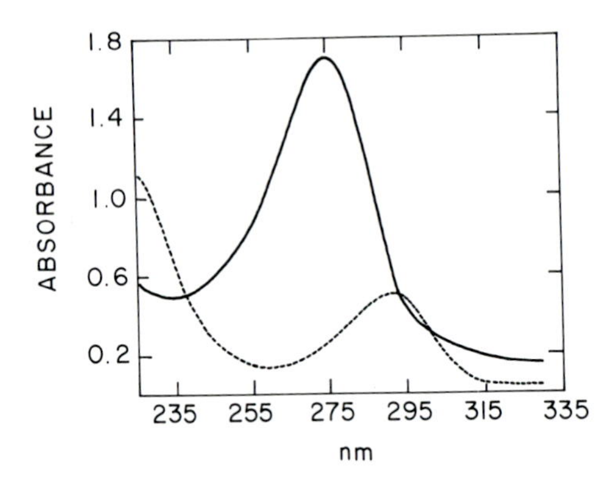

FIG. 4.22. Absorption spectra of coenzyme Q. Oxidized (——); reduced (-----).

oped for determining their oxidation–reduction states. Before considering the redox potentials of the mitochondrial carriers, it may be useful to describe how such measurements are made.

The electrical potential of a compound capable of accepting or donating electrons is always related to the electrical potential of the hydrogen couple

$$H^+ + 1e^- \rightleftharpoons \tfrac{1}{2}H_2 \qquad [4.1]$$

which is arbitrarily said to be 0 volts when the pressure of hydrogen gas is 1 atmosphere and the concentration of H^+ is 1 M (pH = 0). For a compound capable of undergoing an oxidation–reduction change,

$$C_{ox} + ne^- \rightleftharpoons C_{red} \qquad [4.2]$$

the electrical potential of the couple is defined by the Nernst equation,

$$E_h = E_0 + \frac{RT}{nF} \ln \frac{(C_{ox})}{(C_{red})} \qquad [4.3]$$

where R is the gas constant, T the absolute temperature, F the Faraday, and n the number of electrons involved in the reaction. At 25° this equation reduces to:

$$E_h = E_0 + \frac{0.059}{n} \log \frac{(C_{ox})}{(C_{red})} \qquad [4.4]$$

It should be noted that when the concentrations of the oxidized and reduced forms of the compound are equal to each other, $E_h = E_0$. The value of E_0 therefore, is referred to as the standard or midpoint potential and is a measure of the tendency of the compound to accept or donate electrons relative to the hydrogen couple. Oxidation–reduction couples with values of E_0 less than 0 have a stronger tendency than hydrogen to donate electrons and are stronger reduc-

TABLE 4.8. Restoration of Respiratory Activities of Pentane-Extracted Mitochondria with Coenzyme Q_{10}[a]

	Respiratory rates (nmoles O_2/min/mg protein)	
	NADH	Succinate
Unextracted mitochondria	389	450
Extracted mitochondria	9	10
Extracted mitochondria plus cytochrome *c* and phospholipids	10	12
Extracted mitochondria plus cytochrome *c*, phospholipids and coenzyme Q_{10}	404	430

[a]From Szarkowska (1966).

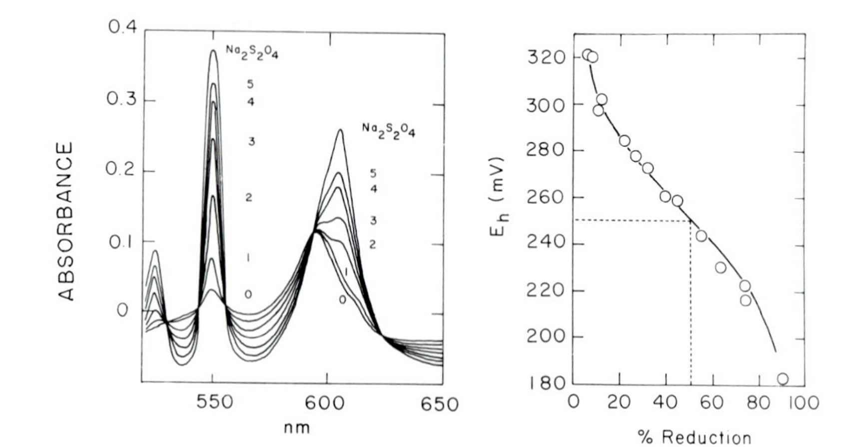

FIG. 4.23. Midpoint potential of cytochrome *a*. Left: A mixture of beef-heart cytochrome oxidase and cytochrome *c* was reacted with carbon monoxide in the absence of oxygen. Under these conditions, the enzyme consists of $a^{3+}a_3^{2+}$-CO (spectrum 0). Graded amounts of ascorbate were added and spectra recorded after a sufficient period of equilibration (spectra 1–5). The mixture was finally reduced completely with dithionite. Right: The percent reduction of cytochrome *a* and cytochrome *c* was measured after each addition of ascorbate. The percent reduction of cytochrome *a* is plotted on the abscissa, and the E_h calculated from the percent reduction of cytochrome *c* is plotted on the ordinate. The midpoint potential of cytochrome *a* corresponding to the E_h at 50% reduction is seen to be 250 mV. (Tzagoloff and Wharton, 1965.)

TABLE 4.9. Midpoint Potentials of Respiratory Carriers

Carrier	E_0, pH 7.2 (mV)
NADH	−320
FAD	−45 (?)
FMN	?
FeS 1	−305
FeS 2	−20
FeS 3	−20
FeS 4	−245
FeS_{s-1}	0 to +30
FeS_{s-2}	−260
FeS_{s-3}	+60 to +120
FeS_{III}	+220 to +280
Cu	+250 to +280
Cytochrome *a*	+200 to +250
Cytochrome a_3	+385
Cytochrome b_K	+30 to +65
Cytochrome b_T	−30 to −50
Cytochrome b_{558}	+120
Cytochrome *c*	+230 to +280
Cytochrome c_1	+225
Coenzyme Q_{10}	0 to +65
O_2	+820

tants. Conversely, couples with E_0 values greater than 0 have a greater tendency to accept electrons and are stronger oxidants.

Two general methods have been used to measure the redox potentials of the electron transfer carriers. In the older method, the reduction state of the carrier (usually determined by optical or EPR spectroscopy) is related to that of some other couple of known potential. It is most convenient if the reference couple can also be assayed spectrophotometrically and has an E_0 close to that of the unknown. An example of this method is shown in Fig. 4.23. Here, the oxidation–reduction state of the cytochrome *a* component of cytochrome oxidase is related to the cytochrome *c* couple. The series of spectra shown represent equilibrium states that are attained after graded additions of the reductant ascorbate. By plotting the observed percent reduction of cytochrome *a* versus the E_h of the cytochrome *c* couple (these values are obtained using the Nernst equation and substituting the known E_0 of cytochrome *c* and the ratios of reduced versus oxidized cytochrome at the different equilibrium states), the E_0 of cytochrome *a* is calculated as the E_h at which 50% of the cytochrome *a* is reduced. The E_0 from this experiment was estimated to be +250 mV.

The second method is a potentiometric one in which the oxidation–reduction state of the carrier is directly related to the electrical potential measured with a metal electrode. This method has been used successfully by Dutton and Wilson who have done extensive studies of the potentials of the various carriers in different energy states of mitochondria.

A list of the carriers and their estimated midpoint potentials is presented in Table 4.9. When plotted on an electropotential scale as shown in Fig. 4.24,

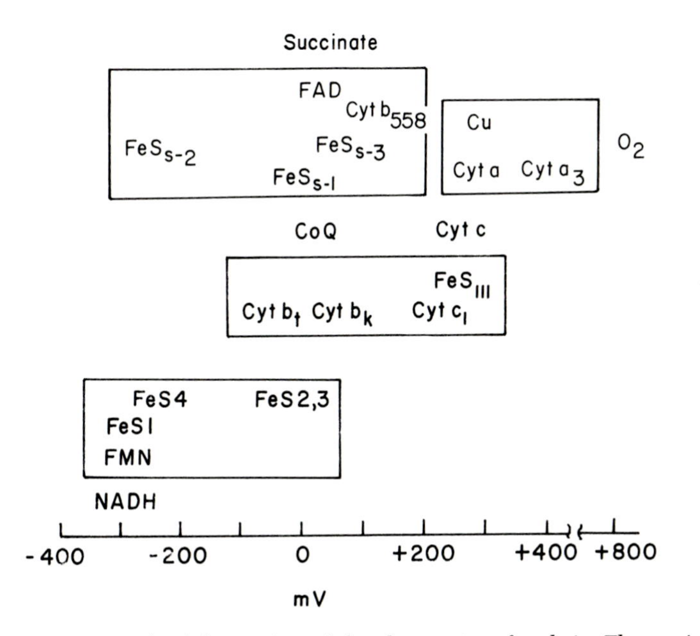

FIG. 4.24. Midpoint potentials of the carriers of the electron transfer chain. The carriers of each respiratory complex are enclosed by the solid lines.

the carriers are seen to fall into clusters that can be correlated with the respiratory complexes with which they are associated. As will be seen in the subsequent section, the sequence of carriers suggested by their midpoint potentials agrees reasonably well with the electron transfer chain sequence established on the basis of kinetic data and inhibitor studies.

There are several additional points to be made concerning the electropotential scale shown in Fig. 4.24. The difference in the midpoint potentials of the $NAD^+/NADH$ couple and the O_2/H_2O couple is equivalent to 1140 mV or 1.14 V. This potential drop can be translated to a free energy change (ΔG^0) by the formula, $\Delta G^0 = -nF\Delta E_0$, where n is the number of electrons involved in the reaction, F is the Faraday (23.062 kcal), and ΔE_0 is the difference in the midpoint potentials of the two couples. The ΔG^0 for the oxidation of NADH calculates to be $-2(23.062)(1.14) = -52.6$ kcal.

The standard free energy change for the esterification of the terminal pyrophosphate bond of ATP has been estimated to be 7.3 kcal. The oxidation of NADH, therefore, produces enough energy to synthesize 7.2 moles of ATP (52.6/7.3). Since it is known that NADH oxidation is coupled to the synthesis of only 3 moles of ATP per mole of NADH oxidized (see Chapter 3, Section VIB), the process has an efficiency of 40%.

Another interesting problem is concerned with the spans that yield sufficient energy for ATP synthesis. For a net transfer of two electrons, any two carrier couples that are separated by a ΔE_0 of 158 mV in theory generate enough energy to synthesize ATP. As can be seen from Fig. 4.24, there are a number of such spans. Let us consider the spans of just those reactions catalyzed by the respiratory complexes. The data summarized in Table 4.10 indicate that three of the four reactions, namely, those catalyzed by complexes I, III, and IV, have ΔG^0 greater than -7.3 kcal and are therefore capable on thermodynamic grounds of generating ATP. The reduction of coenzyme Q by succinate, on the other hand, falls short of the energy requirement, its ΔG^0 being only -2.9 kcal. In Chapter 6 we will see that in fact, the energetic considerations fit with the postulated coupling sites of oxidative phosphorylation determined by completely different means.

TABLE 4.10. Standard Free Energy Changes of the Reactions Catalyzed by the Respiratory Complexes

Complex	Reaction	ΔE_0 (mV)[a]	ΔG_0 (kcal)
I	NADH–CoQ	385	-17.7
II	Succinate–CoQ	65	-2.9
III	$CoQH_2$–cyt. c	195	-9.0
IV	Cyt. c–O_2	520	-23.9

[a]The following midpoint potentials were used to calculate the ΔE_0 of the reactions: NAD/NADH = -320 mV; succinate/fumarate = 0 mV; $CoQ/CoQH_2$ = $+65$ mV; cyt. c_{ox}/cyt. c_{red} = $+260$ mV; O_2/H_2O = $+820$ mV.

III. Sequence of Electron Transfer Carriers Based on Their Rates of Oxidation and Reduction

Kinetic measurements of the relative and absolute rates with which respiratory carriers become oxidized or reduced permit the sequence in which electrons are transferred from substrates to oxygen to be determined. This approach was pioneered in the laboratory of Chance who developed refined methods for the analysis of cytochromes and other spectroscopically detectable components. Later, the kinetic approach was used to study iron–sulfur centers by the EPR technique.

A. Cytochrome Chain

Figure 4.25 shows the difference spectrum of reduced versus oxidized mitochondria in the visible region. The following components of the chain can be discerned in the spectrum: cytochromes *a* and a_3 with their α bands at 605 nm and γ bands at 445 nm; cytochrome *b* with the α band as a shoulder at 564 nm; cytochromes *c* and c_1 with α bands at 550 nm and γ bands at 430 nm; NADH with a maximum at 340 nm; and flavins which appear as a broad trough in the 420- to 480-nm region. Taking advantage of the different spectral properties of the carriers, Chance and Williams were able to measure the rates of reduction and oxidation of flavin and cytochromes. The various carriers were measured at wavelengths corresponding to the maxima at which they absorb. In order to increase the sensitivity and accuracy of the method, Chance and

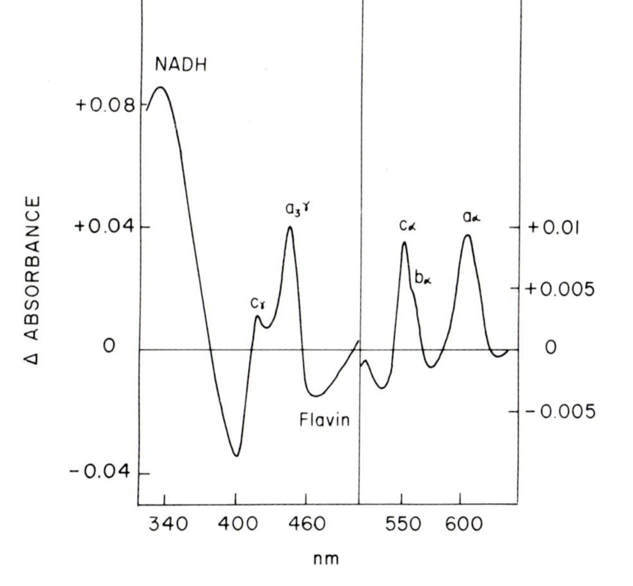

FIG. 4.25. Reduced-versus-oxidized difference spectrum of rat liver mitochondria. The mitochondria were reduced with substrate under anaerobic conditions. Note the difference in scale in the α and γ regions of the spectrum. (Chance and Williams, 1955.)

Williams recorded the changes in absorbancy with time against another nearby wavelength at which the carrier has nearly identical absorbance in the oxidized and reduced form (isosbestic wavelength). This eliminates the problem of spectral artifacts caused by changes in the optical properties of the mitochondria. For example, the difference spectrum of Fig. 4.25 shows that cytochrome *c* has an α band at 550 nm but that at 540 nm there is virtually no difference in absorbance in the spectrum of oxidized versus reduced mitochondria. The wavelength pair 550/540 nm, therefore, gives a suitable estimate of the state of reduction of cytochrome *c*. The wavelength pairs used for the different carriers are indicated in Table 4.11.

In establishing the sequence of the oxidation–reduction events, the assumption is made that under aerobic steady-state conditions, i.e., when the substrate is present, carriers closest to oxygen will be more oxidized relative to other carriers that are further removed along the chain. Similarly, if a pulse of oxygen is introduced into an anaerobic suspension of mitochondria that are fully reduced, the rates with which the carriers become oxidized will depend on how close they are to oxygen in the sequence. The results of the Chance and Williams experiment are shown in Table 4.11. From these data it was concluded that the *a* cytochromes are closest to oxygen, followed by cytochrome *c* and then cytochrome *b*. At the time the experiments were done, there was a problem in establishing the position of cytochrome c_1 because of the similarity of its spectrum to cytochrome *c*. By extracting most of the cytochrome *c* from mitochondria, Chance overcame this problem and was able to locate cytochrome c_1 between cytochromes *b* and *c*.

There are at least three spectral species of cytochrome *b* in the respiratory chain. Considerable work has been done in recent years to establish where these newly discovered carriers fit in the electron transfer sequence. In oxygen pulse experiments analogous to those described above, it has been shown that cytochrome b_K becomes more rapidly oxidized than cytochrome b_T. This, however, is only true of well-coupled mitochondria. In damaged mitochondria or in submitochondrial particles, the kinetics of oxidation of both components are the same. There obviously are still uncertainties in the placement of these

TABLE 4.11. Rates of Oxidation or Reduction of Cytochromes and Flavin in Rat Liver Mitochondria[a]

	Carrier					
	a_3	*a*	*c*	*b*	Flavin	NAD
Rate of reduction (μM/sec) of carrier at 4° after addition of substrate	—	—	0.015	0.019	0.07	0.064
Rate of oxidation of carrier (sec^{-1}) after addition of oxygen pulse to anaerobic mitochondria	158	120	39	20	13	—
Wavelength pairs used	445 455	605 630	550 540	564 575	465 510	340 374

[a]After Chance and Williams (1955).

cytochromes, but it is presently assumed that at least in coupled mitochondria, cytochrome b_K is on the oxygen side of b_T. The third component, cytochrome b_{558}, appears to be part of the succinate-coenzyme Q reductase complex, and its status is not clear at present. It has a low midpoint potential, is not reducible by succinate, and probably does not function in the main respiratory pathways for succinate or NADH oxidation.

The sequence of the cytochrome carriers as established by kinetic data can be written as follows:

$$\text{NADH} \rightarrow \text{flavin} \rightarrow \text{cytochrome } b_T \rightarrow b_K \rightarrow c_1 \rightarrow c \rightarrow a \rightarrow a_3 \rightarrow O_2 \quad [4.5]$$

B. Placement of the Iron–Sulfur Centers

These carriers have been studied by EPR spectroscopy in different laboratories, and in some cases it has been possible to locate them along the main sequence of the chain on the basis of kinetic and titration (redox potential) data. Orme-Johnson and Beinert have examined the centers of NADH-coenzyme Q reductase both in the isolated complex and in submitochondrial particles. From the relative rates of reduction of the centers associated with this segment of the chain, as well as their equilibrium levels of reduction in titration experiments, the deduced order is:

$$\text{FMN} \rightarrow \text{FeS 1} \rightarrow \text{FeS 4} \rightarrow \text{FeS 3} \rightarrow \text{FeS 2} \quad [4.6]$$

The kinetic behavior of the iron–sulfur centers of succinate-coenzyme Q reductase have been examined by Beinert and others. It has not proven possible to see any significant differences in the rates of reduction of FeS_{s-3} and FAD, suggesting that the two are in rapid electronic equilibrium. The essentiality of FeS_{s-3}, however, is supported by the experiments of Hatefi and of Ohnishi and King who observed that in the soluble succinate dehydrogenase, the absence or destruction of the center is correlated with a loss in the ability of the enzyme to reconstitute succinate oxidase of alkali-extracted particles (see Section I). The roles of FeS_{s-1} and FeS_{s-2} are even more ambiguous. These centers are only poorly reduced by succinate. The highly electronegative potential of FeS_{s-1} probably excludes this component from being on the main pathway of the succinate oxidase chain.

The iron–sulfur center of coenzyme QH_2–cytochrome *c* reductase cannot be distinguished from cytochrome c_1 in titration experiments. This is expected in view of their redox potentials which are nearly identical. From kinetic data, Rieske concluded that FeS_{III} is on the oxygen side of cytochrome *b* but could not resolve it from cytochrome c_1. It is therefore still not known if this component is on the main electron transfer pathway.

C. Placement of Copper

Van Gelder and Beinert have done an extensive study of the copper component of cytochrome oxidase. Both kinetic and titration experiments indicate that there are two distinct coppers in the enzyme. One has the EPR signal at g = 2.03, whereas the other copper is not detectable by EPR. The EPR-silent copper accepts reducing equivalents first and titrates in parallel with cytochrome *a*. The copper with the EPR signal, on the other hand, is reduced with the same kinetics as cytochrome a_3. Each copper has been interpreted to be in close electronic equilibrium with one of the two cytochromes.

D. Placement of Coenzyme Q

Early evidence from reconstitution experiments indicated that coenzyme Q is an obligatory carrier on the main pathway of the respiratory chain. This was supported by kinetic data from Klingenberg's laboratory. In a Chance and Williams type of experiment, Klingenberg was able to show that the rates of reduction of coenzyme Q or the rates of its oxidation after an oxygen pulse were compatible with a function in electron transport. Furthermore, from the determined kinetic constants, coenzyme Q was found to function on the substrate side of the cytochromes.

E. Deduced Sequence of Electron Transfer Carriers Based on Their Kinetic Properties

The sum total of the evidence discussed above permits a more or less complete sequence of the carriers to be presented. This is shown in Fig. 4.26. Those components whose roles in electron transfer are still to be assessed are placed in equilibrium with components that have similar midpoint potentials. The relationship of the carriers to the electron transfer complexes is also indicated in this scheme.

IV. Inhibitors of Electron Transfer

There are a number of reagents that inhibit electron transfer at specific sites of the chain. These have been very useful for functional dissections of the

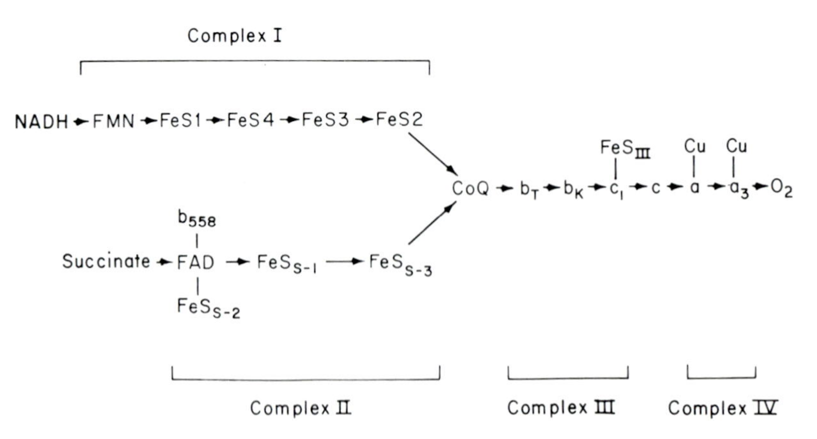

FIG. 4.26. The electron transfer chain. (After Hatefi, 1976.)

$SH_2 \rightarrow a_{red} \rightarrow b_{red} \rightarrow c_{red} \rightarrow d_{red} \rightarrow O_2$	All carriers partially or completely reduced
Inhibitor (between b_{red} and c_{ox}) $SH_2 \rightarrow a_{red} \rightarrow b_{red} \rightarrow c_{ox} \rightarrow d_{ox} \rightarrow O_2$	Only carriers a and b are reduced
$SH_2 \rightarrow a_{red} \rightarrow b_{red} \rightarrow c_{red} \rightarrow d_{red}$ (anaerobic) Inhibitor (between b_{red} and c_{ox}) $SH_2 \rightarrow a_{red} \rightarrow b_{red} \rightarrow c_{ox} \rightarrow d_{ox} \rightarrow O_2$	Carriers c and d become oxidized after oxygen pulse

FIG. 4.27. Effect of an electron transfer inhibitor on the oxidation–reduction state of a hypothetical sequence of carriers.

chain and also for studying the sequence of the carriers. The use of inhibitors for sequence studies is illustrated in Fig. 4.27. Here, a hypothetical oxidation–reduction pathway is blocked by the inhibitor at a site located between two carriers, B and C. In such a sequence, the addition of substrate under aerobic conditions should result in the reduction of components A and B but not C and D. Similarly, if the chain is reduced with the substrate under anaerobic conditions, followed by the addition of inhibitor, an oxygen pulse should cause components C and D to become oxidized and A and B to remain reduced.

An actual example is shown in Fig. 4.28 in which the spectra of mitochondria are compared after reduction with substrate alone and in the presence of

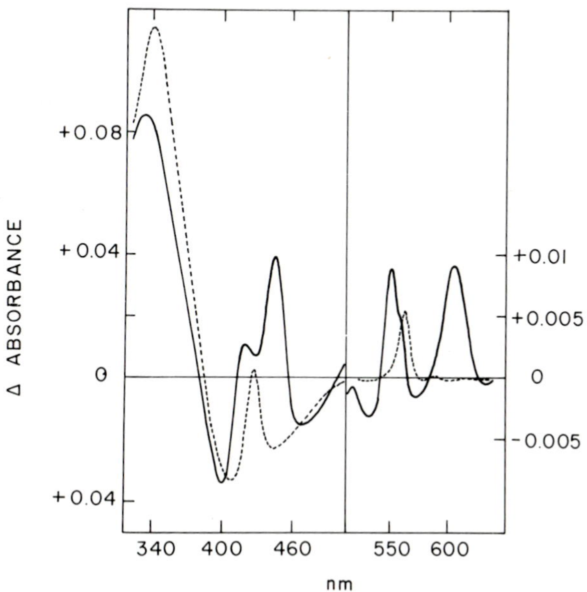

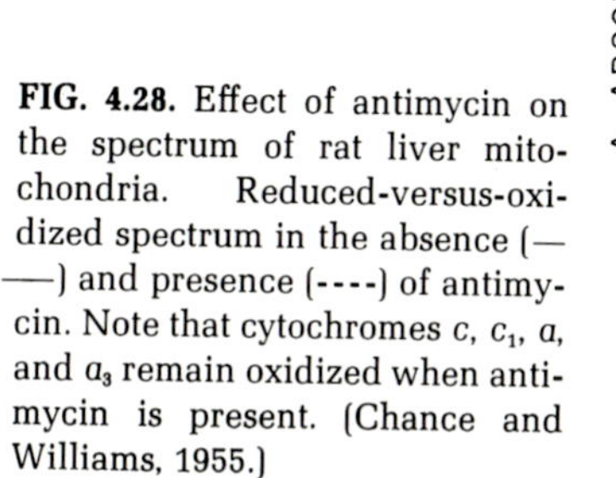
FIG. 4.28. Effect of antimycin on the spectrum of rat liver mitochondria. Reduced-versus-oxidized spectrum in the absence (——) and presence (----) of antimycin. Note that cytochromes c, c_1, a, and a_3 remain oxidized when antimycin is present. (Chance and Williams, 1955.)

the electron transfer inhibitor antimycin. It is seen that when the antimycin block is introduced, some of the carriers, namely, cytochromes *c*, c_1, *a*, and a_3, remain oxidized while flavin and cytochrome *b* are reduced. Thus, the antibiotic dissects the chain into two groups of carriers, those that are on the oxygen side of cytochrome b (*c*, c_1, *a*, and a_3) and flavin which is on the substrate side.

Some of the best-studied inhibitors of respiration and their effects on the reducibility of the electron transfer chain carriers in the NADH → O_2 branch are listed in Table 4.12. Since these inhibitors will be mentioned again in other chapters, it may be worthwhile to review some of their properties and known sites of action.

A. Cyanide and Carbon Monoxide

Both of these compounds are inhibitors of cytochrome oxidase and have long been known to react specifically with cytochrome a_3, the terminal carrier of the chain. Cyanide and carbon monoxide exert their inhibitory effect by competing with oxygen for the sixth coordinate of the heme *a* iron.

B. Antimycin A

This antibiotic (Fig. 4.29) is a very potent inhibitor of the coenzyme QH_2–cytochrome *c* reductase. Complete inhibition of enzymatic activity in mitochondria and in the purified complex is observed at concentrations of antimycin that are stoichiometric with cytochrome c_1. In the presence of the inhibitor, the cytochrome b_K absorption bands are shifted towards the red by about 2 nm, and its reducibility by substrate is enhanced. Both cytochrome *b* and the g = 1.90 iron–sulfur center have been considered possible targets of antimycin, but

TABLE 4.12. Reducibility by NADH of Electron Transfer Carriers in the Presence of Various Inhibitors

	Inhibitor		
Carrier	Rotenone	Antimycin	Cyanide
FAD	Yes	Yes	Yes
FeS 1	Yes	Yes	Yes
FeS 2	Yes	Yes	Yes
FeS 3	Yes	Yes	Yes
FeS 4	Yes	Yes	Yes
CoQ	No	Yes	Yes
Cyt b_T	No	Yes	Yes
Cyt b_K	No	Yes	Yes
FeS_{III}	No	Slow	Yes
Cyt c_1	No	No	Yes
Cyt *c*	No	No	Yes
Cyt *a*	No	No	Yes
Cyt a_3	No	No	Yes[a]

[a]Modified spectrum.

the evidence is only circumstantial. The site of inhibition has been studied by Rieske and co-workers who have synthesized a deformoamido derivative of antimycin and used it as an affinity label. The radioactive compound was found to bind covalently to a 12,500-dalton polypeptide in complex III. Since pretreatment of complex III with antimycin abolished the subsequent reaction of the affinity label, the participation of the low-molecular-weight protein in the binding is strongly indicated by these experiments. Whether or not this protein has any electron carrier function is not known at present.

C. Rotenone, Piericidin, Amobarbital

These antibiotics and barbiturates inhibit the NADH–coenzyme Q reductase segment. Rotenone (Fig. 4.29) has been most extensively used, but all three are probably closely related in terms of their sites of action. This is supported by the observation that the binding of rotenone is reduced in the presence of either amytal or piericidin. Palmer has shown that all three inhibitors probably act between the iron–sulfur centers of complex I and coenzyme Q. The EPR signals at $g = 1.94$ corresponding to the centers of complex I are reducible by substrate to 75% of their normal levels even when the enzymatic activity is more than 99% inhibited. It is interesting to note that the structure of piericidin is similar to that of coenzyme Q. This inhibitor binds very tightly to mitochondria, probably at the site where coenzyme Q normally reacts.

V. Properties of the Respiratory Complexes

The preceding discussion of the chain has dealt with the individual carriers, their functional groups, and the sequence of oxidation–reduction reactions that occur during electron transfer. Most of this information has been obtained from studies of complex systems such as mitochondria, submitochon-

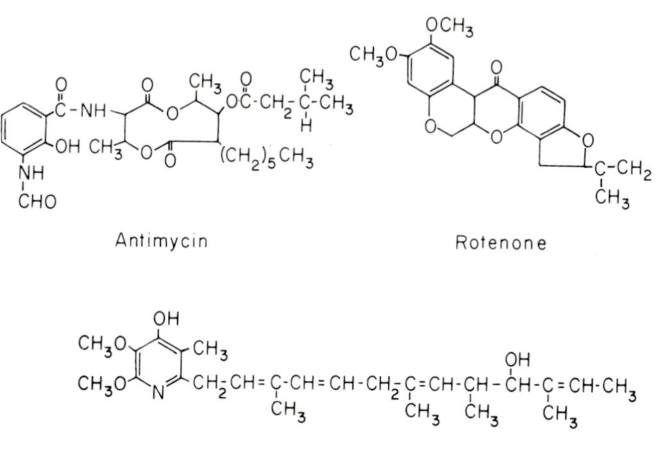

FIG. 4.29. Structural formulas of electron transfer inhibitors.

drial particles, and in some instances even whole cells. It was, nonetheless, recognized that like other metabolic pathways, the respiratory chain must be composed of enzymes whose physical and catalytic properties could be studied once they were isolated. This proved to be very difficult because the proteins involved were highly insoluble in water and classical procedures of protein fractionation could not be used in most instances. By the 1960s, however, there was a substantial backlog of practical experience in dealing with membrane lipoproteins, and Hatefi and co-workers succeeded in purifying four respiratory complexes that appeared on the basis of their properties to be natural segments of the electron transfer chain.

There are two extreme situations one can envision of how the electron transfer carriers are organized in the membrane. On the one hand, we can think of the chain as a composite of protein carriers, each being a self-contained unit of enzymatic activity. According to this view, it is not necessary to postulate any specific physical associations of the enzymes except that they all be present in the membrane where by virtue of spatial proximity and translational motion, electron transport can proceed at an efficient rate. At the other extreme, the chain can be visualized as a highly integrated macromolecular assembly in which the ability of any single carrier to function as an electron transfer catalyst depends on its physical association with the other proteins and carriers of the chain. In fact, we know that the real situation is somewhere in between. The chain has been separated into four complexes, each with a well-defined set of carriers and capable of carrying out part of the overall electron transfer process. All evidence at present points to the respiratory complexes as being the minimal functional units of electron transfer since further fragmentation invariably leads to a loss of enzymatic activity and modification of the catalytic properties of the individual carriers. This is illustrated with cytochrome oxidase which contains the carriers cytochromes a and a_3 and copper. The purified enzyme catalyzes the oxidation of ferrocytochrome c by oxygen, a reaction that can also be demonstrated in whole mitochondria. Attempts at further fractionation of this respiratory complex have been totally unsuccessful and have resulted in loss of enzymatic activity and modification of the cyto-

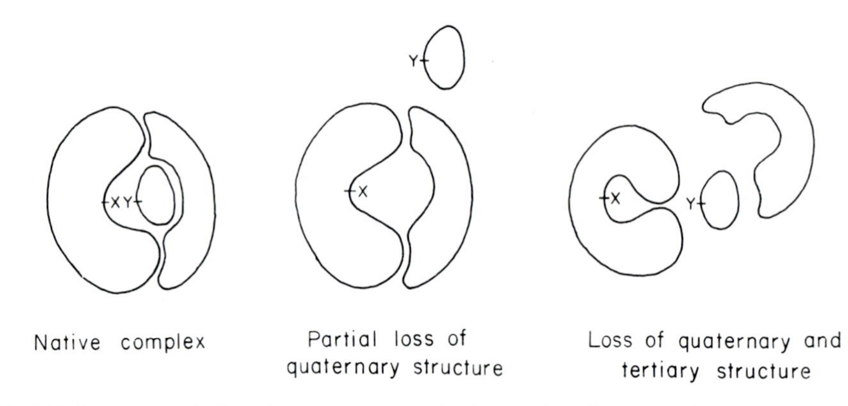

FIG. 4.30. Requirement of quaternary structure for interaction of two prosthetic groups in a multiprotein enzyme complex.

chromes. The catalytic properties of the cytochromes a and a_3 carriers, therefore, appear to depend on the highly ordered arrangement in which they exist in the complex.

Some examples of the way the catalytic function of two carriers could be determined by their association in a complex is shown in Fig. 4.30. The first example shows the requirement of quaternary structure for a proper juxtaposition of prosthetic groups located in two different carrier proteins. Disruption of the quaternary structure leads to a new arrangement of the carrier proteins such that the prosthetic groups are no longer capable of interacting. In the second example, the quaternary structure of the complex assures a proper tertiary structure of the proteins. Separation of the carriers from neighboring proteins in the complex causes a new tertiary structure in which the prosthetic groups become masked.

A. Purification of Respiratory Complexes

The respiratory complexes are integral constituents of the mitochondrial inner membrane, and their solubilization and purification depend on the use of detergents or surface-active reagents that disrupt the hydrophobic protein–protein and protein–lipid interactions of the membrane. Ideally, the detergent and conditions under which it is used should release the enzyme from the membrane but should not affect the internal structure of the complex. The

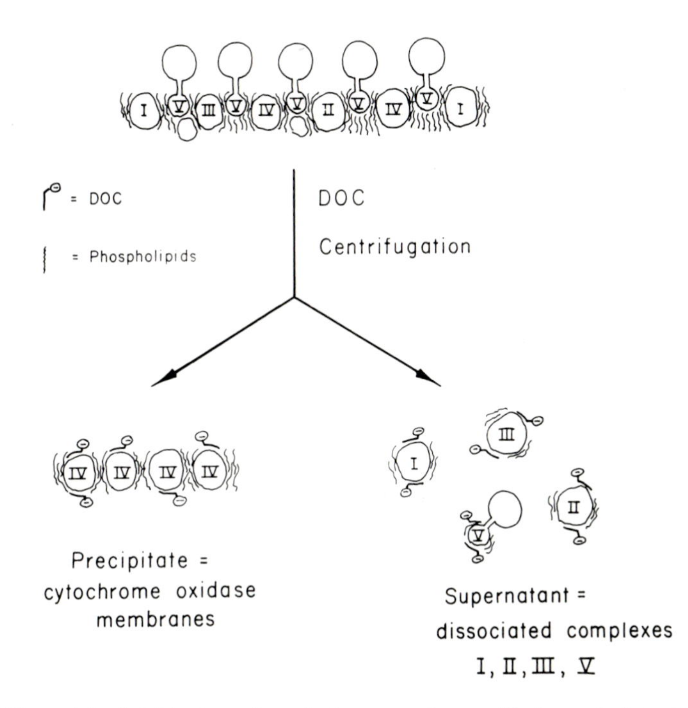

FIG. 4.31. Differential solubilization of respiratory complexes with deoxycholate. At low concentrations of DOC, complexes I, II, III, and V dissociate from the membrane, whereas complex IV remains in a membrane form. See also fractionation scheme of Fig. 4.32.

choice of detergent is therefore of critical importance and is usually governed by the empirical criterion of maximal solubilization and minimal inactivation. In the case of the mitochondrial respiratory complexes, the most useful solubilizing agents have turned out to be the two bile acids, cholate and deoxycholate (DOC).

All detergents are amphipathic molecules; i.e., they exhibit both water and lipid solubility and tend to orient themselves at a water–organic solvent interphase. The amphipathic properties of bile acids arise from the asymmetric distribution of the hydroxyl and carboxylic groups which are on one side of the molecule and give it a polar character while the other side is highly apolar. These reagents intercalate with the membrane lipid and protein constituents and in the process make them soluble in water. There are probably several factors involved in the solubilization of membrane enzymes. By coating the apolar regions of the proteins, bile acids decrease the hydrophobic interactions that hold the membrane together. At the same time, the bile acids because of their carboxylic group confer a net negative charge on the proteins which will tend to disaggregate the membrane because of charge-repulsion forces. The trick most often employed in the purification of membrane enzymes is to find a detergent that at a certain concentration effects a limited solubilization of only certain components (Fig. 4.31) which can then be further purified by salt fractionation and other methods.

A flow sheet summarizing the procedures that have been used to purify the complexes of the inner membrane is shown in Fig. 4.32. Mitochondria or submitochondrial particles are first treated with a low concentration of deoxycholate to solubilize respiratory complexes I, II, III, and the ATPase. Cytochrome oxidase remains associated with the insoluble membrane residue and is purified by solubilization at higher deoxycholate concentrations and salt fraction-

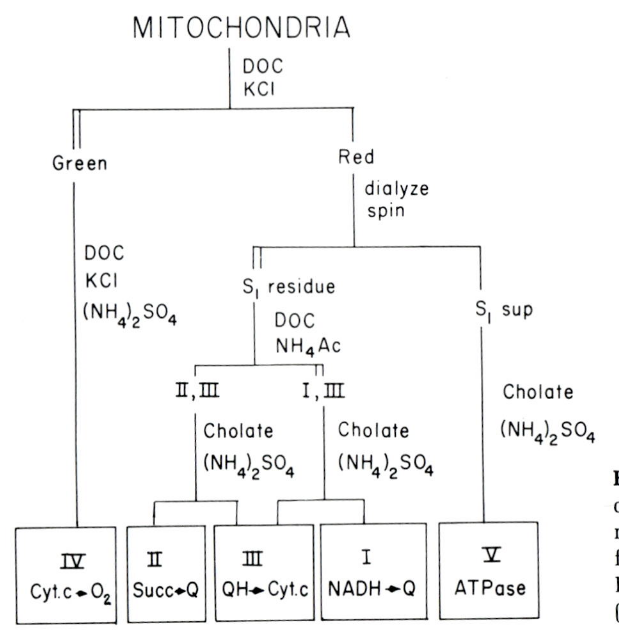

FIG. 4.32. Summary of the procedures used to purify the four respiratory and ATPase complexes from beef heart mitochondria. Double lines indicate precipitates. (Hatefi, 1976.)

ation. Complexes I, II, and III are separated from the ATPase after partial removal of the bile acid from the initial extract. This leads to a preferential aggregation of the respiratory complexes which are further purified by repeated solubilization with deoxycholate or cholate and precipitation with ammonium sulfate and ammonium acetate. The purified complexes form optically clear solutions as long as there is bile acid present but aggregate readily if the detergent is removed.

B. Compositions and Enzymatic Properties of the Respiratory Complexes

The respiratory complexes function both as catalysts of the electron transfer chain and as structural elements of the mitochondrial inner membrane. From the standpoint of their catalytic functions, each complex is different, having a unique set of carriers and capable of catalyzing a part of the overall process. As structural units, however, they share certain common properties, the most important being the ability to combine with phospholipids and to form membranes. Our discussion of the complexes here will be restricted to those aspects that are relevant to their electron transfer activity. The structural aspect will be dealt with in later chapters when we consider in more detail the cytochrome oxidase and ATPase complexes.

1. NADH-Coenzyme Q Reductase (Complex I)

This enzyme consists of some ten different subunit proteins some of which have been identified with the NADH dehydrogenase and nonheme iron protein. In addition, the complex contains coenzyme Q and about 20% by weight phospholipids. The contents of the various prosthetic groups are presented in Table 4.13. The nonheme iron of the complex can be accounted for by the four iron-sulfur centers FeS 1, 2, 3, and 4 which have been estimated to be present in about equimolar amounts. The stoichiometry of the carriers, assuming that the centers are of the bacterial ferredoxin type with four irons and four sulfurs per center, is 1 FMN/1FeS 1/1FeS 2/1FeS 3/1FeS 4.

Purified NADH-coenzyme Q reductase catalyzes the reduction of coenzyme Q_1 and Q_6 with high specific activities (Table 4.13). The natural homologue, coenzyme Q_{10}, is also reduced but only at a relatively slow rate—this has been attributed to assay problems caused by the highly lipophilic properties of coenzyme Q_{10}. Of particular significance is the fact that the reduction of coenzyme Q_1 and Q_6 is completely sensitive to rotenone and amobarbital which also inhibit coenzyme Q reduction in mitochondrial and submitochondrial particles. The preservation of the sensitivity to these inhibitors is a strong indication that the essential catalytic properties of the enzyme have not been modified during the course of the isolation. The complex also reduces ferricyanide at a high rate, but dyes such as dichloroindophenol and menadione are poor acceptors. The reduction of these artificial acceptors is not sensitive to rotenone or amobarbital, probably because they bypass the site at which the inhibitors block electron transport.

In recent studies, Hatefi and co-workers have found that complex I has two

TABLE 4.13. Composition and Enzymatic Properties of Complex I[a]

Component	Concentration (per mg protein)	
FMN	1.4–1.5 nmoles	
Nonheme iron	23–26 ng-atoms	
Acid-labile sulfur	23–26 nmoles	
Coenzyme Q_{10}	4.2–4.5 nmoles	
Phospholipids	0.22 mg	
Reaction	**Specific activity (μmoles/min per mg)**	**Inhibition by rotenone or amobarbital**
NADH → coenzyme O_1	25	yes
NADH → coenzyme Q_6	21.5	yes
NADH → $K_3Fe(CN)_6$	685	no
NADH → menadione	1.9	no
NADH → 2,6-dichloroindophenol	1.5	no
NADPH → coenzyme Q_1	yes[b]	yes
NADPH → $K_3Fe(CN)_6$	0.9	no

[a]From Hatefi and Stempel (1969).
[b]Specific activity not reported.

other activities which had not been detected before, (1) a transhydrogenation reaction in which the hydrogen of NADPH is transferred to NAD^+ and (2) an oxidation of NADPH to $NADP^+$. Both reactions have a similar pH optimum which is different from the optimum pH for NADH oxidation. In both cases also, it is the 4β hydrogen of NADPH that is removed. These findings suggest that a common component may be involved in the transhydrogenation and dehydrogenation reactions. It is also of interest that the oxidation of NADPH appears to proceed through some of the same carriers that participate in the oxidation of NADH. For example, the iron–sulfur centers FeS 2 and FeS 3 but not FeS 1 and FeS 4 are reduced by NADPH.

Complex I has been resolved into three fractions: (1) a soluble NADH

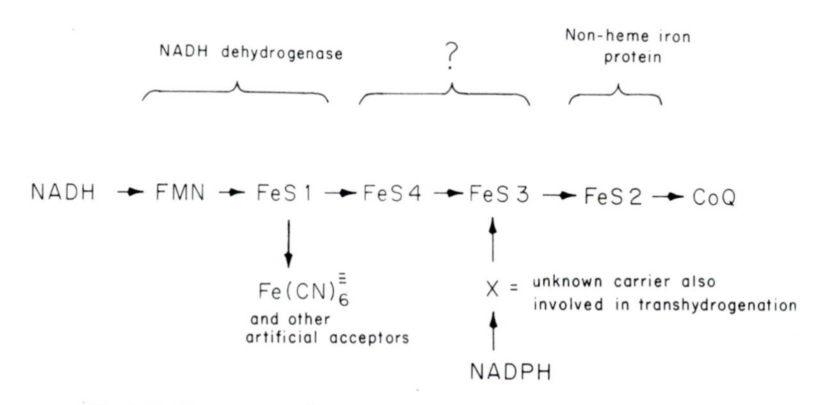

FIG. 4.33. Electron transfer activities of NADH–coenzyme Q reductase.

dehydrogenase of molecular weight 70,000, (2) a soluble nonheme iron protein with center FeS 2, and (3) a crude particulate fraction with centers FeS 3 and FeS 4. The properties of these components have already been described in Section I of this chapter. Based on the information currently available, the various activities of complex I may be related to its known carriers as shown in Fig. 4.33.

2. Succinate–Coenzyme Q Reductase (Complex II)

Although this enzyme is considered a respiratory complex, it should be recalled that it is responsible for catalyzing step 6 of the TCA cycle (see Chapter 3, Section VA). The purified complex consists of 7–8 nonidentical subunit proteins whose molecular weights range from 70,000 to 12,000.

The concentration of the various oxidation–reduction groups in, and the enzymatic reactions catalyzed by complex II are summarized in Table 4.14. In addition to flavin and nonheme iron, complex II contains cytochrome b_{558} whose function is not known at present. Based on the currently believed number of irons in the different FeS centers and the nonheme iron content of the enzyme, the stoichiometry of FAD:FeS_{s-1}:FeS_{s-2}:FeS_{s-3} is probably 1:1:1:1.

Complex II catalyzes the reduction of various artificial acceptors as does the soluble succinate dehydrogenase isolated from the complex. In contrast to the dehydrogenase, however, complex II is also capable of reducing coenzyme Q_1, Q_2, and higher analogues. The reduction of coenzyme Q is inhibited by 2-thenoyltrifluoroacetone (TTFA), an iron-chelating agent. This inhibitor probably acts on some FeS center that is involved in the reduction of coenzyme Q. Since the reduction of artificial acceptors by complex II or by succinate dehydrogenase is not affected by TTFA, these compounds probably become reduced by some carrier group that is on the succinate side of the block.

TABLE 4.14. Composition and Enzymatic Properties of Complex II[a]

Component	Concentration (per mg protein)
FAD	4.5–5.0 nmoles
Nonheme iron	32–40 ng-atoms
Acid-labile sulfur	32–40 nmoles
Cytochrome b_{558}	4.5–5.0 nmoles
Phospholipids	0.2 mg
Reaction	**Specific activity (μmoles/min per mg)**
Succinate → coenzyme Q_2	56
Succinate → coenzyme Q_{10}	12
Succinate → $K_3Fe(CN)_6$	48
Succinate → phenazine methosulfate	5
Succinate → 2,6-dichloroindophenol	24

[a]From Ziegler and Doeg (1962) and Baginsky and Hatefi (1969).

A soluble and "reconstitutively" active succinate dehydrogenase has been purified from complex II by extraction with the chaotrope $NaClO_4$. The pure succinate dehydrogenase contains FAD and the iron–sulfur centers but does not contain cytochrome b_{558}. It consists of two subunits, one a flavoprotein with covalently bound FAD and centers FeS_{s-1} and FeS_{s-2} and a nonheme iron protein with only FeS_{s-3}. The properties of these proteins have been described in Section I.

3. Coenzyme QH_2-Cytochrome *c* Reductase (Complex III)

This respiratory complex catalyzes the reduction of cytochrome *c* by coenzyme QH_2 in an antimycin-sensitive fashion. The same reaction has been shown to occur in mitochondria and submitochondrial particles. The purified complex has been determined to have a molecular weight of 260,000–280,000 by light scattering which is in reasonably good agreement with the minimum molecular weight based on the content of cytochrome c_1 as well as its size in the electron microscope. Negatively stained preparations of complex III indicate that it is a spherical particle with a diameter of 80–90 Å.

The composition of complex III is indicated in Table 4.15. It contains cytochrome *b*, cytochrome c_1, and nonheme iron in a 2:1:2 stoichiometry. Only 50% of the cytochrome *b* has been found to be reducible by coenzyme QH_2. The component reduced by substrate has an α band maximum at 562 nm and has been identified as cytochrome b_K. The remaining cytochrome *b* absorbs light at 565 nm and is reducible by dithionite but not by substrate and is probably cytochrome b_T. Since the two *b* cytochromes are present in equal amounts, the stoichiometry of the carriers in the complex is probably 1 cytochrome b_K:1 cytochrome b_T:2 FeS_{III}.

In addition to the above carriers, complex III contains other proteins whose

TABLE 4.15. Composition and Enzymatic Properties of Complex III[a]

Component	Concentration (per mg protein)	
Cytochrome *b*	8.2 nmoles	
Cytochrome c_1	4.1 nmoles	
Nonheme iron	8.2 ng-atoms	
Acid-labile sulfur	8.2 nmoles	
Coenzyme Q_{10}	1 nmole	
Phospholipids	0.16 mg	
Reaction	**Specific activity (μmoles/min per mg)**	**Inhibition by antimycin**
Reduced coenzyme Q_1 → cytochrome *c*	90–180	yes
Reduced coenzyme Q_2 → cytochrome *c*	300–600	yes
Reduced coenzyme Q_{10} → cytochrome *c*	30–60	yes

[a]Adapted from Rieske (1976).

functions are yet to be determined. Almost 50% of the mass of the complex is contributed by two proteins (core proteins) with no detectable prosthetic groups. Also present in the complex are some low-molecular-weight proteins, one of which has been found to bind antimycin. Several laboratories have attempted to relate the different components of complex III to the subunit proteins that are seen in polyacrylamide gels of the dissociated complex. Although there are inconsistencies in some of the results, most of the subunits of the complex have been identified. The subunit composition of the enzyme is summarized in Table 4.16. It should be noted that the identification of cytochrome *b* with a 15,000-dalton subunit is probably incorrect. The gene sequences of the human, bovine, and yeast cytochrome *b*s indicate the true molecular weight to be 44,000 (see Chapter 11, Section VIIC). Moreover, recent genetic and biochemical data suggest that b_K and b_T are associated with a single polypeptide chain with two separate heme-binding sites. The spectral and potentiometric differences attributed to the two cytochromes can best be rationalized in terms of different environments conferred on the hemes by the same protein.

Antimycin completely inhibits the enzymatic activity of coenzyme QH_2–cytochrome *c* reductase at a concentration equimolar with the cytochrome c_1 component. The inhibitor also induces some changes in the physical and spectral properties of the complex. Normally, cholate and salt or guanidine hydrochloride cause a cleavage of the complex into two fractions: one is an insoluble fraction containing cytochrome *b* and the iron protein, and the other is a soluble fraction consisting of cytochrome c_1 and core proteins. This cleavage reaction is blocked by antimycin, a finding that has been interpreted as indicating that antimycin protects some sites in the complex against the dissociating action of bile acids and chaotropes. Antimycin has also been shown to have several effects on the cytochrome *b* component of complex III. The visible absorption bands of cytochrome b_K are shifted towards the red region of the spectrum by about 2 nm, and there is a greater extent of reduction of the *b* cytochromes. Under anaerobic conditions only 30% of the cytochrome *b* of the complex is

TABLE 4.16. Subunit Composition of Complex III[a]

Subunit	Molecular weight	Identity with known components	Number of subunits	Total molecular weight
1	46,000	Core protein I	1	46,000
2	43,000	Core protein II	2	86,000
3	29,000	Cytochrome c_1	1	29,000
4	25,000	Nonheme iron protein	1	25,000
5	15,000	Cytochrome *b*	2	30,000
6	14,000	?	1	14,000
7	11,500	Antimycin protein	1	11,500
8	8,000	?	1	8,000
		Composite molecular weight		249,500

[a]Coenzyme QH_2–cytochrome *c* reductase of beef heart mitochondria. Based on data of Das Gupta and Rieske (1973).

reduced by substrate. When antimycin is added, however, virtually all of the cytochrome *b* becomes reduced provided that an oxidant such as ferricyanide is present.

This interesting observation has stimulated several proposals for the catalytic mechanism of the enzyme. One of these is a model devised by Mitchell which attempts to explain the observed transport by coenzyme QH_2-cytochrome *c* reductase of two protons for a net single electron transfer. This mechanism, shown in Fig. 4.34, also accounts for the induced reducibility of cytochrome *b* in the presence of antimycin and an oxidant. The scheme involves a cycle in which reduced coenzyme Q is first oxidized to the semiquinone by cytochrome b_T and then to the fully oxidized quinone by cytochrome c_1. In the reductive part of the cycle, coenzyme Q is first reduced by cytochrome b_K to the semiquinone and then to fully reduced coenzyme QH_2 by substrate (the electron and hydrogens in this last reaction would come from complex I or II). Since one of the electrons is cycled between cytochromes b_K and b_T, there is a net transfer of only one electron per cycle and a transport of two protons across the membrane. Antimycin is postulated to inhibit the reduction of coenzyme Q to the semiquinone. In the presence of the oxidant, the effect of antimycin would be to prevent the reoxidation of cytochrome b_T and an enhancement in the reduction of cytochrome b_K by coenzyme QH_2.

4. Cytochrome Oxidase (Complex IV)

Only a few properties of this complex will be mentioned here since it will be discussed in more detail in the next chapter.

Cytochrome oxidase has been obtained in highly purified form from different sources including beef-heart, human, yeast, and *Neurospora* mitochondria. These preparations have similar catalytic and physical properties and are also alike in their subunit protein compositions. The purified complex has a molecular weight of 200,000–250,000 and consists of seven nonidentical subunits. It contains the carriers cytochromes *a* and a_3 and copper in a 1:1:2 stoi-

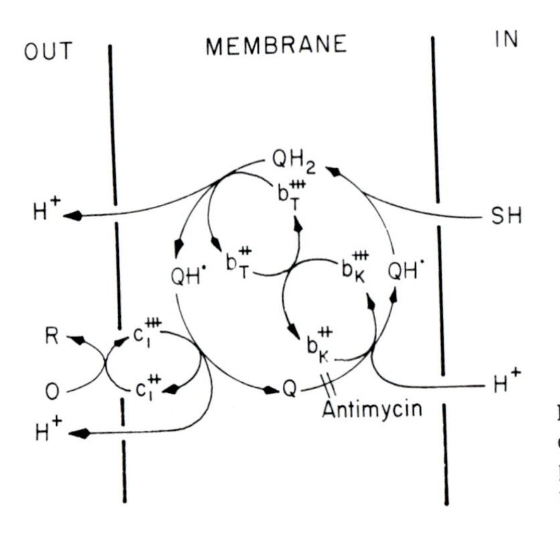

FIG. 4.34. Electron transfer mechanism of coenzyme QH_2–cytochrome *c* reductase proposed by Mitchell. (Drawn after Rieske, 1976.)

TABLE 4.17. Composition of Complex IV[a]

Component	Concentration (per mg protein)
Cytochrome *a*	4.5–5.0 nmoles
Cytochrome a_3	4.5–5.0 nmoles
Copper	10–12 nmoles
Phospholipids	0.3–0.4 mg

Reaction	Specific activity (μmoles/min per mg)
Ferrocytochrome *c* → Ferricytochrome *c*	25–50

[a]Beef-heart cytochrome oxidase.

chiometry. The specific contents of these components in beef cytochrome oxidase are shown in Table 4.17.

VI. Stoichiometry of the Electron Transfer Carriers

The concentration of the electron transfer carriers and their ratios have been studied in different types of mitochondria. The procedures used in such analyses are quite elaborate and are probably not completely accurate. This is especially true of cytochrome components whose visible absorption bands overlap in narrow regions of the spectrum (e.g., cytochromes *b*, c_1, and *c*) and therefore require special correction for their estimations. Flavoproteins also present problems, since their quantitation depends on a selective extraction of FMN or FAD and must be corrected for the presence of other flavoproteins of mitochondria with the same prosthetic groups.

In spite of these difficulties, the data obtained with different mitochondria are in fairly good agreement and indicate a constancy in the stoichiometry of the carriers. The results presented in Table 4.18 compare the concentrations of the various components present in beef heart mitochondria and in the isolated complexes. There are several points to be made from these data. It is obvious that the "mobile" carriers NAD^+ and coenzyme Q are present in considerable excess over the other carriers that are constituents of the respiratory complexes. Not all of the extra NAD^+ and coenzyme Q is necessary for optimal respiratory activity. For example, it is possible to deplete mitochondria of most of their endogenous NAD^+ without significantly affecting the rate of oxidation of NAD-linked substrates such as malate or glutamate.

The concentration of the other components that are constituents of the complexes, when normalized to that of cytochrome c_1, are almost the same. The exceptions are the cytochrome oxidase carriers, cytochromes *a* and a_3 and Cu which are approximately three times as abundant as cytochrome c_1. The electron transfer chain of beef-heart mitochondria, therefore, appears to be composed of one unit each of complexes I, II, and III and three units of complex IV. The reason why there is more cytochrome oxidase is not clear, particularly since the rate-limiting reactions in the chain occur at the flavoenzyme steps.

VII. Other Respiratory Chain-Linked Dehydrogenases

The electron transfer chain serves as a funnel for many oxidation reactions in which oxygen is the final acceptor of the hydrogens and electrons. So far we have considered those reactions that feed into the chain through NAD^+. These include the various NAD-linked dehydrogenases of the TCA cycle and other enzymes such as glutamate, pyruvate, and β-hydroxybutyrate dehydrogenases. There are other dehydrogenases, however, that connect with the respiratory chain through coenzyme Q or cytochrome *c*. This was seen to be true of the TCA cycle enzyme, succinate–coenzyme Q reductase. The coenzyme Q pathway is also involved in the oxidation of glycerol-3-phosphate and of fatty acyl-CoAs.

Some mitochondria, e.g., those of yeast and insect flight muscle, are capable of catalyzing the oxidation of glycerol-3-phosphate to dihydroxyacetone phosphate. The enzyme responsible for this reaction is known as α-glycerophosphate dehydrogenase (also L-glycerol-3-phosphate dehydrogenase) and is to be distinguished from another enzyme that catalyzes the same reaction but is present in the cytoplasm. The cytoplasmic dehydrogenase is a soluble protein that does not contain flavin and uses NAD^+ as the acceptor. The mitochondrial

TABLE 4.18. Stoichiometry of Respiratory Carriers[a]

	Mitochondria		Complex			
Carrier	Concentration (nmoles/mg)	Proportion relative to cytochrome c_1	I	II	III	IV
NAD	7.5	35				
FMN	0.15	1	1			
FAD	0.20	1		1		
FeS 1	↑	↑	1			
FeS 2			1			
FeS 3			1			
FeS 4	Total nonheme iron	Total nonheme iron	1			
FeS_{s-1}	6.4	30		1		
FeS_{s-2}				1		
FeS_{s-3}				1		
FeS_{III}	↓	↓			1	
Cyt. b_K	↑	↑			1	
	Total cyt. b	Total cyt. b				
Cyt. b_T	0.68	3.2			1	
Cyt. b_{558}	↓	↓		1		
Cyt. c_1	0.21	1			1	
Cyt. *c*	0.45	2.1				
Cyt. *a*	0.65	3.1				1
Cyt. a_3	0.65	3.1				1
Cu	1.47	7.0				2
Coenzyme Q	4.0	19				

[a]Based on analyses of beef-heart mitochondria and respiratory complexes. From Green and Wharton (1963).

α-glycerophosphate dehydrogenase is an insoluble inner membrane component that has been solubilized and partially purified from pig brain mitochondria. It contains FAD and nonheme iron. Several lines of evidence suggest that the oxidation of glycerol-3-phosphate is linked to the electron transfer chain through coenzyme Q. (1) Glycerol-3-phosphate oxidation is inhibited by antimycin and cyanide but not by amobarbital and rotenone; (2) there is a substrate-dependent reduction of coenzyme Q; and (3) the oxidation of glycerol-3-phosphate decreases in the presence of succinate, suggesting that there is a competition between the two substrates.

In our previous discussion of β-oxidation (Chapter 3, Section IIB), it was indicated that the conversion of fatty acyl-CoAs to the α,β-unsaturated fatty acyl-CoAs is catalyzed by a soluble flavoenzyme, fatty acyl-CoA dehydrogenase, present in the mitochondrial matrix. The two hydrogens extracted from the fatty acid are first used to reduce the FAD of the fatty acyl-CoA dehydrogenase. The reduced FAD is reoxidized by another FAD-containing flavoenzyme called electron transfer flavoprotein (ETF). The ETF transfers the reducing equivalents to the respiratory chain via a pathway that is thought to involve coenzyme Q. In analogy to glycerol-3-phosphate, the oxidation of fatty acyl-CoAs is inhibited by antimycin but not by rotenone. Fatty acyl-CoAs have also been shown to reduce coenzyme Q as do succinate and glycerol-3-phosphate.

Evidence that the chain may be tapped at the level of cytochrome *c* comes from studies of the L-lactate dehydrogenase system of yeast. L-Lactate can support the aerobic growth of yeast such as *Saccharomyces cerevisiae*. The enzyme responsible for the oxidation of L-lactate has been obtained in crystalline form and shown to be a tetrameric protein consisting of four identical subunits each with FMN and protoheme. The enzyme, therefore, is a cytochrome as well as a flavoenzyme and is known as cytochrome b_2. During the oxidation of L-lactate, the reducing equivalents pass from flavin to cytochrome b_2. It is thought that cytochrome b_2 is linked to cytochrome oxidase. This is based on the observation that cytochrome *c* acts as an electron acceptor of L-lactate dehydrogenase and that growth of yeast on L-lactate is inhibited by cyanide but not by antimycin.

The various branches that feed into the electron transfer chain are shown in Fig. 4.35.

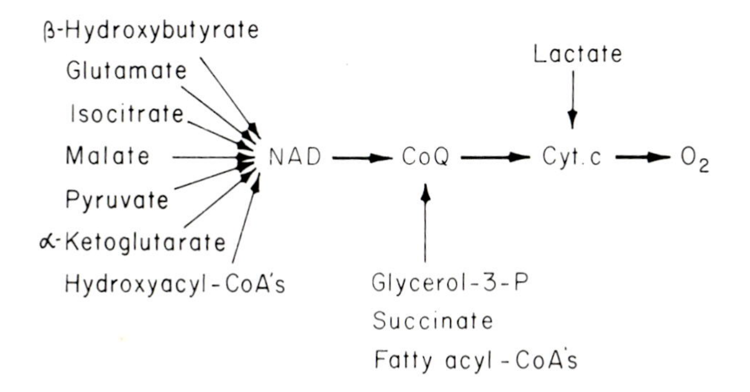

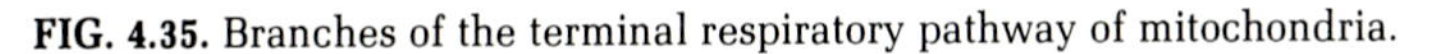
FIG. 4.35. Branches of the terminal respiratory pathway of mitochondria.

Selected Readings

Baginsky, M.L., and Hatefi, Y. (1969) Reconstitution of succinate-coenzyme Q reductase and succinate oxidase activities by a highly purified reactivated succinate dehydrogenase, *J. Biol. Chem.* **244:**5313.

Beinert, H., and Palmer, G. (1965) Contribution of EPR spectroscopy to our knowledge of oxidative enzymes, *Adv. Enzymol.* **27:**105.

Beinert, H., Ackrell, B. A. C., Kearney, E. B., and Singer, T. P. (1974) EPR studies on the mechanism of action of succinate dehydrogenase in activated preparations, *Biochem. Biophys. Res. Commun.* **58:**564.

Chance, B., and Williams, G. R. (1955) A method for the localization of sites for oxidative phosphorylation, *Nature* **176:**63.

Chance, B., Wilson, D. F., Dutton, P. L., and Erecinska, M. (1970) Energy-coupling mechanisms in mitochondria: Kinetic, spectroscopic and thermodynamic properties of an energy-transducing form of cytochrome *b*, *Proc. Natl. Acad. Sci. U.S.A.* **66:**1175.

Crane, F. L. (1961) Isolation and characterization of the coenzyme Q (ubiquinone) group and plastoquinone, in *Quinones in Electron Transport* (G. E. W. Wolstenholme and C. M. O'Connor, eds.), Little, Brown, Boston, pp. 36–78.

Das Gupta, U., and Rieske, J. S. (1973) Identification of a protein component of the antimycin-binding site of the respiratory chain by photoaffinity labeling, *Biochem. Biophys. Res. Commun.* **54:**1247.

Davis, K. A., and Hatefi, Y. (1971) Succinate dehydrogenase. I. Purification, molecular properties and substructure. *Biochemistry* **10:**2509.

Dickerson, R. E., and Timkovich, R. (1975) Cytochrome *c*, in *The Enzymes* (P. D. Boyer, ed.), Vol. XI, part A, Academic Press, New York, pp. 397–547.

Dutton, P. L., and Wilson, D. F. (1974) Redox potentiometry in mitochondrial and photosynthetic bioenergetics, *Biochim. Biophys. Acta* **346:**165.

Green, D. E., and Wharton, D. C. (1963) Stoichiometry of the fixed oxidation-reduction components of the electron transfer chain of beef heart mitochondria, *Biochem. Z.* **338:**335.

Hatefi, Y. (1976) The enzymes and the enzyme complexes of the mitochondrial oxidative phosphorylation system, in *The Enzymes of Biological Membranes* (A. Martonosi, ed.), Vol. 4, Plenum Press, New York, pp. 3–41.

Hatefi, Y., and Stempel, K. E. (1969) Isolation and enzymatic properties of the mitochondrial reduced diphosphopyridine nucleotide dehydrogenase, *J. Biol. Chem.* **244:**2350.

Hatefi, Y., Haavik, A. G., and Griffiths, D. E. (1962) Studies on the electron transfer system XLI. Reduced coenzyme Q (QH_2)-cytochrome *c* reductase, *J. Biol. Chem.* **237:**1681.

Keilin, D., and Hartree, E. F. (1939) Cytochrome and cytochrome oxidase, *Proc. R. Soc. Lond.* [*Biol.*] **127:**167.

King, T. E. (1963) Reconstitution of respiratory chain enzyme systems XII. Some observations on the reconstitution of the succinate oxidase system from heart muscle, *J. Biol. Chem.* **238:**4037.

Klingenberg, M., and Kroger, A. (1967) On the role of ubiquinone in the respiratory chain, in *The Biochemistry of Mitochondria* (E. C. Slater, Z. Kaninga and L. Wojtczak, eds.), Academic Press, New York, pp. 11–27.

MacLennan, D. H., and Tzagoloff, A. (1965) The isolation of a copper protein from cytochrome oxidase, *Biochim. Biophys. Acta* **96:**166.

Mitchell, P. (1975) Protonmotive mechanism of the cytochrome b–c_1 complex in the respiratory chain: Protonmotive ubiquinone cycle, *FEBS Lett.* **56:**1.

Nelson, B. D., and Gellerfors, P. (1974) The redox properties of cytochromes in purified complex III, *Biochim. Biophys. Acta* **357:**358.

Ohnishi, T. (1979) Mitochondrial iron-sulfur flavodehydrogenases in *Membrane Proteins in Energy Transduction* (R. A. Capaldi, ed.) Marcel Dekker, New York, pp. 1–87.

Ohnishi, T., Winter, D. B., Lim, J., and King, T. E. (1973) Low temperature electron paramagnetic resonance studies on two iron-sulfur centers in cardiac succinate dehydrogenase, *Biochem. Biophys. Res. Commun.* **53:**231.

Orme-Johnson, N. R., Hansen, R. E., and Beinert, H. (1974) Electron paramagnetic resonance detectable electron acceptors in beef heart mitochondria. Reduced diphosphopyridine nucleotide ubiquinone reductase segment of the electron transfer system, *J. Biol. Chem.* **249:**1922.

Palmer, G. (1975) Iron–sulfur proteins, in *The Enzymes* (P. D. Boyer, ed.), Vol. XII, part B, Academic Press, New York, pp. 2–56.

Palmer, G., Horgan, D. J., Tisdale, H., Singer, T. P., and Beinert, H. (1968) Studies on the respiratory chain linked reduced nicotinamide adenine dinucleotide dehydrogenase XIV. Location of the site of inhibition of rotenone, barbiturates and piericidin by means of electron magnetic resonance spectroscopy, *J. Biol. Chem.* **243**:844.

Rieske, J. S. (1967) Preparation and properties of coenzyme Q-cytochrome *c* reductase (complex III of the respiratory chain), *Methods Enzymol.* **10**:239.

Rieske, J. S. (1976) Composition and function of complex III of the respiratory chain, *Biochim. Biophys. Acta* **456**:195.

Salach, J., Walker, W. H., Singer, T. P., Ehrenberg, A., Hemmerich, P., Ghisla, S., and Hartman, U. (1972) Studies on succinate dehydrogenase. Site of attachment of the covalently bound flavin to the peptide chain, *Eur. J. Biochem.* **26**:267.

Schatz, G., and Ross, E. (1976) Cytochrome c_1 of bakers' yeast, *J. Biol. Chem.* **251**:991.

Singer, T. P., Kearney, E. B., and Ackrell, W. C. (1973) Succinate dehydrogenase, *Adv. Enzymol.* **37**:189.

Slater, E. C. (1973) The mechanism of action of the respiratory inhibitor antimycin, *Biochim. Biophys. Acta* **301**:129.

Szarkowska, L. (1966) The restoration of DPNH-oxidase activity by coenzyme Q (ubiquinone), *Arch. Biochem. Biophys.* **113**:519.

Tzagoloff, A., and Wharton, D. C. (1965) Studies on the electron transfer system LXII. The reaction of cytochrome oxidase with carbon monoxide, *J. Biol. Chem.* **240**:2628.

VanGelder, B. F., and Beinert, H. (1969) Studies on the heme components of cytochrome oxidase by EPR spectroscopy, *Biochim. Biophys. Acta* **189**:1.

Weiss, H., and Ziganke, B. (1974) Cytochrome *b* in *Neurospora crassa* mitochondria, *Eur. J. Biochem.* **41**:63.

Warburg, O., and Negelein, E. (1929) Uber das Absorptionsspectrum des Atmungsferment, *Biochem. Z.* **214**:64.

Yonetani, T. (1963) The *a*-type cytochromes, in *The Enzymes* (P. D. Boyer, H. Lardy and K. Myrback, eds.), Vol. 8, Academic Press, New York, pp. 41–95.

Yu, L., and King, T. E. (1972) Preparation and properties of cardiac cytochrome c_1, *J. Biol. Chem.* **247**:1012.

Ziegler, D. M., and Doeg, K. A. (1962) Studies on the electron transport system XLII. The isolation of a succinic-coenzyme Q reductase from beef heart mitochondria, *Arch. Biochem. Biophys.* **97**:41.

5

Cytochrome Oxidase
Model of a Membrane Enzyme

The purpose of this chapter is to describe in more detail what is known of the enzymatic and structural properties of cytochrome oxidase. There are several reasons for emphasizing this particular respiratory complex. Cytochrome oxidase is an insoluble lipoprotein complex that can be isolated in high yields by relatively simple procedures.

Since there is a fair amount of information about its composition, structure, and function, it is well suited for studying questions that are generally relevant to membrane enzymes as well as specific questions that have to do with the mechanism of electron transfer. There is yet another reason why cytochrome oxidase has become important in mitochondrial studies. It was one of the first complexes shown to be capable of forming membranes. The reconstituted cytochrome oxidase membranes have been found to carry out energy-dependent processes such as ion transport and oxidation phosphorylation and are currently being used in a number of laboratories as a model for studying energy-coupling mechanisms (Chapter 8).

I. Purification and Assay of Cytochrome Oxidase

Cytochrome oxidase has been isolated from mammalian, yeast, *Neurospora crassa*, and other mitochondria. It makes a substantial contribution to the mass of the inner membrane (15–20% in beef-heart mitochondria). In practical terms, this means that only a five- to sevenfold purification is needed to obtain a homogeneous enzyme. Most procedures rely on solubilization of the complex with bile acids or nonionic detergents (Triton) and separation from other membrane components by salt fractionation or adsorption on hydrophobic or ion-exchange columns.

The composition and spectral properties of purified cytochrome oxidase have already been described (see Chapter 4, Section IVB). Depending on the method of isolation, the enzyme contains either very small amounts or as much as 30% by weight of phospholipids. Since the phospholipids are required for

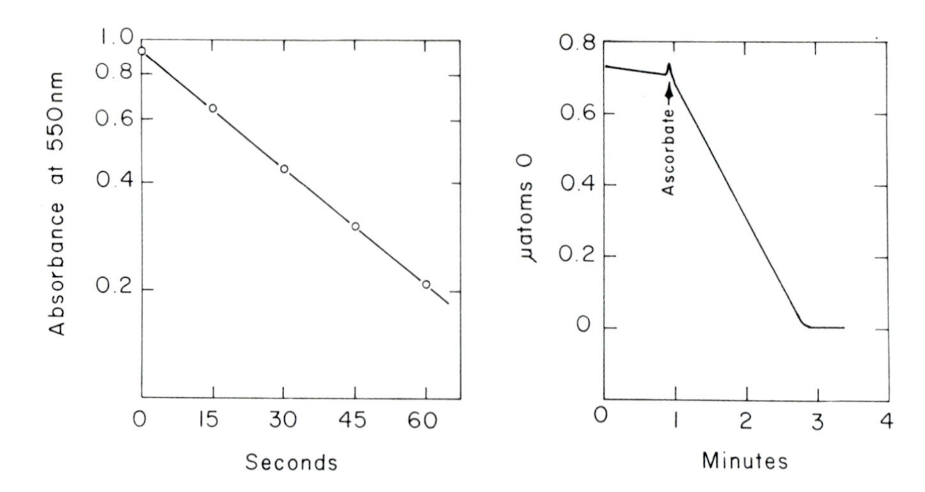

FIG. 5.1. Two assays of cytochrome oxidase. Left: Semilogarithmic plot of ferrocytochrome *c* oxidation versus time. The reaction mixture contains phosphate buffer, ferrocytochrome *c*, and enzyme. Right: Trace showing oxygen consumption when ascorbate serves as the reductant for cytochrome *c*. The reaction mixture contains phosphate buffer, TMPD, catalytic amounts of cytochrome *c*, and enzyme. The reaction is started with ascorbate (arrow). Note that the rate of oxygen disappearance from the solution is linear with time.

optimal enzymatic activity, low-lipid preparations are usually assayed in the presence of exogenous phospholipids.

There are several convenient ways to assay cytochrome oxidase. One method relies on measuring the rate of oxidation of ferrocytochrome *c* by oxygen. This reaction can be followed spectrophotometrically at 550 nm, which corresponds to the absorption maximum of the α band of ferrocytochrome *c* (see Chapter 4, Section IB). The reaction is first order with respect to substrate and therefore becomes exponentially slower with time. When the decrease in absorbance at 550 nm is plotted on a semilogarithmic scale (Fig. 5.1), a straight line is obtained whose slope is equal to the first-order velocity constant (k). The initial velocity of the reaction is calculated by multiplying the first-order velocity constant by the starting concentration of ferrocytochrome *c*. An alternative method of assaying the enzyme is to measure the rate at which oxygen is consumed during the oxidation of cytochrome *c*. In this assay, a catalytic concentration of cytochrome *c* is kept in a reduced form by the presence of excess ascorbate and a small amount of tetramethylphenylene diamine (TMPD) which is a redox dye that mediates the reduction of cytochrome *c* by ascorbate. Since the concentration of ferrocytochrome *c* does not change during the assay, the rate of oxygen uptake proceeds at a linear rate with time (Fig. 5.1).

Cytochrome oxidase has a high turnover number (350–500 sec^{-1} based on cytochrome a or a_3) and specific activities as high as 100 μmoles cytochrome *c* oxidized per minute per milligram protein have been reported. The enzyme is effectively inhibited by low concentrations of cyanide, azide, and sulfide which act as ligands of the heme in the cytochrome a_3 component and prevent the binding of oxygen.

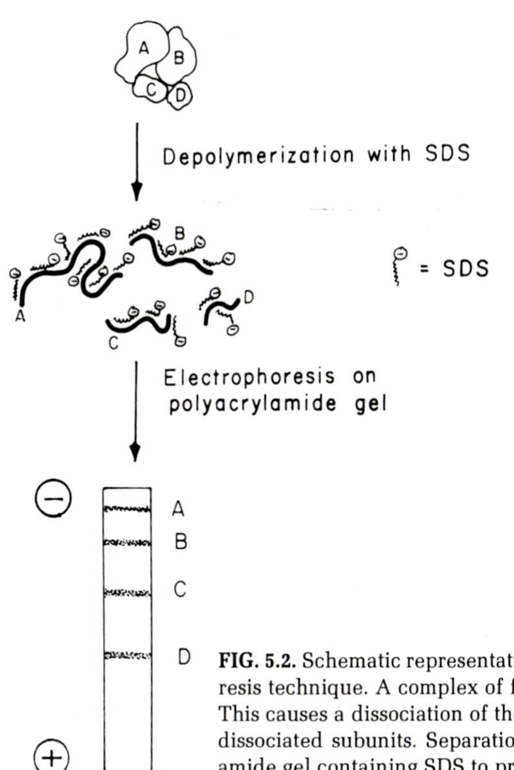

FIG. 5.2. Schematic representation of the SDS polyacrylamide gel electrophoresis technique. A complex of four proteins (A, B, C, D) is treated with SDS. This causes a dissociation of the complex and loss of tertiary structure in the dissociated subunits. Separation of the subunits is effected on a polyacrylamide gel containing SDS to prevent reassociation of the subunits.

II. Structure of Cytochrome Oxidase

A. Subunit Composition

The subunit proteins of cytochrome oxidase have been studied by electrophoresis on polyacrylamide gels after dissociation of the enzyme with sodium dodecyl sulfate (SDS). This technique is very useful for determining the number of different proteins in polymeric enzymes, viral particles, membranes, etc. Sodium dodecyl sulfate is a powerful ionic detergent that binds in high molar excess to proteins and weakens the hydrophobic and electrostatic interactions involved in the tertiary and quaternary conformation of the proteins. When a membrane enzyme or for that matter whole membranes are treated with SDS, all the protein components are depolymerized to their monomeric form and can be separated by electrophoresis on polyacrylamide gels (Fig. 5.2). Because of the negative charges of the bound detergent molecules, the proteins migrate toward the anode in this electrophoretic system, and their separation is determined not by differences in charge but rather by the size of the proteins because of the sieving properties of the polyacrylamide gel—the larger the protein, the greater its retardation in the gel. By appropriate calibration of the gel, this technique can be used to estimate molecular weights of monomeric proteins.

The subunits of yeast cytochrome oxidase separated on a 7.5% polyacrylamide gel are shown in Fig. 5.3. The enzyme is seen to consist of seven distinct subunit proteins with a molecular weight range of 40,000 to 9500. Similar subunit compositions have been found for the cytochrome oxidase complexes of beef-heart and *Neurospora crassa* mitochondria (Table 5.1). The apparent absence of subunit 2 from beef-heart cytochrome oxidase has recently been shown to be because it migrates with the same mobility as subunit 3.

The relative proportions of the subunits have been determined in the yeast and *Neurospora* enzymes. Based on the *in vivo* incorporation of radioactive leucine into the different subunits and their known molecular weights and amino acid compositions, Sebald and co-workers have concluded that they are present in equimolar concentrations. From the data of Table 5.1, it can be calculated that a cytochrome oxidase enzyme consisting of one polypeptide chain of each of the seven subunits would have a composite molecular weight of 140,000 to 150,000. Since the purest preparations of yeast and *Neurospora* cytochrome oxidase have a total heme *a* (cytochromes *a* and a_3) content of 13–15 nmoles per milligram protein, the minimum molecular weight, assuming one cytochrome *a* or a_3 per unit of enzyme, can be calculated to be 133,000–153,000. Although these numbers are in good agreement with each other, they are inconsistent with the molecular weight values of 200,000–250,000 that have been measured by sedimentation and light scattering. The discrepancy can be rationalized either by assuming that the active enzyme is a dimer or that the molecular weights determined by the physical procedures are incorrect. In view of the known tendency of the enzyme to aggregate and the problems inherent in measuring physical constants of lipoprotein complexes, the latter explanation appears more likely at present.

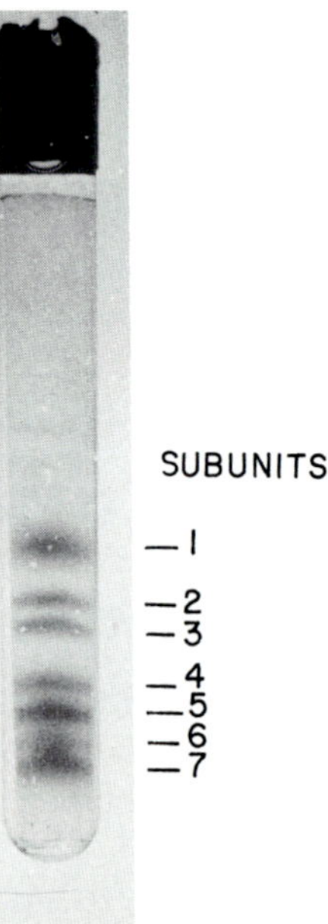

FIG. 5.3. Subunits of yeast cytochrome oxidase. The purified enzyme was dissociated in 1% SDS and separated on a 7.5% polyacrylamide gel containing 0.1% SDS. The gel was stained with Amido Black. The molecular weights of the subunits are listed in Table 5.1.

B. Purification of Subunits and Identity with Known Carriers

The subunit proteins of yeast, *Neurospora*, and bovine cytochrome oxidase have been purified, and most of them, particularly those of the bovine enzyme, are now sequenced. As can be seen from the amino acid compositions of bovine cytochrome oxidase (Table 5.2), the three largest-molecular-weight polypeptides (these are the mitochondrially synthesized subunits) have a considerably higher content of nonpolar amino acids. Similar results have been obtained with the subunits of *Neurospora* and yeast cytochrome oxidase. The hydrophobicity of the large subunits is also evident by their insolubility in water—this is in contrast to the other four subunits which, after separation from the complex, are reasonably soluble in water.

Most of the procedures used for the purification of the subunits have employed sodium dodecyl sulfate to depolymerize the complex followed by separation of the monomeric proteins either by preparative polyacrylamide gel electrophoresis or by chromatography on molecular sieves. Unfortunately, these conditions displace the prosthetic groups from the enzyme and, in the case of the cytochromes, cause an actual chemical destruction of heme *a*. In view of this, it has been difficult to identify conclusively the subunits with electron carrier functions. There is good circumstantial evidence, however, that the largest polypeptide (subunit 1) is the hemoprotein component of the complex. This protein has been purified after depolymerization of yeast and bovine cytochrome oxidase with guanidine thiocyanate, a reagent that does not cause destruction of heme.

In this procedure, all the heme *a* is associated with subunit 1. The purified protein has a heme *a* content of 25–30 nmoles per milligram protein, but its spectral properties are different from those of the native enzyme: the visible absorption bands are shifted by 7 nm toward shorter wavelengths, and the dis-

TABLE 5.1. Molecular Weights of Subunit Proteins of Yeast, Bovine, and *Neurospora crassa* Cytochrome Oxidase

	Molecular weights		
Subunit	Yeast[a]	Beef heart[b]	*Neurospora crassa*[c]
1	40,000[d]	36,000	41,000
2	27,300	21,000	28,500
3	25,000	19,000	21,000
4	13,800	14,000	16,000
5	13,000	12,500[e]	14,000
6	10,200	11,000	11,500
7	9,500	10,000	10,000
8	absent	6,000	absent

[a]Based on separations in 10% polyacrylamide gels (data of Rubin and Tzagoloff, 1973).
[b]Estimated from 15% polyacrylamide gels (Steffens and Buse, 1976).
[c]Estimated from 15% polyacrylamide gels (data of Sebald *et al.*, 1973).
[d]The real molecular weight of this subunit is probably 56,000.
[e]In the bovine cytochrome oxidase, there are two subunits with nearly identical molecular weights.

tinction between cytochromes a and a_3 is abolished. A second line of evidence suggesting the association of heme with subunit 1 comes from studies of *Paracoccus denitrificans* cytochrome oxidase. The *Paracoccus* enzyme is similar to mitochondrial cytochrome oxidase, having both the a and a_3 components and copper. This bacterial oxidase, however, consists of only two subunits. Based on SDS gel electrophoresis, the two proteins appear to correspond to subunits 1 and 2 of mitochondrial cytochrome oxidase. Since subunit 2 is probably the copper protein of the complex (see below), these interesting results imply that in *Paracoccus* and, by analogy, in mitochondrial cytochrome oxidase, there is only one hemoprotein constituent with two heme a binding sites.

Although the molecular weight of subunit 1 has been estimated to be 40,000, this is probably incorrect based on recent information about the gene sequence of the protein (see Chapter 11, Section VIIC). The amino acid sequences of yeast and human subunit 1 deduced from the corresponding nucleotide sequences of the two genes indicate the molecular weight to be 56,000.

The primary structure of bovine subunit 2 has been completely determined by Steffens and Buse. These authors have found substantial sequence homology between subunit 2 and two copper-binding proteins, azurin and plastocyanin (Fig. 5.4). Most significant is the conservation of two histidines at residues 37 and 87, a cysteine at residue 84, and a methionine at residue 92 of plastocyanin. From X-ray crystallographic studies, these residues have been determined to

TABLE 5.2. Amino Acid Compositions of Bovine Cytochrome Oxidase Subunits[a]

Amino acid	Cytochrome oxidase	Subunit						
		1	2	3	4	5 + 6	7	8
Lysine	4.9	2.6	3.3	3.8	10.3	6.2	7.0	7.9
Histidine	3.9	3.8	3.5	4.9	2.7	2.7	3.0	2.8
Arginine	4.3	2.6	3.1	3.6	4.4	5.6	7.1	3.8
Aspartic acid	7.4	7.3	7.3	6.6	9.3	10.9	9.4	7.3
Threonine	7.2	7.7	7.8	7.9	4.6	5.5	5.7	6.0
Serine	7.0	6.7	9.8	7.3	7.8	4.8	5.4	6.3
Glutamic acid	7.5	4.7	9.7	8.5	11.6	11.9	9.3	8.1
Proline	5.5	5.9	6.3	5.3	5.2	7.5	5.6	7.8
Glycine	7.8	9.3	5.1	8.8	5.6	7.7	8.3	8.0
Alanine	8.1	8.1	4.6	8.4	9.4	8.9	10.8	9.2
Cysteine	0.7	n.d.	n.d.	n.d.	n.d.	n.d.	n.d.	n.d.
Valine	6.0	6.2	4.8	5.7	6.4	6.5	5.4	5.4
Methionine	4.5	5.8	7.1	3.4	2.5	1.6	2.1	1.8
Isoleucine	4.6	5.8	5.0	4.1	3.4	4.9	3.8	4.1
Leucine	11.1	12.2	14.0	11.1	8.1	9.3	7.6	11.3
Tyrosine	4.0	3.9	5.1	3.8	4.4	3.4	3.7	4.0
Phenylalanine	6.1	7.5	3.7	6.6	4.5	3.0	5.8	6.4
Percent polarity	42.2	35.4	44.5	42.6	50.7	47.6	46.9	42.2

[a]The values are reported as mole percent. The polarity was calculated as mole percent of lysine, histidine, arginine, aspartic acid, threonine, serine, and glutamic acid. (From Steffens and Buse, 1976.)

be involved in the copper-binding function of plastocyanin. The homology in the primary structures of the proteins supports the idea that the copper of cytochrome oxidase is linked to subunit 2 in a manner similar to that found in other copper proteins. In earlier studies, subunit 2 was isolated by succinylation of cytochrome oxidase. The purified succinylated protein was found to have two coppers per polypeptide chain, indicating that all of the copper of cytochrome oxidase is probably associated with this subunit.

The role of the other subunits of mitochondrial cytochrome oxidase is unclear at present and will require the development of less destructive isolation procedures in order for the functional groups and tertiary structures of the proteins to be studied.

C. Quaternary Structure

Very little is known about the arrangement of the subunits in the complex. A possible approach to this problem is suggested by studies of Schatz and coworkers who have used several different types of surface-probing reagents to examine the quaternary structure of yeast cytochrome oxidase. One of the probes was bovine serum albumin that had been derivatized with tolylene 2,4-diisocyanate (TC-BSA). Since this is a high-molecular-weight probe, it was reasoned that the conjugation would occur only with the lysine residues (the isocyanate reacts specifically with the ε-amino group of lysine) of subunits on the surface of the complex (Fig. 5.5). Another advantage of the probe is that the extent of conjugation of the TC-BSA to the different subunits could be quantitated in a relatively simple way by SDS gel electrophoresis of the dissociated complex. This was possible because of the increased molecular weight and slower mobility of the conjugates. The results of such experiments are shown in Fig. 5.6 and Table 5.3.

The protein profile of cytochrome oxidase treated with TC-BSA reveals a decrease in the peaks corresponding to subunits 3, 4, 5, 6, and 7 but virtually no change in subunits 1 and 2. The greatest decrease is seen in subunits 5 and 6 which appear to be especially accessible to the probe. In this type of experiment, the failure of a subunit to react with the TC-BSA could mean either that it is buried and inaccessible to the probe or that it does not have exposed lysine

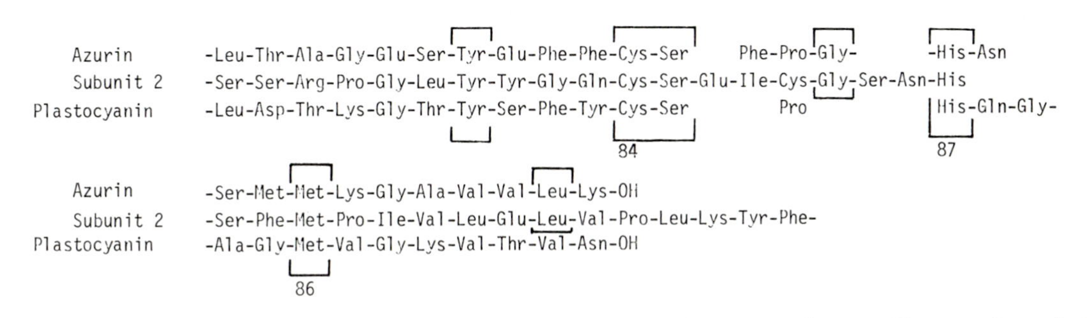

FIG. 5.4. Amino acid homologies in the primary structures of subunit 2 of bovine cytochrome oxidase, azurin, and plastocyanin. (Data of Steffens and Buse, 1979.)

residues. The latter possibility cannot entirely be excluded but is unlikely in view of the fact that qualitatively similar results were obtained with another probe, diazonium benzene sulfonate, which reacts with a wider range of amino acids including lysine, histidine, cysteine, and tyrosine. Although these are still beginning efforts, they suggest that the hydrophobic proteins may form a core that is surrounded by the more hydrophilic subunits. Obviously, further refinements of this as well as other approaches are needed to fill in the existent gaps in our understanding of how the subunits of the complex are organized.

III. Formation of Cytochrome Oxidase Membranes

Cytochrome oxidase is freely soluble in buffers containing bile acids or synthetic detergents. When these dispersing agents are removed or when their concentrations are effectively lowered, a rapid polymerization of the enzyme ensues. This phenomenon was first studied by McConnell who made the important discovery that the polymerization of cytochrome oxidase is a self-assembly process that leads to the formation of membranes. McConnell also showed that the reconstitution of membranes from cytochrome oxidase depends only on the presence of the purified enzyme and phospholipids and that no other factors are required.

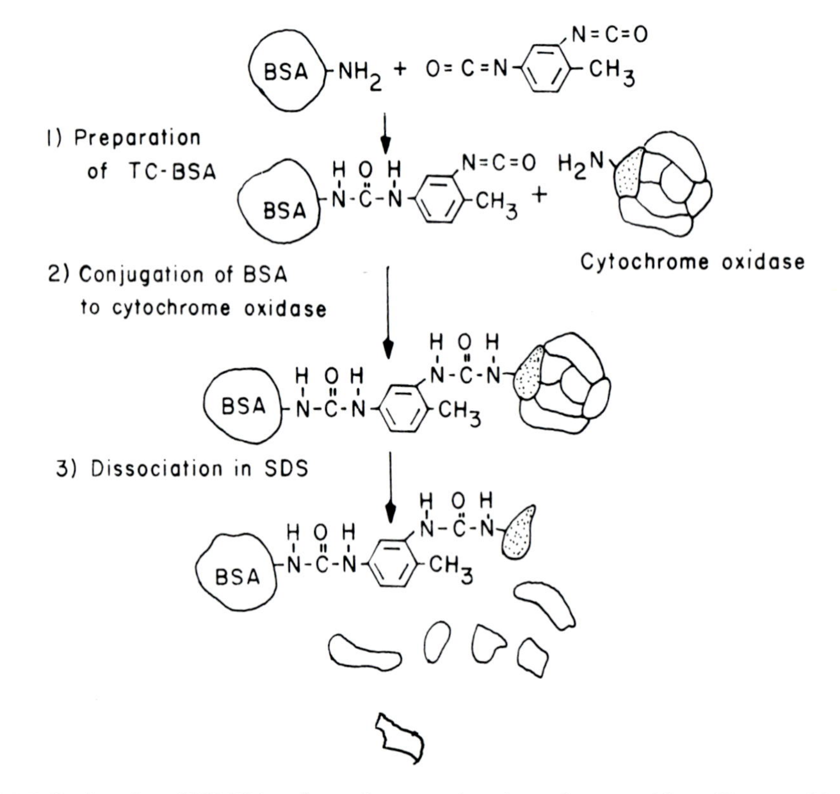

FIG. 5.5. Conjugation of TC-BSA to the surface proteins of cytochrome oxidase. (Eytan and Schatz, 1975.)

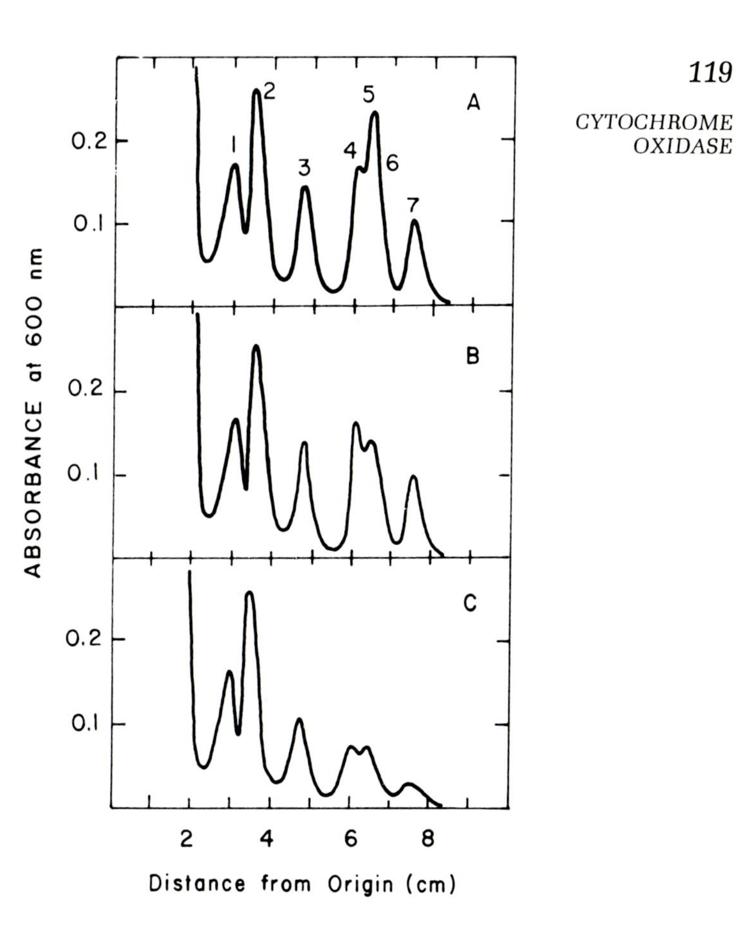

FIG. 5.6. SDS polyacrylamide gels of cytochrome oxidase after conjugation with TC-BSA. A: Untreated yeast cytochrome oxidase. B: Enzyme reacted with 1 mg TC-BSA per mg protein. C: Enzyme reacted with 2 mg TC-BSA per mg protein. The gels were stained, and the amount of each subunit estimated from the areas under the peaks. See Table 5.3 for the quantitative results. (Eytan and Schatz, 1975.)

Cytochrome oxidase prepared from beef-heart mitochondria by fractionation with deoxycholate and salt usually contains about 30% by weight of phospholipids and an equal amount of bile acid. The deoxycholate is reduced to less than 1% of the protein weight without affecting the phospholipid content by dialysis or passage of the enzyme through a molecular sieving column. Electron

TABLE 5.3. Conjugation of TC-BSA with Subunits of Yeast Cytochrome Oxidase[a]

	Percent loss of subunit	
Subunit	1 mg TC-BSA per mg enzyme	2 mg TC-BSA per mg enzyme
1	1	0
2	10	0
3	0	32
4	5	57
5 + 6	60	73
7	21	60

[a]Data of Eytan and Schatz (1975).

micrographs of these preparations negatively stained with phosphotungstate indicate that removal of the bile acid results in the formation of membrane vesicles whose substructure is a particle with the same dimensions as the dispersed enzyme (Fig. 5.7). The simplest interpretation of these findings is that the reconstituted membranes consist of cytochrome oxidase particles embedded in a phospholipid bilayer.

The requirement of phospholipid for the *in vitro* assembly of cytochrome oxidase membranes can be demonstrated with preparations of enzyme depleted of its endogenous phospholipids. Repeated precipitation of cytochrome oxidase with ammonium sulfate in the presence of high concentrations of cholate reduces the phospholipid content to values as low as 2% by weight. When such low-lipid-containing preparations are dialyzed, the enzyme particles polymerize into large bulk-phase aggregates instead of membrane monolayers. The addition of phospholipids to the lipid-depleted enzyme once again restores its capacity to form membranes (Fig. 5.8).

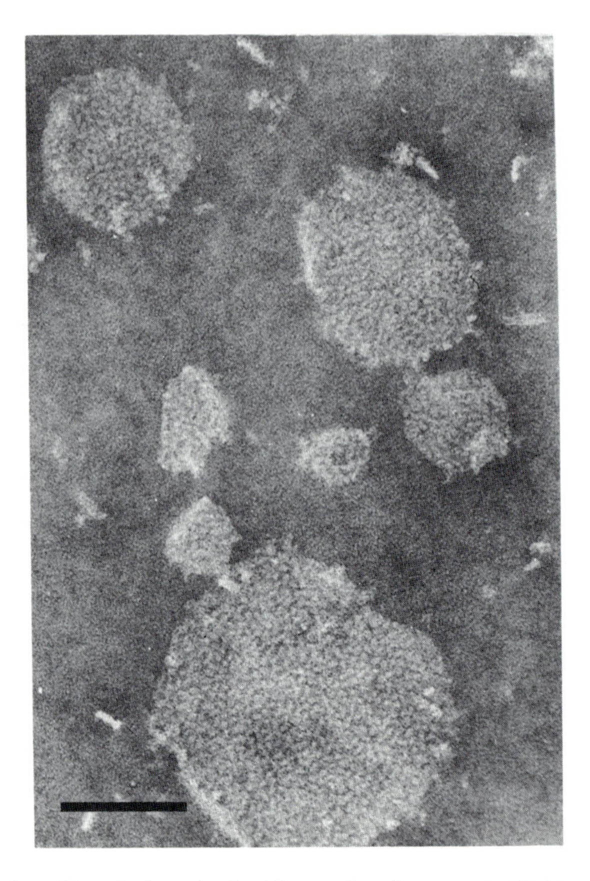

FIG. 5.7. Formation of membranes from purified beef heart cytochrome oxidase. The enzyme used contained 30% by weight of phospholipids. Left: Electron micrograph of cytochrome oxidase dispersed in approximately 1% deoxycholate. Negatively stained with 1% phosphotungstate. Right: Electron micrograph of the enzyme after passage through a column of Sephadex G-50 to remove the deoxycholate. The scale bar represents 0.1 μm.

The above observations have been extended to the other complexes of the respiratory chain and have been helpful in interpreting the structure of the mitochondrial inner membrane. The ability of the purified enzymes to form membranes in the absence of any structural elements other than phospholipids indicates that they are important structural units of the membrane. This is supported by two lines of evidence. The first has to do with the fact that the respiratory chain components account for most of the protein mass of the inner membrane—this is particularly true of mitochondria with high oxidative activity. Second, the purified complexes have been shown to be globular particles whose dimensions are consistent with the particles seen in fracture faces and in negatively stained preparations of the inner membrane. It is therefore reasonable to assume that the majority of the globular particles embedded in the lipid bilayer of the membrane must correspond to the electron transfer chain complexes.

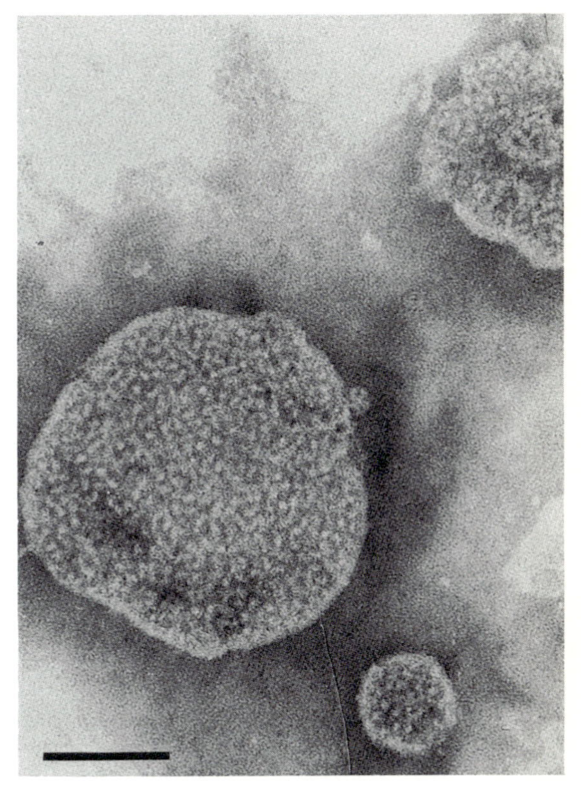

FIG. 5.8. Dependence of membrane formation on phospholipids. The enzyme used contained 4% by weight of phospholipids. Left: Electron micrograph of cytochrome oxidase after passage through a Sephadex G-50 column to remove endogenous deoxycholate. Negatively stained with 1% phosphotungstate. Note the highly aggregated nature of the preparation and the absence of vesicular membranes. Right: The enzyme was mixed with mitochondrial phospholipids in a 1:1 (w/w) ratio in the presence of 1% deoxycholate. The mixture was then passed through Sephadex G-50 and examined in the microscope after staining with phosphotungstate. Note the presence of vesicular membranes. The scale bar represents 0.1 μm.

IV. Shape of Cytochrome Oxidase

There have been several reports of crystalline cytochrome oxidase. These have not been confirmed, and X-ray crystallographic data have not been obtained on the crystals. Cytochrome oxidase, however, can be induced to form two-dimensional crystals with a sufficiently regular lattice structure to collect X-ray diffraction data. Henderson and Capaldi have deduced the general shape of the cytochrome oxidase molecule from three-dimensional reconstructions of several different types of crystalline arrays. Since the maximal resolution attained so far is of the order of 25 Å, the analysis does not give any information about the substructure or topology of the subunits in the complex.

Two types of preparations have been studied. Membranes of cytochrome oxidase prepared in the presence of Triton X-100 have numerous patches with crystalline structure. Although X-ray diffraction analyses have been done on this preparation, they have been difficult to interpret because the data are

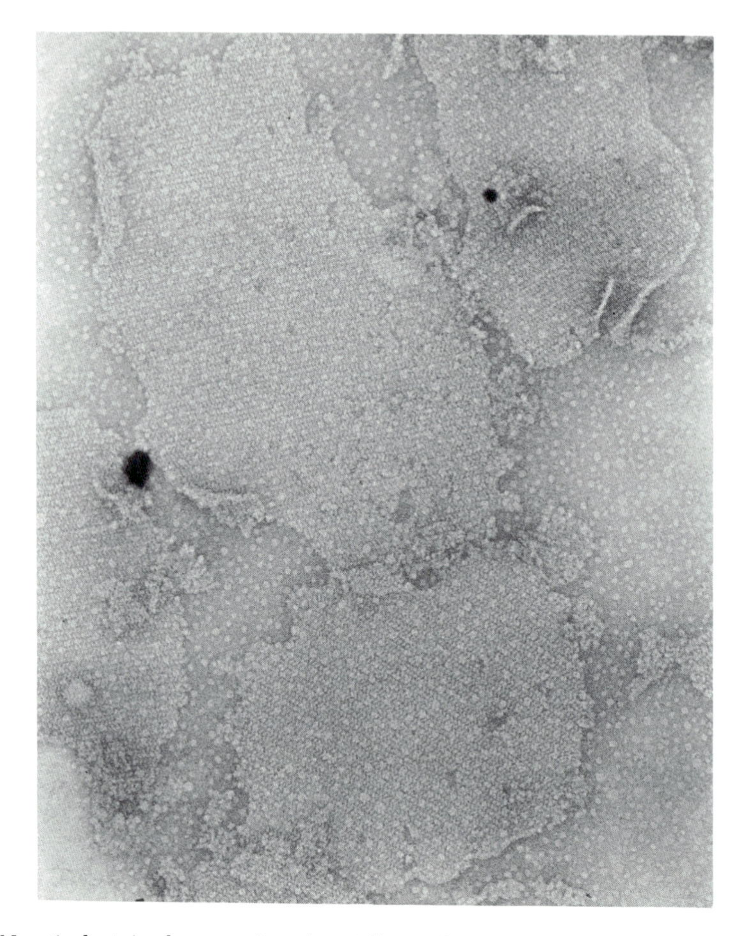

FIG. 5.9. Negatively stained preparation of partially purified cytochrome oxidase from bovine heart mitochondria. The enzyme was stained with 1% phosphotungstic acid. Magnification ×90,000. (Courtesy of Dr. Ronald Capaldi.)

obtained on two superimposed membranes collapsed on each other. Furthermore, the enzyme in the crystalline areas is arranged as a dimer whose boundaries are not clearly defined because of the low resolution of the electron micrographs.

A more suitable preparation consists of the partially purified cytochrome oxidase fraction remaining after extraction of mitochondrial membranes with deoxycholate (see Fig. 4.31). When this particulate fraction is negatively stained with phosphotungstate or uranyl acetate, the enzyme is seen to be packed as a single-layered crystal with lattice dimensions consistent with a monomer repeat unit (Fig. 5.9). Apparently, complete aggregation of the enzyme into closed membrane vesicles is prevented by the residual deoxycholate present in the preparation. Three-dimensional reconstructions of the diffraction patterns indicate that the monomer is packed in rows with opposite orientation (Fig. 5.10). Each monomer consists of three parts. There are two arms approximately 50 Å in length that project into the matrix side and a stem 55 Å long at the opposite end of the molecule facing the cytoplasmic side of the membrane. The molecule has an overall Y shape (Fig. 5.10). Higher resolution is presently limited by the use of negative stains. The method, however, is potentially capable of achieving a resolution of 7 Å when the diffraction analysis is done on unstained wet specimens. If the regular lattice structure of cytochrome oxidase can be preserved in the absence of stains, it may be possible in the future to deduce the subunit arrangement and to some extent the tertiary structures of the polypeptides.

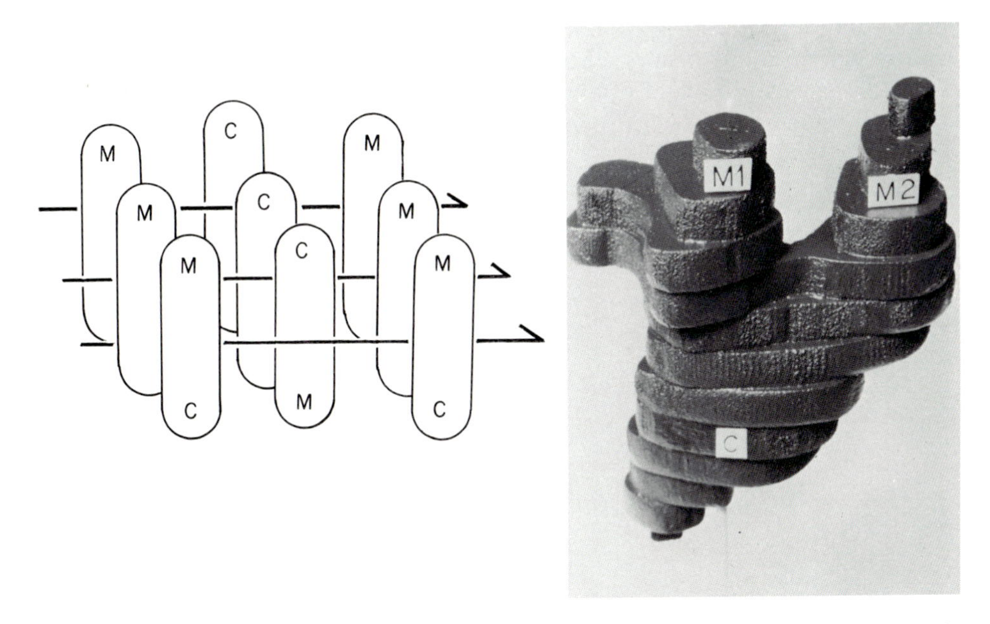

FIG. 5.10. Models of cytochrome oxidase based on reconstruction of two-dimensional crystals. Left: Arrangement of cytochrome oxidase monomers in the crystalline array. Right: Shape of a single molecule. The M1 and M2 arms face the matrix side of the membrane, whereas the C base lies on the cytoplasmic side. (Taken from Fuller *et al.*, 1979.)

V. Phospholipid Activation of Enzymatic Activity

Most membrane enzymes require phospholipids for enzymatic activity. Usually, this requirement can be satisfied with crude mixtures or with individual phospholipids. In the case of mitochondrial enzymes, phospholipids have been shown to activate cytochrome oxidase and the other respiratory complexes. The titrations of Fig. 5.11 indicate that maximal activation of cytochrome oxidase occurs at about 30% by weight of phospholipid.

There are several ways in which phospholipids might be expected to exert a stimulatory effect on the catalytic activity of cytochrome oxidase. Normally, the assay is done by adding a small amount of enzyme to a large excess of reaction buffer. Under these conditions, the endogenous bile acid or detergent becomes diluted, and the enzyme polymerizes into membranes or bulk aggregates depending on how much lipid is present. Since a membrane represents a more dispersed state than a bulk aggregate, it may be anticipated a priori that cytochrome oxidase would be more active in the membrane form because of greater accessibility of substrate, more efficient diffusion of oxygen to the active sites, etc. The activation by phospholipids may therefore simply reflect their effect on the state of dispersion of the enzyme. An alternative explanation is that phospholipids play a role in the catalytic mechanism itself.

The activation of cytochrome oxidase based on the dispersive effect of phospholipids alone is excluded from the following experimental observations. It is possible to chemically modify cytochrome oxidase to a water-soluble form in the absence of detergents. This can be done by acetylation of the ϵ-amino groups of the lysine residues. The masking of the positively charged amino groups confers a net negative charge on the protein and prevents the polymerization that normally occurs when detergents are removed. The experiment of Table 5.4 shows the effect of chemical modification on the enzymatic activity of cytochrome oxidase with high and low lipid contents. Although both types

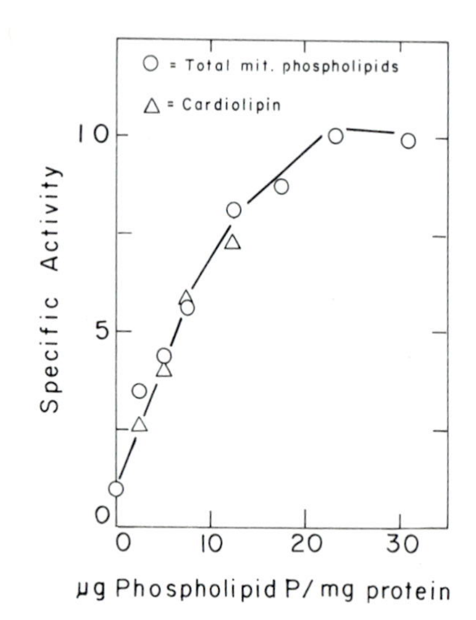

FIG. 5.11. Activation of beef heart cytochrome oxidase with phospholipids. A low-lipid-containing cytochrome oxidase was mixed with the indicated amounts of total mitochondrial phospholipids or with purified cardiolipin and assayed for enzymatic activity. The specific activity refers to μmoles of ferrocytochrome c oxidized per min per mg protein. A microgram of phosphorus is equivalent to approximately 0.03 mg of phospholipid. (Tzagoloff and MacLennan, 1965.)

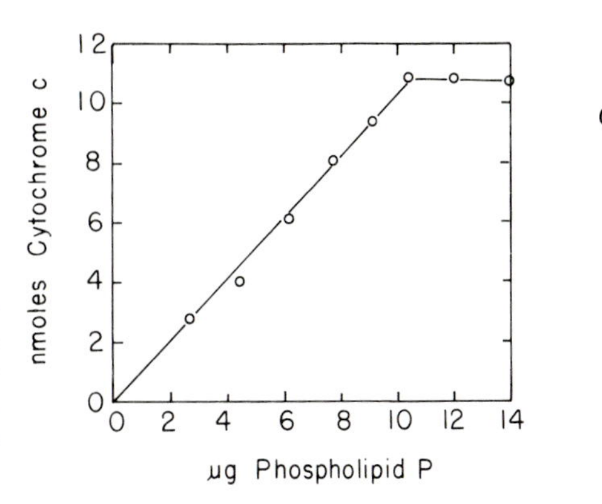

FIG. 5.14. Stoichiometry of cytochrome *c* and phospholipids for optimal cytochrome oxidase activity. The endpoints from the titrations shown in Fig. 5.10 are plotted to show the ratio of cytochrome *c* to phospholipids at which maximal enzymatic activity is observed. (Tzagoloff and MacLennan, 1965.)

The traditional view has been that there is an obligatory sequence of electron transfer from cytochrome a to a_3 and that the ultimate electron donor for oxygen is cytochrome a_3. This interpretation arose from Keilin's notion of two separate heme carriers (cytochrome a and a_3) in the terminal oxidase and was supported by later kinetic and potentiometric evidence indicating that the a_3 component is closest to oxygen (Chapter 4, Section IV).

Although there have been many sequential schemes proposed involving

CONCERTED MECHANISM

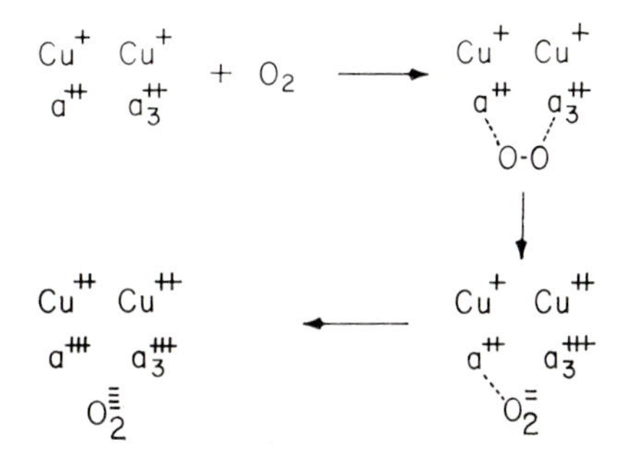

SEQUENTIAL MECHANISM

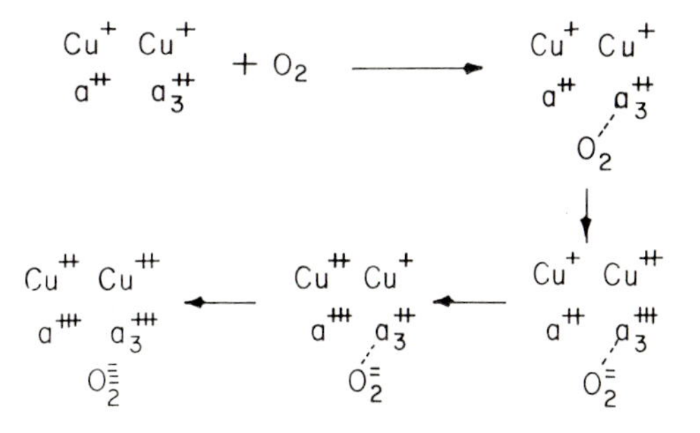

FIG. 5.15. Two possible mechanisms of oxygen reduction by cytochrome oxidase.

different intermediate states of the enzyme and of oxygen, these have been largely speculative and based on analogies with other reactions involving oxygen and heme iron catalysts. Recently, however, Chance and co-workers have claimed spectroscopic evidence of transient intermediates of cytochrome a_3 and oxygen in the oxidase reaction. These intermediates were trapped by allowing oxygen to react with reduced mitochondria at very low temperatures and were measured by observing changes in the visible absorption bands of cytochrome a_3 and Cu. Two intermediates were found: the first was stable in the temperature range of −125° to −100°, and the second in the range of −100° to −60°. The properties of the two intermediates and the reactions proposed by Chance and co-workers are summarized in Table 5.5. The first intermediate which was only stable at temperatures below −100° had an absorption spectrum similar to the cytochrome a_3-CO adduct and was postulated to be cytochrome a_3^{2+}-O_2. In the higher temperature range of −100° to −60°, a second intermediate was detected by spectral changes suggesting an oxidation of cytochrome a_3 and Cu. The two electrons extracted from the a_3-Cu pair were assumed to be used in the reduction of oxygen to peroxide. Hence, this intermediate was postulated to be a_3^{2+}-$Cu^{2+} \cdot O_2$. No other intermediates could be detected at temperatures above −60°, and the sequence of Table 5.5 after step 3 is hypothetical. It should be pointed out that this reaction mechanism is not significantly different from other schemes that have usually also involved superoxide and peroxide intermediates.

Even though most of the existing evidence tends to favor an $a \rightarrow a_3 \rightarrow O_2$ sequence, there are some observations that are more consistent with a concerted type of mechanism. In a sense, this problem is closely related to the issue of the difference between cytochromes a and a_3. Let us recall that the two cytochromes are distinguished on the basis of their reactivity or lack of reactivity with ligands such as CO, cyanide, etc. (see Chapter 4, Section IB). The differential reactivities have led to the generally accepted interpretation of two functionally distinct cytochromes. There are some investigators, however, who believe that there is only one cytochrome in cytochrome oxidase. This so called "unitarian" viewpoint has been most strongly championed by Okunuki.

TABLE 5.5. Intermediates in the Reduction of Oxygen[a]

Reaction	Intermediate detected	Temperature stability	Spectral properties
1. $a_3^{2+}Cu^+ + O_2 \rightarrow a_3^{2+}Cu^+ \cdot O_2$	Yes	−125° to −100°	Similar to CO adduct
2. $a_3^{2+}Cu^+ \cdot O_2 \rightarrow a_3^{3+}Cu^+ \cdot O_2^-$	No	—	—
3. $a_3^{3+}Cu^+ \cdot O_2^- \rightarrow a_3^{3+}Cu^{2+} \cdot O_2^{2-}$	Yes	−100° to −60°	Spectrum of oxidized a_3 and Cu
4. $a_3^{3+}Cu^{2+} \cdot O_2^{2-} + a^{2+}Cu^+ \rightarrow a_3^{3+}Cu^+ \cdot O_2^{2-} + a^{2+}Cu^{2+}$	No	—	—
5. $a_3^{3+}Cu^+ \cdot O_2^{2-} + a^{2+}Cu^{2+} \rightarrow a_3^{2+}Cu^+ \cdot O_2^{2-} + a^{3+}Cu^{2+}$	No	—	—
6. $a_3^{2+}Cu^+ \cdot O_2^{2-} + a^{2+}Cu^+ \rightarrow a_3^{2+}Cu^+O_2^{4-} + a^{3+}Cu^{2+}$	No	—	—
7. $a_3^{2+}Cu^+ \cdot O_2^{4-} + 4H^+ \rightarrow a_3^{2+}Cu^+ + 2H_2O$	No	—	—

[a] Data of Chance *et al.* (1975).

Okunuki and co-workers first discovered that when reduced cytochrome oxidase reacts with oxygen, a new spectral species is formed which is distinct from the reduced or oxidized enzyme. The oxygenated enzyme is characterized by a γ band with a maximum at 428 nm, a position intermediate between the γ band of the oxidized (418 nm) and that of reduced (445 nm) enzyme. Oxygenated cytochrome oxidase is quite stable and becomes oxidized only after relatively long (30 sec) incubation in the presence of oxygen. Okunuki has postulated that oxygenated cytochrome oxidase is the first product of the reaction with oxygen. The stability of this intermediate was explained by the fact that it is formed in the absence of substrate that would ordinarily cause a reduction of oxygen.

The oxygenated compound has been difficult to explain in terms of two separate heme carriers, since its spectrum does not show any evidence of unreacted cytochrome *a*, as is the case in the carbon monoxide adduct. This puzzling observation has been interpreted to indicate the presence of only one species of cytochrome *a* in the enzyme. If we go back to our model of cytochrome oxidase with two heme–Cu pairs, according to the unitarian concept, both pairs are equivalent and can interact with oxygen (Fig. 5.16). In contrast to oxygen, carbon monoxide can only react with one of the heme–Cu pairs—the reason for the asymmetric reaction with carbon monoxide need not be specified but could reflect steric problems if the two pairs are closely spaced or if there is a redistribution of electrons that decreases the affinity of the second heme for carbon monoxide. Accordingly, the spectral and potentiometric properties of the carbon monoxide adduct which have been interpreted in terms of two functionally different cytochromes may be the consequence of, rather than the reason for, the asymmetric reaction with carbon monoxide. Such a model of cytochrome oxidase would tend to favor a concerted type of catalytic mechanism in which both heme–Cu pairs can deliver electrons to the dioxygen molecule.

Despite the fact that cytochrome oxidase was the first respiratory complex to be isolated and studied, its mechanism of action is still poorly understood. It may turn out that answers to some of the existent questions will emerge from the structural studies that are being actively pursued at present.

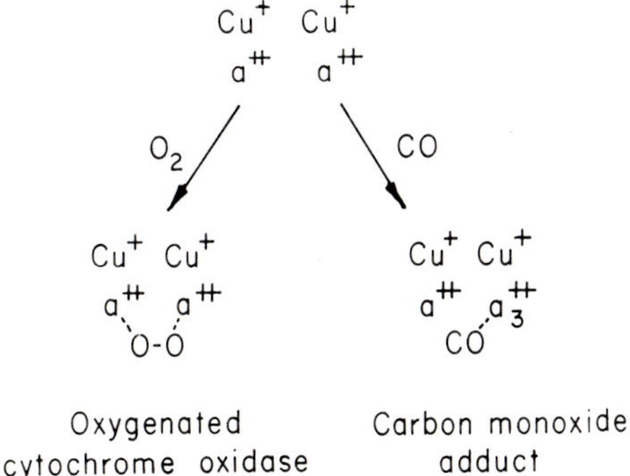

FIG. 5.16. Unitarian model of cytochrome oxidase. (After Okunuki.)

Selected Readings

Brierley, G. P., Merola, A. J., and Fleischer, S. (1962) Studies of the electron transfer chain. XLIX. Sites of phospholipid involvement in the electron transfer chain, *Biochim. Biophys. Acta* **64**:218.

Caughey, W. S., Wallace, W. J., Volpe, J. A., and Yoshikawa, S. (1976) Cytochrome *c* oxidase, in *The Enzymes* (P. D. Boyer, ed.), Vol. XIII, Academic Press, New York, pp. 299–344.

Chance, B., Saronio, C., and Leigh, J. S., Jr. (1975) Functional intermediates in the reaction of cytochrome oxidase with oxygen, *Proc. Natl. Acad. Sci. U.S.A.* **72**:1635.

Eytan, G. D., and Schatz, G. (1975) Cytochrome *c* oxidase from baker's yeast V. Arrangement of the subunits in the isolated and membrane-bound enzyme, *J. Biol. Chem.* **250**:767.

Fowler, L. R., Richardson, S. W., and Hatefi, Y. (1962) A rapid method for the preparation of highly purified cytochrome oxidase, *Biochim. Biophys. Acta.* **64**:170.

Fuller, S. D., Capaldi, R. A., and Henderson, R. (1979) Structure of cytochrome *c* oxidase in deoxycholate-derived two-dimensional crystals, *J. Mol. Biol.* **134**:304.

King, T. E., Kuboyama, M., and Takemori, S. (1964) On cardiac cytochrome oxidase. A cytochrome *c*-cytochrome oxidase complex, in *Oxidases and Related Redox Systems* (T. E. King, H. S. Mason, and M. Morrison, eds.), Wiley, New York, pp. 707–736.

Lemberg, M. R. (1969) Cytochrome oxidase, *Physiol. Rev.* **49**:48.

Ludwig, B., and Schatz, G. (1980) A two subunit cytochrome oxidase (cytochrome aa_3) from *Paracoccus denitrificans, Proc. Natl. Acad. Sci. U.S.A.* **77**:196.

Mason, T. L., Poyton, R. O., Wharton, D. C., and Schatz, G. (1973) Cytochrome *c* oxidase from baker's yeast I. Isolation and properties, *J. Biol. Chem.* **248**:1346.

McConnell, D. G., Tzagoloff, A., MacLennan, D. H., and Green, D. E. (1966) Studies of the electron transfer system. The formation of membranes by purified cytochrome oxidase, *J. Biol. Chem.* **241**:2373.

Okunuki, K. (1966) Cytochromes and cytochrome oxidase, in *Comprehensive Biochemistry. Biological Oxidations* (M. Florkin and E. Slotz, eds.), Vol. 14, Elsevier, Amsterdam, pp. 232–308.

Poyton, R. O., and Schatz, G. (1975) Cytochrome *c* oxidase from baker's yeast III. Physical characterization of isolated subunits and chemical evidence for two different classes of polypeptides, *J. Biol. Chem.* **250**:752.

Rubin, M., and Tzagoloff, A. (1973) Assembly of the mitochondrial membrane system IX. Purification, characterization, and subunit structure of yeast and beef cytochrome oxidase, *J. Biol. Chem.* **248**:4269.

Sebald, W., Machleidt, W., and Otto, J. (1973) Products of mitochondrial protein synthesis in *Neurospora crassa.* Determination of equimolar amounts of three products in cytochrome oxidase on the basis of amino acid analysis, *Eur. J. Biochem.* **38**:311.

Smith, L. (1955) Spectrophotometric assay of cytochrome *c* oxidase in *Methods in Biochemical Analysis* (D. Glick, ed.), Interscience, New York, pp. 427–434.

Steffens, G., and Buse, G. (1976) Studien an Cytochrom *c* Oxidase, I. Reinigung und Charackterisierung des Enzyms aus Rinderhersen und Identifizierung der im komplex enthaltenen Peptidketten, *Hoppe Seylers Z. Physiol. Chem.* **357**:1125.

Steffens, G. J., and Buse, G. (1979) Studies on cytochrome *c* oxidase, IV. Primary structure and function of subunit II. *Hoppe Seylers Z. Physiol. Chem.* **360**:613.

Tzagoloff, A., and MacLennan, D. H. (1965) Studies of the electron-transfer system LXIV. Role of phospholipid in cytochrome oxidase, *Biochim. Biophys. Acta* **99**:476.

Tzagoloff, A., and Rubin, M. S. (1974) Mitochondrial products of yeast ATPase and cytochrome oxidase, in *The Biogenesis of Mitochondria* (A. Kroon and M. Saccone, eds.), Academic Press, New York, pp. 405–421.

Tzagoloff, A., and Wharton, D. C. (1965) Studies of the electron-transfer system LXII. The reaction of cytochrome oxidase with carbon monoxide, *J. Biol. Chem.* **240**:2628.

Yonetani, T. (1963) The "*a*" type cytochrome, in *The Enzymes* (P. D. Boyer, H. Lardy, and K. Myrback, eds.), Vol. 8, Academic Press, New York, pp. 41–79.

6

Oxidative Phosphorylation

Oxidative phosphorylation is the single most important function of mitochondria. As the term implies, it is the process whereby the energy released from the oxidation reactions of the electron transfer chain is used for the synthesis of ATP. Because of its central role in aerobic metabolism, biochemists have invested substantial time and effort in studies of the mechanism of oxidative phosphorylation. Although numerous solutions to this problem have been proposed, they usually had to be abandoned in light of later evidence—the regularity with which this has happened has caused the remark to be made that no worse fate could befall anyone working on oxidative phosphorylation than to solve it.

Although there are still many aspects of oxidative phosphorylation that are not understood, significant advances have been made over the years, and the broad outlines of the mechanism are known. It is not possible in a short space to deal adequately with all or even a fraction of the phenomenology and details that have emerged from some 40 years of experimentation. The discussion will therefore be restricted to only the most salient findings and some of the more tenable mechanistic theories. It is hoped that this will give the reader, if not a good, at least a general feeling for the problem.

I. The Concept of Coupling

Mitochondria are often referred to as being tightly or loosely coupled. What do these terms signify? The concept of the coupled state is best illustrated by a simple experiment in which the respiratory activity of mitochondria is measured in the presence or absence of the phosphate acceptor, ADP. The rate of oxygen consumption can be followed either manometrically in a Warburg apparatus or polarographically with a platinum or Clark electrode. The latter method monitors the concentration of oxygen in the assay solution and is more convenient if many different additions have to be made during the course of the reaction.

The oxygen uptake trace shown in Fig. 6.1 is typical of measurements made with a Clark oxygen electrode. Mitochondria are added to a reaction mixture containing phosphate buffer, Mg^{2+}, and an NAD-linked substrate such as glutamate. The oxygen uptake under these conditions is called *state 4 respiration* and is the rate at which mitochondria oxidize substrates in the absence of a phosphate acceptor. In state 4, there is no phosphorylation, and if the mitochondria are well coupled, the respiratory rate is low. When ADP is added, the rate of oxygen uptake increases, and this new rate is called *state 3 respiration.* The ratio of the state 3 to state 4 respiration is the *respiratory control quotient (R.C.)* which provides a measure of how tightly coupled the mitochondria are. In poorly coupled or uncoupled mitochondria, state 4 respiration is high, and the R.C. approaches values of 1. The dependence of substrate oxidation on the presence of a phosphate acceptor indicates that, in coupled mitochondria, electron transfer and ATP synthesis are geared to each other. Unless the energy released during electron transport is used in some energy-dependent process such as ATP synthesis or ion transport, oxidation of substrate is limited. The true index of coupling, therefore, is the state 4 respiration which shows how readily the oxidative energy can be dissipated when it is not being utilized, and in perfectly coupled mitochondria, it might be expected that the rate would be zero.

II. Uncouplers and Inhibitors of Oxidative Phosphorylation

There are three general classes of reagents that prevent the coupled synthesis of ATP by mitochondria: uncouplers, inhibitors, and ionophores. Ionophores are compounds that facilitate the uptake of ions—they will be discussed later in the context of mitochondrial ion transport.

As a rule, uncouplers are lipophilic compounds that have properties of weak acids. The list of such reagents is very long, but the most commonly used are 2,4-dinitrophenol (DNP) and carbonyl cyanide *m*-chlorophenylhydrazone

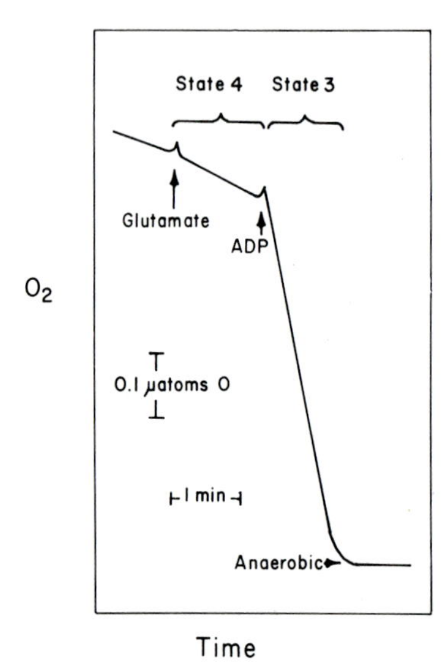

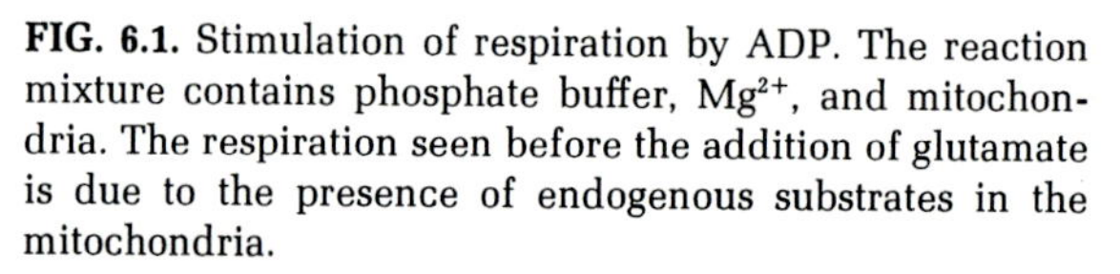
FIG. 6.1. Stimulation of respiration by ADP. The reaction mixture contains phosphate buffer, Mg^{2+}, and mitochondria. The respiration seen before the addition of glutamate is due to the presence of endogenous substrates in the mitochondria.

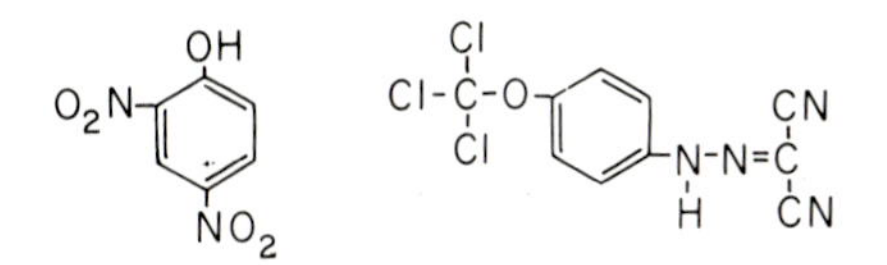

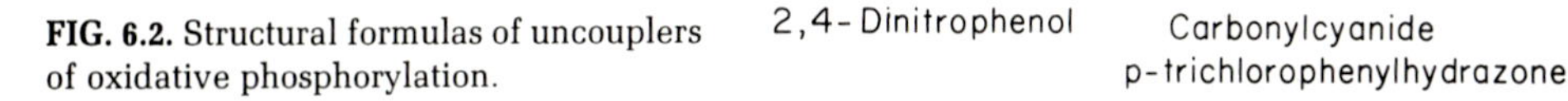

FIG. 6.2. Structural formulas of uncouplers of oxidative phosphorylation.

(*m*-CCCP) (Fig. 6.2). The latter is an especially potent uncoupler that stops phosphorylation at concentrations of 10^{-7} M. Uncouplers are characterized by their ability to release state 4 respiration in the absence of phosphate acceptor. This is shown in Fig. 6.3 where it is seen that after addition of dinitrophenol, there is an immediate increase in the rate of respiration which continues until all the oxygen is exhausted. The rate in the presence of the uncoupler is usually the same as the state 3 rate. Reagents such as DNP and *m*-CCCP are thought to prevent ATP synthesis by promoting the dissipation of the energy produced from electron transport.

Inhibitors of oxidative phosphorylation also prevent ATP synthesis, but their mode of action is entirely different from that of uncouplers. The best-known inhibitors are oligomycin, tributyltin chloride, and aurovertin (Fig. 6.4). Although inhibitors are also lipid-soluble compounds, there are no obvious chemical or structural properties that unify them as a group. The effect of inhibitors on the respiratory activity of coupled mitochondria is easily distinguished from that of uncouplers. This is shown in Fig. 6.5. Here, the state 4 respiration is released with ADP, and after a state 3 rate is established, the inhibitor oligomycin is added. The inhibitor causes the rate to return to the state 4 level. When the experiment is repeated and state 4 respiration is released with an uncoupler instead of ADP, oligomycin has no further effect. Oligomycin and other inhibitors are thought to prevent the utilization of the oxidative energy for ATP synthesis by blocking some terminal phosphorylation step.

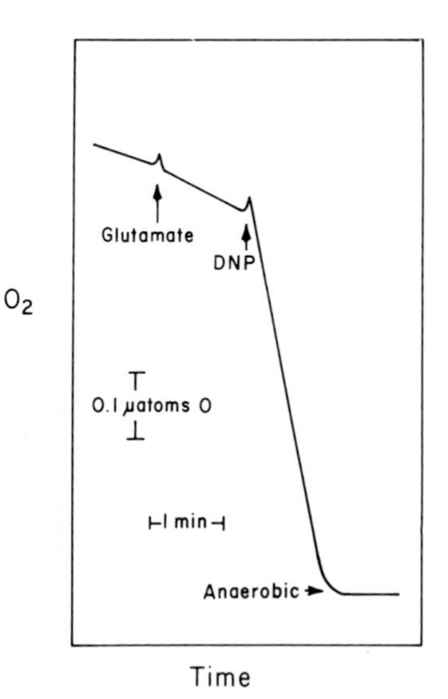

FIG. 6.3. Stimulation of respiration by dinitrophenol. The reaction mixture contains phosphate buffer, Mg^{2+}, and mitochondria. The substrate is glutamate.

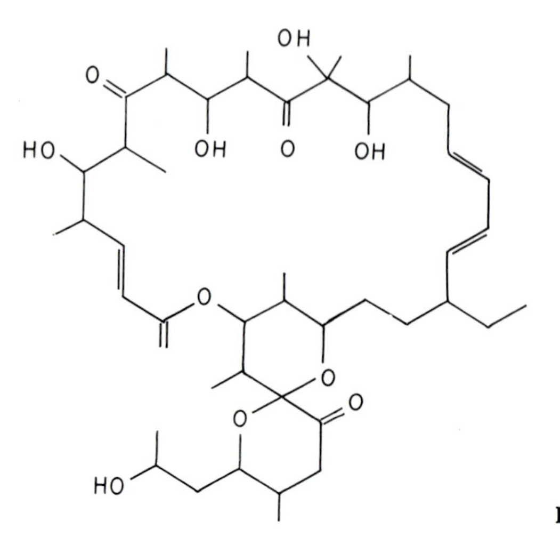

FIG. 6.4. Structural formula of oligomycin B.

III. Efficiency of Phosphorylation

A standard way of expressing the efficiency of phosphorylation is the P/O ratio or the mole equivalents of phosphate esterified per oxygen consumed. The P/O values can be determined by one of several methods. In the older method, oxygen uptake was measured manometrically in a Warburg respirometer, and phosphate esterification was assayed by the disappearance of inorganic phosphate from the reaction mixture. In recent years it has become more popular to determine the ATP formed by a radiochemical assay of the incorporation of $^{32}P_i$ into organic phosphate.

An alternative method of measuring the phosphorylation efficiency of

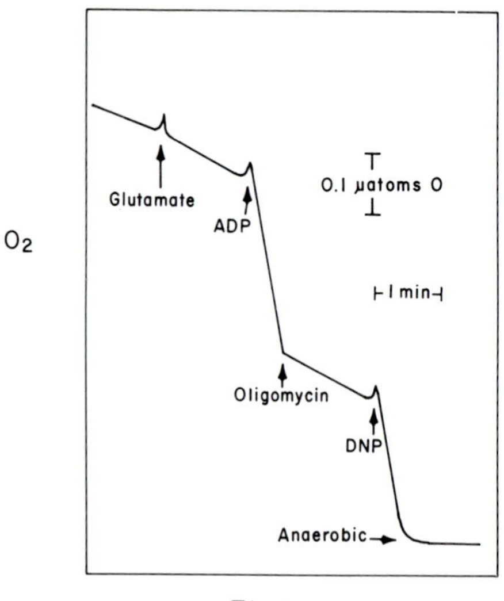

FIG. 6.5. Inhibition of respiration by oligomycin. Same conditions as in Fig. 6.1.

mitochondria is based on the release of state 4 respiration by ADP. In the experiment of Fig. 6.1, the amount of ADP added was in excess of the total oxygen present in the reaction mixture, and state 3 respiration proceeded until all the oxygen was used up from the solution. If instead, a limiting amount of ADP is added, the rate of state 3 respiration will be maintained until all the ADP has been converted to ATP, at which point the initial state 4 rate will be reestablished. This can be repeated several times by successive additions of graded amounts of ADP as shown in Fig. 6.6. The efficiency of phosphorylation is calculated from the oxygen trace and is equal to the ratio of the known amount of ADP added to the amount of oxygen consumed during the burst of the state 3 respiration. The calculated ADP/O ratio for the example of Fig. 6.6 is 3.0. It has been experimentally verified that the ADP/O and the P/O values are identical. This is understandable, since in coupled mitochondria, all externally added ADP is quantitatively converted to ATP and therefore represents the amount of phosphate esterified.

The yields of ATP with different substrates have been carefully studied in many laboratories. Mitochondria that exhibit high respiratory control also have high P/O values. This also applies to mitoplasts made by the digitonin procedure (see Chapter 2, Section II.B). In both types of preparations, the P/O values approach 3 with NAD-linked substrates and 2 with succinate. Submitochondrial particles obtained by sonic disruption of mitochondria, however, have very high rates of oxidation of NADH and succinate in the absence of ADP and show no respiratory control. Nonetheless, when properly made, such particles are capable of efficient phosphorylation with P/O values that are almost as high as those of coupled mitochondria. These observations indicate that the submitochondrial vesicles, even though poorly coupled, are still able to trap oxidative energy. The trapping or coupling system therefore appears to be able to tolerate a certain degree of "uncoupling."

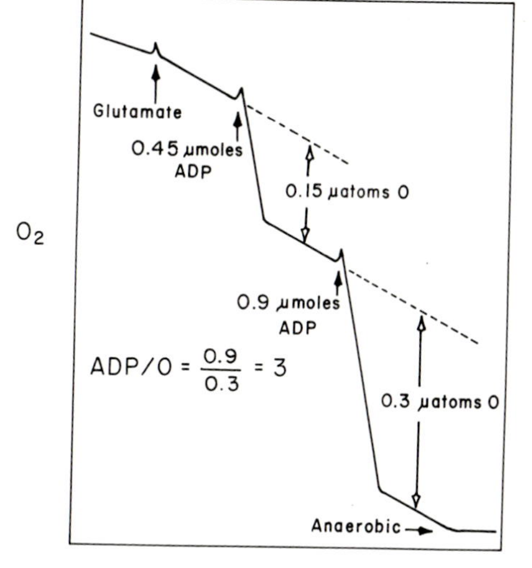

FIG. 6.6. Estimation of the efficiency of phosphorylation. The ADP/O ratio is calculated from the amount of oxygen consumed in the burst of state 3 respiration that follows the addition of ADP. Same conditions as in Fig. 6.1.

IV. Phosphorylation Sites

We have seen that the oxidation of NADH and succinate is coupled to the synthesis of 3 and 2 mole equivalents of ATP, respectively. Several approaches have been used to locate the sites in the electron transfer chain where the energy conservation events occur.

A. Crossover Points

The phosphorylation sites have been studied by Chance and Williams by a spectrophotometric method in which the oxidation–reduction states of the different components of the chain are measured in the transition of state 4 to state 3 respiration.

During state 3 respiration, the steady-state levels of reduction of the electron transfer carriers are determined by their positions relative to substrate and oxygen. The carriers closest to substrate are more reduced than those closer to oxygen (Table 6.1). Let us recall that when an inhibitor of electron transport such as antimycin is added, all the components of the chain on the substrate side of and including cytochrome *b* become completely reduced, whereas components that are on the oxygen side of cytochrome *b* become oxidized (see Fig. 4.28). This experiment tells us that antimycin blocks the oxidation of cytochrome *b* and localizes its site of action either at the level of cytochrome *b* (this could be conceived in terms of an antimycin–cytochrome *b* complex that can be reduced but not oxidized) or between cytochrome *b* and cytochrome c_1. Chance and Williams have defined the *crossover point* to be the site where a large difference in the oxidation–reduction levels of neighboring carriers is observed. In the case of antimycin-inhibited mitochondria, the crossover occurs between cytochromes *b* and c_1.

Crossover points have proven to be of general usefulness in identifying the reactions in a metabolic sequence that exert a regulatory influence on the overall rate. For example, in the case of the respiratory chain sequence, the transition of state 3 to state 4 respiration is accompanied by an inhibition of electron transport. This inhibition may be visualized as resulting from the accumulation of one or more of the carriers in some energized form that cannot be oxidized unless the energy of oxidation is used for ATP synthesis. It should, therefore, be possible to observe crossover points between those components of the chain where energy conservation events occur, and the number of such crossovers should be equal to the number of energy conservation steps or phosphorylation sites.

TABLE 6.1. Crossover Points during State 4 Respiration in Rat Liver Mitochondria[a]

State	O_2	ADP Level	Respiration rate	Rate limiting	Steady-state percent reduction				
					a	*c*	*b*	Flavoprotein	NADH
4	>0	low	slow	ADP	0	14	35	40	>99
3	>0	high	fast	Resp. chain	4	6	16	20	53

[a]The percent reduction of the carriers was estimated spectrophotometrically as in Table 4.11 (Chance and Williams, 1956).

From the steady-state values of Table 6.1, it is seen that in state 4 respiration, a crossover occurs between cytochrome *c* and cytochrome *a*. Chance and Williams proposed this to be one of the phosphorylation sites. It is also evident that the largest differences in the oxidation–reduction levels of the other components examined are between NADH and flavoprotein and between cytochromes *b* and c_1. The differences were made more evident when the rate of substrate oxidation was retarded by partially inhibiting cytochrome oxidase with a low concentration of azide. At very low concentrations of azide, a clear crossover was observed between cytochrome *b* and *c*, and at higher concentrations between NADH and flavin. These experiments suggested that there are three phosphorylation sites in the NADH oxidase chain, the first at the level of flavin, the second in the cytochrome *b*, and the third in the cytochrome oxidase regions.

B. Site-Specific Assays

The phosphorylation sites have also been determined from the known P/O values with different substrates and by dissection of the chain with the combined use of electron transfer inhibitors and artificial electron donors and acceptors. Since some of the substrates and artificial electron donors are not accessible to the inner compartment of the mitochondrion, most of these studies have been done with submitochondrial particles where there are no permeability barriers. The assays used to study the phosphorylation associated with the spans NADH → coenzyme Q, NADH → cytochrome *c*, and cytochrome *c* → O_2 are shown in Fig. 6.7.

The oxidation of NADH by coenzyme Q in the presence of an antimycin

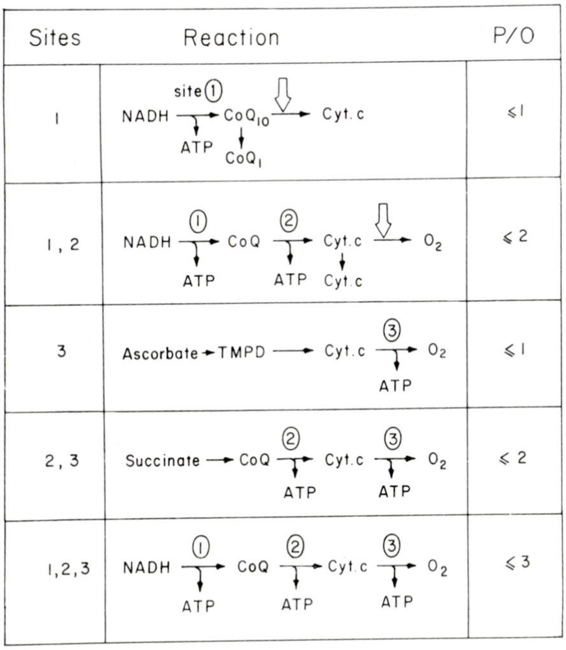

Sites	Reaction	P/O
1	NADH →(site ①, ATP) CoQ_{10} ⇩→ Cyt. c; CoQ_{10} ↓ CoQ_1	≤1
1, 2	NADH →(①, ATP) CoQ →(②, ATP) Cyt. c ⇩→ O_2; Cyt. c ↓ Cyt. c	≤2
3	Ascorbate→TMPD → Cyt. c →(③, ATP) O_2	≤1
2, 3	Succinate → CoQ →(②, ATP) Cyt. c →(③, ATP) O_2	≤2
1,2,3	NADH →(①, ATP) CoQ →(②, ATP) Cyt. c →(③, ATP) O_2	≤3

FIG. 6.7. Site-specific assays of oxidative phosphorylation. The assay of site 1 is done in the presence of antimycin (clear arrow) to block the oxidation of coenzyme Q. In the sites 1 plus 2 assay, cyanide (clear arrow) is used to block oxidation of cytochrome *c*.

block has been shown to yield close to 1 mole equivalent of ATP per mole of NADH oxidized, thus establishing this span to have a phosphorylation site. When the chain is blocked with cyanide and cytochrome *c* is used as the electron acceptor, the yield of ATP is greater than 1 but less than 2, indicating that there is a second site between coenzyme Q and cytochrome *c*. The third site occurs at the level of cytochrome oxidase. This can be shown in an assay that utilizes ascorbate and TMPD to reduce cytochrome *c* and antimycin to block any contribution from the first two sites. Under these conditions, the oxidation of cytochrome *c* by oxygen is coupled to the synthesis of ATP with P/O values of approximately 1. No synthesis of ATP has been found to occur in the succinate → coenzyme Q span. This accounts for the observed P/O values of 3 with NADH and 2 with succinate.

The localization of the phosphorylation sites by the crossover method and by the site-specific assays points to each respiratory complex of the NADH oxidase chain as being a unit of energy conservation. It should also be recalled that each of these complexes of the electron transfer chain represents a sufficient electropotential drop to permit the synthesis of at least 1 mole of ATP for each net transfer of two electrons (see Chapter 4, Section II).

V. Evidence for the Existence of a Nonphosphorylated Intermediate

A substantial body of experimental evidence revolving around the effects of uncouplers and inhibitors on various energy-coupled processes of mitochondria has been rationalized in terms of a generalized scheme of oxidative phosphorylation as shown in Fig. 6.8. The scheme incorporates two different events. The first is a conservation step in which the energy derived from the oxidation of substrate is coupled to the formation of an intermediate denoted by the "squiggle" symbol. As will be seen subsequently, the nature of this intermediate is a crucial issue in understanding the coupling mechanism; however, it need not concern us here and for convenience will be referred to as the nonphosphorylated high-energy intermediate without the implication that it is a real chemical species. A second feature of the scheme is that the common nonphosphorylated intermediate can be used in various ways either for the synthesis of ATP or for other work performances of the mitochondrion.

There are several other aspects of the mechanism to be noted. As indicated in Fig. 6.8, both the coupling reaction that leads to the formation of the nonphosphorylated intermediate and its utilization in the phosphorylation of ADP

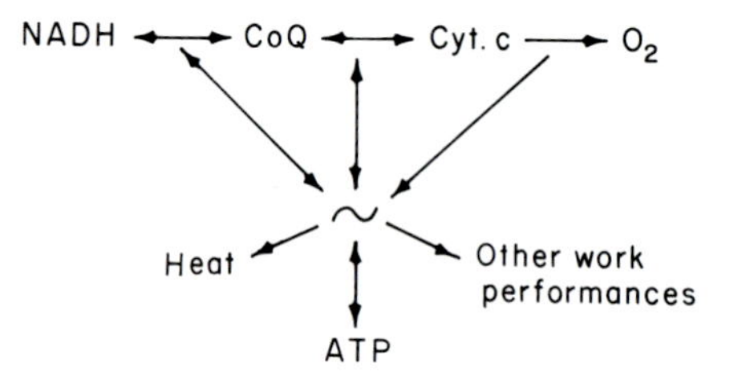

FIG. 6.8. Formation of a common high energy intermediate of oxidative phosphorylation. The high energy intermediate (squiggle) can be used for ATP synthesis as well as other energy dependent functions according to this scheme.

are reversible events. This means that the squiggle can be generated either from the oxidation of substrates or from the hydrolysis of ATP. The scheme also illustrates the essential difference in the way uncouplers and inhibitors of oxidative phosphorylation exert their effects. The uncouplers act by facilitating a breakdown or discharge of the nonphosphorylated intermediate, whereas the inhibitors prevent the terminal reactions involved in the phosphorylation of ADP. Based on the postulated sites and mode of action, uncouplers should affect energy-dependent processes when the nonphosphorylated intermediate is generated either from the substrate or the ATP side. On the other hand, inhibitors should only affect the formation of the intermediate from the ATP side.

We can now proceed to examine how the general mechanism outlined above is useful in explaining a large number of findings related to the energy-coupling functions of mitochondria.

1. Respiratory Control

The tight coupling between the energy-producing and requiring reactions is evidenced by the phenomenon of respiratory control. This phenomenon has been interpreted to be an expression of the stability of the nonphosphorylated intermediate which creates a back pressure preventing further electron flow through the chain. The back pressure arises from the postulated reversibility of the coupling reaction.

The scheme of Fig. 6.8 shows three ways in which respiration can be stimulated by creating a drain on the intermediate: phosphorylation of ADP, dissipation to heat by uncouplers, and utilization for other energy-dependent reactions. We have already seen that both ADP and chemical uncouplers are capable of markedly stimulating the rates of substrate oxidation. Enhancement of respiration has also been shown to occur when mitochondria are engaged in other work functions. For example, mitochondria can accumulate Ca^{2+} against a concentration gradient. This transport is dependent on an energy source and is thought to involve the participation of the same intermediate that is used for ATP synthesis. When Ca^{2+} is added to mitochondria under state 4 conditions, the rate of oxygen uptake increases and is equal to that obtained in the presence of an uncoupler (Fig. 6.9). The release of state 4 respiration by Ca^{2+} or ADP probably occurs through a discharge of the nonphosphorylated intermediate, in one case because of its utilization for transport and in the other for the esterification of ADP and phosphate.

2. ATPase Activity

The ATPase activity of well-coupled mitochondria is very low. An active ATPase is elicited by uncouplers or treatments that cause mitochondrial damage. Submitochondrial particles that have no respiratory control and are only poorly coupled also exhibit a high ATPase activity (Table 6.2). Both the uncoupler-stimulated ATPase of mitochondria and the overt ATPase activity of submitochondrial particles are inhibited by oligomycin and other inhibitors of oxidative phosphorylation. These observations have been interpreted to indicate that the latent ATPase is part of the phosphorylation mechanism, and according to the scheme of Fig. 6.8, the hydrolytic reaction must represent a reversal of the terminal phosphorylation reactions. Under normal circumstances, hydrolysis of ATP is prevented because of the stability of the nonphosphorylated intermediate. By introducing a leak in the system, e.g., with uncouplers, the energetically favored hydrolytic reaction can proceed at a high rate.

3. Utilization of the Nonphosphorylated Intermediate for Energy-Requiring Functions

The strongest argument for the participation of a nonphosphorylated intermediate in oxidative phosphorylation comes from studies of various energy-dependent processes of mitochondria such as the reduction of NAD^+ by succinate, the transhydrogenation of pyridine nucleotides, and active transport of monovalent and divalent metals.

Chance first reported that mitochondria catalyze an energy-dependent reduction of NAD^+ by succinate through a reversal of electron flow in the respiratory chain. This reaction has been extensively studied in submitochondrial particles and is now known to utilize the energy of the nonphosphorylated intermediate formed either by the oxidation of substrate or the hydrolysis of ATP.

Figure 6.10 outlines three ways of generating the intermediate in order to drive the reduction of NAD^+. In method 1, succinate acts both as a source of the reducing equivalents for NAD^+ and of energy for the formation of the intermediate at sites 2 and 3 of the chain. In method 2, the oxidation of succinate is prevented by antimycin. The substrate in this assay is ascorbate plus TMPD which can lead to the formation of nonphosphorylated intermediate at site 3. The reduction of NAD^+ in methods 1 and 2 is blocked by uncouplers but not by inhibitors of oxidative phosphorylation. This indicates that the energy for

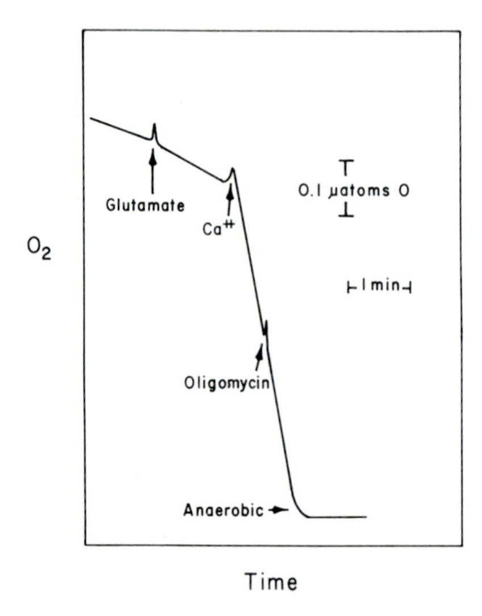

FIG. 6.9. Stimulation of respiration by Ca^{2+}. Same conditions as in Fig. 6.1.

TABLE 6.2. Stimulation of ATPase by Dinitrophenol[a]

	μmoles ATP hydrolyzed per min per mg protein	
	−DNP	+DNP
Mitochondria	0.26	1.25
Submitochondrial particles	1.57	1.50

[a]Beef heart mitochondria and submitochondrial particles were assayed for ATPase in the presence of Mg^{2+} and ATP.

the reaction comes not from ATP but from some uncoupler-sensitive intermediate generated during respiration. That this intermediate can also be formed from ATP is shown in the third method (3) where the oxidation of succinate is again inhibited with antimycin, but no other substrate is added. The reduction of NAD^+ under these conditions can be brought about by ATP. In this case, however, the reaction is sensitive both to uncouplers and to inhibitors, suggesting that the intermediate is formed from the energy of hydrolysis of ATP. The reduction of NAD^+ has been shown to be sensitive to inhibitors of the NADH-coenzyme Q reductase complex (e.g., amobarbital or rotenone) and therefore appears to involve a reversal of electron flow in this segment of the chain. This reaction is energetically unfavorable because of the negative ΔE_0 (~350 mV) which explains the requirement of an energy input.

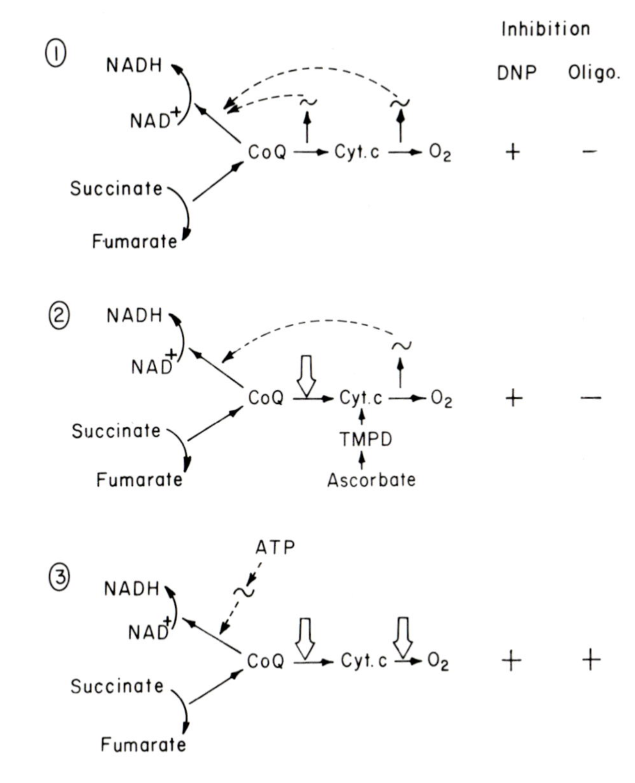

FIG. 6.10. Three assays for the energy-dependent reversal of electron transfer in the NADH–coenzyme Q span. 1: The reduction of NAD^+ is driven by high-energy intermediates generated at sites 2 and 3 from the oxidation of succinate. 2: NAD^+ reduction is driven by the intermediate generated at site 3 only. The clear arrow indicates an antimycin block. 3: NAD^+ reduction is driven by a high-energy intermediate generated from ATP. Electron transfer can be inhibited either by antimycin or cyanide (clear arrows) in this assay.

Submitochondrial particles contain a pyridine nucleotide transhydrogenase that catalyzes the reduction of $NADP^+$ by NADH.

$$\mathrm{NADH} + \mathrm{NADP}^+ \rightleftharpoons \mathrm{NAD}^+ + \mathrm{NADPH} \qquad [6.1]$$

The equilibrium constant of this reaction is approximately 1, and in the presence of excess NADH, the reduction of $NADP^+$ occurs in the absence of any energy input. Ernster and co-workers discovered that the rate of transhydrogenation is enhanced by oxidizable substrates or by ATP. The energized and nonenergized reactions are catalyzed by the same enzyme, but its kinetic properties are different under energized conditions. The participation of the nonphosphorylated intermediate in the energized transhydrogenation is supported by the same type of evidence that has been adduced for the energy-dependent reduction of NAD^+ by succinate. This is shown in Fig. 6.11 where it is seen that the rate of $NADP^+$ reduction by NADH is increased when succinate or ATP is added to the reaction assay. Although both substrate- and ATP-driven transhydrogenation are sensitive to uncouplers, inhibition by oligomycin is observed only in the case of the ATP-driven reaction.

The transport of metals serves as the third example of the utilization of the nonphosphorylated intermediate for a mitochondrial energy-coupled function. Although the ion transport systems of the mitochondrion will be discussed more fully in Chapter 9, it may be mentioned here that the uptake of monovalent and divalent metals shows the same substrate or ATP dependence as reverse electron flow and transhydrogenation. This is illustrated in Table 6.3 for the active transport of Ca^{2+}. The accumulation of Ca^{2+} by mitochondria can be supported by energy derived from a substrate or from ATP. The effects of uncouplers and inhibitors, again, are consistent with the idea that active transport of metals is driven by a nonphosphorylated intermediate of oxidative phosphorylation.

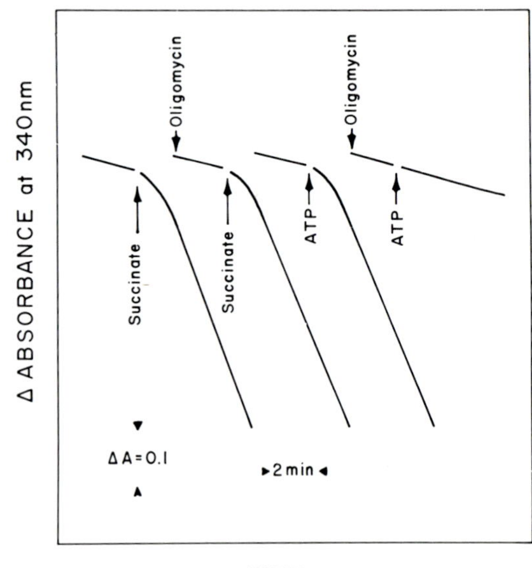

FIG. 6.11. Energy-dependent pyridine nucleotide transhydrogenase. The reaction mixture contains buffer, Mg^{2+}, submitochondrial particles, rotenone (to block NADH oxidation), NADH, $NADP^+$, glutathione, and glutathione reductase (to keep NAD^+ reduced). The downward trace indicates a reduction of $NADP^+$ as measured at 340 nm. (From Lee and Ernster, 1964.)

4. Competition of Oxidative Phosphorylation and Other Coupled Reactions for the Nonphosphorylated Intermediate

The participation of a common nonphosphorylated intermediate in the synthesis of ATP and other energy-dependent functions of mitochondria has been confirmed by experiments in which the rate of a reaction such as energized transhydrogenation is measured under phosphorylating and nonphosphorylating conditions. In each case, it has been found that the phosphorylation of ADP is a competing reaction that reduces the efficiency of the other coupled process.

5. Are the Nonphosphorylated Intermediates Generated at the Three Coupling Sites Identical?

This is an important question in considering possible mechanisms of oxidative phosphorylation, and although an unqualified answer cannot be given at present, most of the available evidence is consistent with at least a functional equivalence of the intermediates formed at the three coupling sites. We have already seen that the reduction of NAD^+ by succinate can be driven by intermediates generated either at sites 2 and 3 (succinate oxidation) or at site 3 alone (ascorbate plus TMPD). Since the reduction of NAD^+ occurs through a reversal of electron flow in the NADH–coenzyme Q segment corresponding to the first coupling site, the above observations indicate that the nonphosphorylated intermediate of sites 2 and 3 can substitute at site 1. This conclusion is also supported by studies of the other energy-dependent functions which in every case can be driven by intermediates generated at any one of the three coupling sites.

VI. Partial Reactions of Oxidative Phosphorylation

Mitochondria catalyze four exchange reactions that are considered to be partial reactions of oxidative phosphorylation.

TABLE 6.3. ATP- and Substrate-Driven Uptake of Ca^{2+} by Beef Heart Mitochondria

	Ca^{2+} (μmoles per mg protein)	
	ATP	ATP + Succinate
Complete system[a]	1.65	2.37
+ antimycin	1.32	1.59
+ oligomycin	0.03	1.20
+ DNP	0.56	0.52

[a]The complete system contained sucrose, Mg^{2+}, phosphate, imidazole buffer, Ca^{2+}, and beef heart mitochondria (Brierley *et al.*, 1964).

A. ATP–P_i Exchange

In this reaction, the phosphate in the γ position of ATP is exchanged with free inorganic phosphate. The exchange is catalyzed by mitochondria or submitochondrial particles and is assayed by measuring the incorporation of $^{32}P_i$ into ATP in the presence of an inhibitor of electron transport in order to prevent any net synthesis of ATP. The ATP–P_i exchange is sensitive to uncouplers and inhibitors of oxidative phosphorylation but is not affected by inhibitors of electron transfer. It has also been shown recently that the exchange reaction can be catalyzed by membranes reconstituted from the ATPase complex (see Chapter 8, Section IV). The ATP–P_i exchange is thought to reflect the reversibility of the the terminal reactions of oxidative phosphorylation.

B. ATP–ADP Exchange

This exchange has been studied by Wadkins and Lehninger. The reaction is catalyzed by mitochondria and is inhibited by uncouplers and inhibitors of oxidative phosphorylation. It has not been observed in submitochondrial particles. Several ATP–ADP exchange enzymes have been purified from mitochondria, but no sensitivity to uncouplers and inhibitors was detected in the isolated proteins. The function of these proteins in oxidative phosphorylation is not known. It is likely, however, that they are not essential to the coupling mechanism but rather play a role in ancillary reactions involving the transfer of high-energy phosphate.

C. ATP–$H_2{}^{18}O$ Exchange

The exchange of ^{18}O of water with ATP is catalyzed by mitochondria and submitochondrial particles and qualifies as a partial reaction of oxidative phosphorylation because of the sensitivity to uncouplers and inhibitors. The exchange has been interpreted by Boyer as indicating that the bridge oxygen in the terminal phosphate of ATP is derived from ADP.

D. P_i–$H_2{}^{18}O$ Exchange

This exchange reaction has also been extensively studied in Boyer's laboratory. The exchange of ^{18}O of water with inorganic phosphate is catalyzed by mitochondria and submitochondrial particles and is sensitive to uncouplers and inhibitors of oxidative phosphorylation. The concentration of uncouplers required to inhibit this exchange reaction is higher than for the other three exchanges. Originally, it was thought that the exchange of the oxygen of water with phosphate occurs through the formation of a high-energy intermediate involving a covalently bound phosphate to a protein. More recently, an alternate explanation has been proposed (see conformational hypothesis in Section VII of this chapter).

VII. Role of Structure in Oxidative Phosphorylation

There is agreement among most people working in the field that certain features of the gross and fine structure of the mitochondrion are essential for energy coupling. The structure and function relationship is an intriguing question that has been difficult to study experimentally; aside from the specific role of the membrane in the chemiosmotic model (see below), current ideas in this area are vague and for the most part founded on correlative observations of the loss of various phenomena in going from the morphologically complex mitochondrion to the simpler membranes of submitochondrial particles.

It is possible to study different expressions of the coupling mechanism in four types of systems of decreasing complexity: (1) mitochondria, (2) mitoplasts, (3) submitochondrial vesicles, and (4) membranes reconstituted from the respiratory and ATPase complexes. Let us confine our attention to the first three systems and consider the more obvious differences in their morphologies and coupling capabilities.

Mitoplasts are of intermediate complexity and differ from mitochondria in the loss of the outer membrane and intermembrane components. Even though the inner membrane of mitoplasts is more extended, and the foldings giving rise to the cristae are absent, its orientation with respect to the outside medium is the same as in mitochondria. The 90-Å ATPase particles are on the interior side of the inner membrane in both mitochondria and mitoplasts. Submitochondrial particles prepared from sonically disrupted mitochondria are virtually free of all the soluble components of the organelle and consist predominantly of sealed inner membrane vesicles. The two faces of the inner membrane, however, have a reverse orientation so that the ATPase is exposed to the solutes in the surrounding medium (Fig. 6.12).

The functional properties of the three preparations (Table 6.4) show essentially no differences between mitochondria and mitoplasts with respect to most of the phenomena usually associated with coupling. The outer membrane and the cristal arrangement of the inner membrane can therefore be excluded as being important in functions such as respiratory control and energy-dependent ion uptake. This, however, does not apply to submitochondrial particles which, despite their ability to carry out efficient phosphorylation, are grossly modified in other respects. Most notably, submitochondrial particles have no respiratory control, exhibit a high ATPase activity, and are deficient in transport functions.

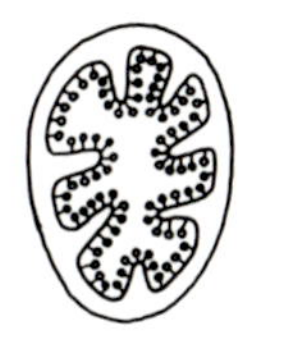

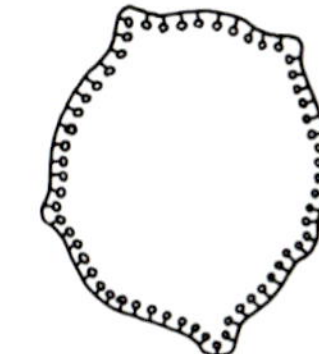

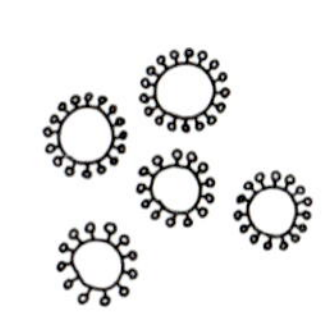

FIG. 6.12. Orientation of inner membrane in mitochondria, mitoplasts, and submitochondrial vesicles.

TABLE 6.4. Energy-Coupled Functions of Mitochondria, Mitoplasts, and Submitochondrial Particles

Function	Mitochondria	Mitoplasts	SMP
P/O	High	High	High
Respiratory control	High	High	Absent
Ion transport	High	High	Absent
Reverse electron flow	High	High	High
ATPase	Low	Low	High
DNP-stimulated ATPase	Yes	Yes	No

Can the loss or acquisition of the new properties in submitochondrial particles be explained by their inverse orientation? The inability of these preparations to perform a vectorial process, e.g., transport, has been ascribed to the wrong sidedness of the membrane. It is unlikely, however, that the membrane orientation per se is related to the loose coupling, since there are conditions under which inverted particles can be induced to be tightly coupled. High concentrations of oligomycin inhibit oxidative phosphorylation and ATP-driven reactions in submitochondrial particles. Ernster and co-workers found that at subinhibitory concentrations, oligomycin actually makes the particles more coupled. For instance, energy-coupled transhydrogenation and NAD^+ reduction by succinate are stimulated in the presence of low concentrations of oligomycin. The coupling effect is also seen in the restoration of respiratory control. Figure 6.13 shows that the oxidation of NADH by submitochondrial particles is inhibited by oligomycin and that the inhibition is relieved by an uncoupler. The coupling effect of oligomycin has been assumed to result from a stabilization of the squiggle or nonphosphorylated intermediate, although how this occurs is not known. Whatever the mechanism, it does point out that loose coupling is not related to the membrane orientation but is probably the result of some more subtle structural changes in the membrane.

The necessity of a membrane structure for oxidative phosphorylation has been a crucial issue on which mechanistic proposals of the coupling process can stand or fall. For example, the chemiosmotic hypothesis which will be discussed in the next section depends on the existence of a membrane that is capable of maintaining a proton gradient generated from electron transport. Although there have been reported claims of oxidative phosphorylation in nonmembranous systems, they have not been confirmed, and it may be assumed that to date even the most resolved coupled systems have depended on the presence of a membrane.

VIII. Current Theories of the Coupling Mechanism

So far this chapter has dealt with the basic phenomenology of oxidative phosphorylation. It was shown that one of the early events in respiratory-chain-linked phosphorylation is an energy conservation step in which the oxidative energy is converted to a form that can be used either for the esterification of ADP and P_i or for a wide range of other energy-demanding functions of the mitochondrion. The identity of the nonphosphorylated intermediate, or squiggle as it is conventionally represented, has been and continues to be the central problem in understanding the coupling mechanism. Some of the hypotheses directed to this question will be reviewed in more or less the same chronological order in which they first appeared.

A. Chemical Hypothesis

In 1953, Slater proposed a mechanism involving chemical intermediates to account for what was then known about oxidative phosphorylation. The intermediates and dismutation reactions were based on the mechanism of ATP synthesis coupled to the oxidation of glyceraldehyde-3-phosphate. Slater's formulation consisted of the following reactions.

$$AH_2 + B + C \rightleftharpoons A{\sim}C + BH_2 \qquad [6.2]$$

$$A{\sim}C + ADP + P_i \rightleftharpoons A + C + ATP \qquad [6.3]$$

$$A{\sim}C + H_2O \rightarrow A + C \qquad [6.4]$$

In the first reaction, the oxidation of the carrier AH_2 is coupled to the synthesis of a high-energy intermediate A~C. This intermediate reacts with ADP and P_i in the second step to form ATP and regenerate A and C. Respiratory control was explained by assuming that in a tightly coupled system, the intermediate A~C is stable and prevents further electron transport by tying up the carrier A. The high respiratory rates observed in the presence of uncouplers or in submitochondrial particles were reasoned to occur through a hydrolytic breakdown of the intermediate as shown in reaction 6.4.

The chemical hypothesis was somewhat modified in subsequent years in

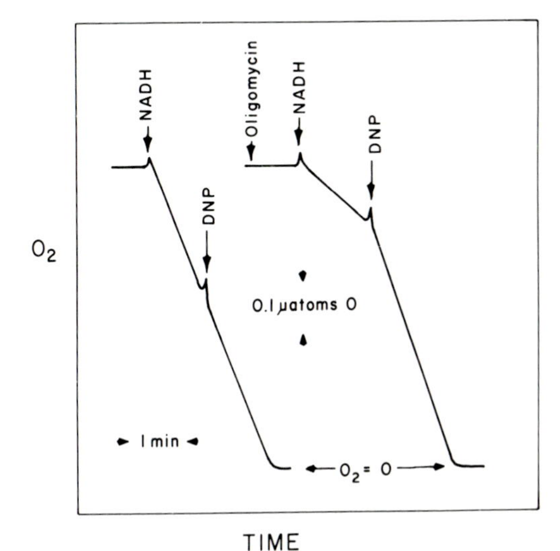

FIG. 6.13. Oligomycin-induced respiratory control in submitochondrial particles. The reaction mixture contains phosphate buffer, Mg^{2+}, and submitochondrial particles. NADH is used as the substrate. Note decrease in respiration when oligomycin is present.

view of new information from studies of partial reactions and evidence about the existence of a common nonphosphorylated intermediate at each of the three phosphorylation sites. The newer schemes have taken the following form.

$$AH_2 + C + B \rightleftharpoons A{\sim}C + BH_2 \quad [6.5]$$

$$A{\sim}C + X \rightleftharpoons A + C{\sim}X \quad [6.6]$$

$$C{\sim}X + P_i \rightleftharpoons X{\sim}P + C \quad [6.7]$$

$$X{\sim}P + ADP \rightleftharpoons ATP + X \quad [6.8]$$

In the original hypothesis, the nonphosphorylated intermediate, A∼C, had to be different for each site, since one of the partners is a carrier of the chain. In the later schemes, this logistic problem was avoided by the introduction of another dismutation reaction leading to the formation of a new chemical species, C∼X in which neither C nor X are carriers. C∼X could, therefore, be common to each of the three phosphorylation sites. A phosphorylated intermediate was included in the scheme in order to account for the exchange of water oxygen with inorganic phosphate. This becomes evident when the reactions are written to indicate the elimination of water.

$$AH_{red} + B_{ox} + C{\cdot}OH \rightleftharpoons A_{ox}{\sim}C + B_{red} + H_2O \quad [6.9]$$

$$A_{ox}{\sim}C + XH \rightleftharpoons X{\sim}C + AH_{red} \quad [6.10]$$

$$X{\sim}C + P{\cdot}OH \rightleftharpoons X{\sim}P + C{\cdot}OH \quad [6.11]$$

Although the chemical hypothesis is consistent with all of the experimental findings, it has fallen into disfavor. The main reason appears to be a general feeling of frustration that has set in as a result of a long and unsuccessful search for evidence of the existence of chemical intermediates.

B. Chemiosmotic Hypothesis

This coupling mechanism was proposed by Mitchell in 1961. The chemiosmotic hypothesis represented a radical departure from the way investigators had been thinking about the problem up to then, and, having gained in stature over the years, it is now the more widely accepted of the different interpretations of oxidative phosphorylation. There are three postulates to the mechanism.

1. The inner mitochondrial membrane has a low conductivity and is impermeable to ionic species including protonated water.
2. The electron transfer chain components are organized in the inner membrane in such a way that during oxidation–reduction, there is an asymmetric translocation of electrons and protons across the membrane.

The immediate consequence of electron transport is the buildup of a pH gradient and of an electrical charge across the membrane. The combined pH and electrical differential is called the protonmotive force which acts as the primary source of energy for ATP synthesis and other energy-coupled reactions.

3. The inner membrane contains a reversible proton-translocating ATPase which can either generate a protonmotive force from the energy of hydrolysis of ATP or, alternatively, can use the potential energy of the protonmotive force to synthesize ATP.

Let us now examine the more detailed features of the last two postulates.

1. Loops of the Respiratory Chain

To account for the three phosphorylation sites, Mitchell proposed three oxidation–reduction loops (O/R loops), each capable of delivering two protons to one side of the membrane per two electrons transported. The principle of an O/R loop is shown in Fig. 6.14. The loop consists of two circuits: one carries protons, and the other electrons. The proton-carrying circuit transfers two H^+ from the substrate SH_2 to one side of the membrane, and the electron circuit transfers the electrons to the next member of the chain and eventually to oxygen. Both flavoproteins and coenzyme Q could be possible candidates for proton carriers and cytochromes; nonheme iron and copper are the obvious choices for electron carriers.

The arrangement of carriers shown in Fig. 6.15 has been proposed as a possible way of translocating six H^+ to the outer side of the inner membrane during the oxidation of substrates. This arrangement consists of three O/R loops, each capable of forming a sufficient protonmotive force to synthesize one ATP molecule. The carriers shown in Fig. 6.15 represent the NADH oxidase chain. The succinate oxidase pathway would correspond to the second and third loops of the diagram. Although the sequence of carriers in Fig. 6.15 is somewhat at variance with the most recent formulations of the chain, this is not a serious criticism, since it is possible to have alternate arrangements of loops

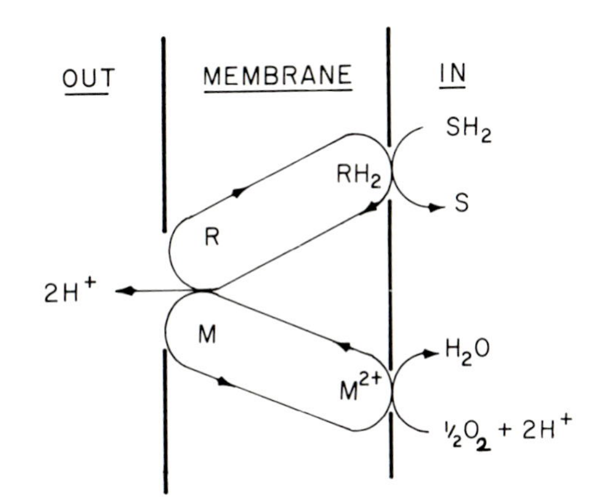

FIG. 6.14. Principle of the O/R loop. The translocation of two protons to the exterior side of the membrane by the hydrogen carrier (R) is coupled to the transfer of two electrons to the inner side where they are used to reduce oxygen. (Mitchell, 1966.)

(see Mitchell's more recent mechanism for the coenzyme QH_2–cytochrome *c* reductase in Chapter 4). Similarly, the model does not rule out other proton-carrying groups in the electron transfer complexes (e.g., SH) that could participate in proton translocation.

2. Proton-Translocating ATPase

At the crux of the chemiosmotic mechanism is the reversible ATPase whose proton-translocating activity discharges the proton gradient generated during electron transport with a simultaneous esterification of ADP and P_i.

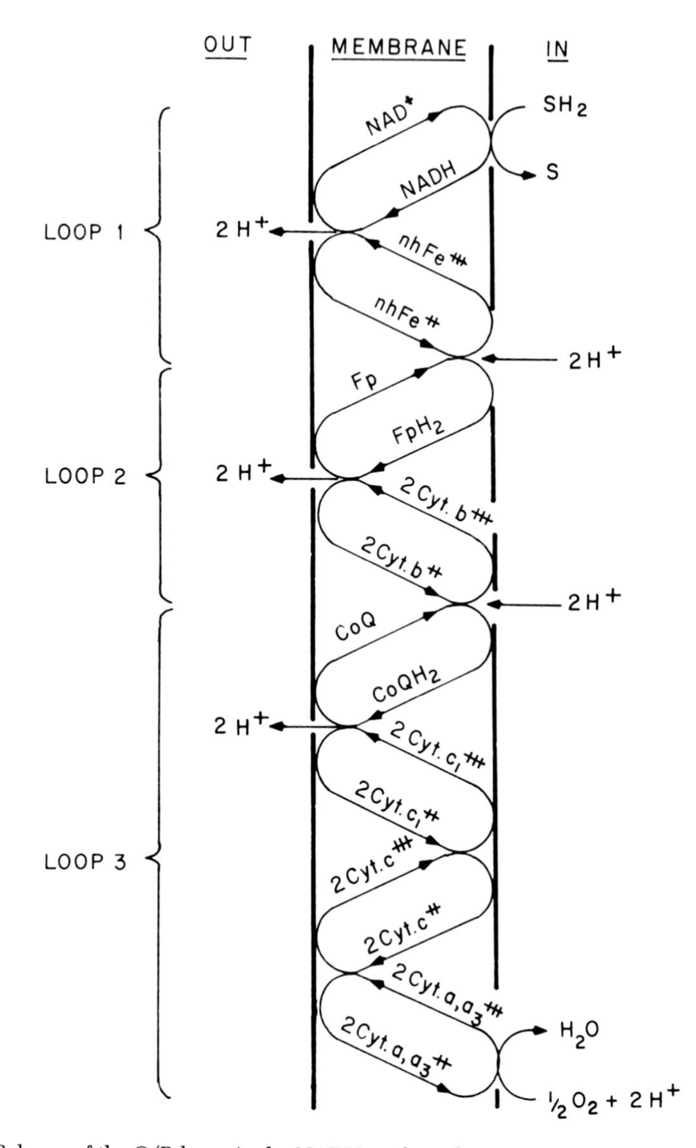

FIG. 6.15. Scheme of the O/R loops in the NADH oxidase chain. A net transfer of six protons from the inner to the outer side occurs during the transfer of two electrons from the substrate (SH_2) to oxygen. The three hydrogen carriers are NAD, flavin (F_p), and coenzyme Q. The electron carriers are nonheme iron (nhFe) and cytochromes. (Mitchell, 1966.)

Two different mechanisms of the ATPase have been considered by Mitchell. In the original version, two proton (or hydroxyl) groups of the enzyme reacted to form a high-energy intermediate which could then be used for ATP synthesis. A more recent and simpler model dispenses with the intermediate and conceives of a direct dehydration of ADP and P_i. In both cases it is assumed that two H^+ are translocated for every ATP formed in a direction opposite to the protons translocated during electron transfer. The two mechanisms are shown in Fig. 6.16.

In the model with the chemical intermediates, the first step is a dehydration reaction in which two groups, X·H and I·OH, form a covalent compound X–I. This reaction is catalyzed by a component of the ATPase located on the outer side of the membrane. The free energy of hydrolysis of X–I is low (about −3 kcal) because of the high H^+ concentration at its site of formation. After translocation to the inner side where the H^+ concentration is high, the free energy of hydrolysis increases (about −10 kcal), and the intermediate can be used in a series of dismutation reactions (not shown) to form ATP.

The second mechanism entails a direct nucleophilic attack by ADP on the phosphate. This model requires that the reactants bind to an active center that exposes phosphate to and insulates ADP from the high-proton side of the membrane. Under these conditions, the negative charges on the phosphate will be neutralized by protonation (further neutralization may occur if the phosphate is complexed to Mg^{2+} or some other groups on the protein), and the ADP will be present as the monoion $ADPO^-$. The reaction sequence shown in Fig. 6.16 involves (1) an initial translocation of the reactants to the active site with a release of two H^+ on the inner side, (2) a nucleophilic attack by $ADPO^-$ on the protonated phosphate group, and (3) a translocation of the product from the active site to the inner compartment. The net result of the esterification is a transfer of two H^+ from the outer to the inner side of the membrane.

The ATPase can function in three different modes. In the one already discussed, the protonmotive force generated by electron transfer is of sufficient magnitude to keep the ATPase poised in favor of ATP synthesis. Alternatively,

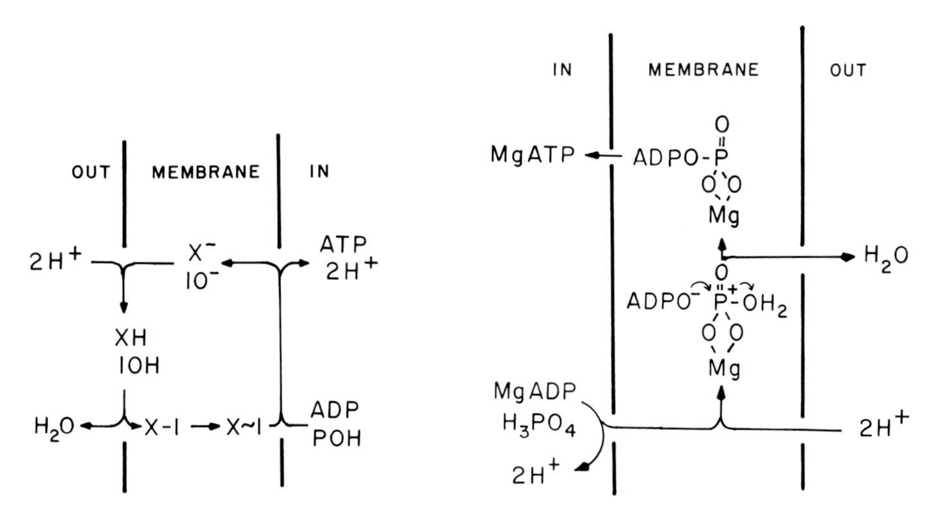

FIG. 6.16. Two mechanisms of the reversible ATPase (ATP synthetase) (Mitchell, 1966, 1974).

when there is no electron transfer, the ATPase can generate a protonmotive force from the coupled hydrolysis of ATP. Finally, when the membrane cannot maintain a pH gradient, the ATPase functions as an uncoupled ATP hydrolase. This occurs in the presence of uncouplers that discharge the gradient by acting as proton carriers or if there are structural changes in the membrane that make it permeable to protons.

3. Evidence for the Chemiosmotic Model

The feasibility of a chemiosmotic coupling mechanism is supported by theoretical considerations of the protonmotive force necessary for ATP synthesis and also by experimental evidence showing that mitochondria are capable of generating a proton gradient from either substrate oxidation or ATP hydrolysis.

The protonmotive force is an electrochemical potential that can be expressed by the formula:

$$\Delta p = \Delta\psi - Z\Delta\text{pH} \qquad [6.12]$$

where $\Delta\psi$ is the membrane potential resulting from the separation of charged species across the membrane, $Z = 2.3RT/F = 59$ mV, and ΔpH is the difference in pH. Mitchell has calculated that at a phosphate concentration of 10 mM, the protonmotive force (in electrical units) necessary to maintain an ATP/ADP ratio of 1 is 210 mV. Under state 4 conditions, the ATP/ADP ratio has been estimated to be close to 100. This would require a protonmotive force of 270 mV or a ΔpH of 4.5 if there is no membrane potential. Actual measurements of the membrane potential and of the pH gradient under conditions imitating state 4 respiration yielded values of the electrochemical potential difference that were approximately 230 mV (Table 6.5) and are reasonably close to the 270 mV required for the high ATP/ADP ratio. Although the ΔpH under the different conditions tried was approximately the same, the relative contribution by the membrane potential and the pH gradient was different depending on whether or not K^+ was present. In the absence of K^+ and other cations, most of the electrochemical potential arose from the pH gradient (Table 6.5). The opposite was true when K^+ was included. Since the experiments were done in the presence of valinomycin which facilitates K^+ transport (see Chapter 9, Sec-

TABLE 6.5. Contributions of the Membrane Potential and the pH Gradient to the Protonmotive Force in Rat Liver Mitochondria[a]

Mitochondria	Medium	$\Delta\psi$	$-Z\Delta$pH	Δp
K^+-depleted	0.2 mM EGTA	199	31	230
K^+-depleted	—	171	58	229
Normal	10 mM KCl	83	150	233

[a]The measurements were made in the presence of valinomycin. Oligomycin was used to prevent the utilization of the pH gradient in ATP synthesis. β-Hydroxybutyrate was the substrate. $Z = -59$ mV (Mitchell and Moyle, 1969).

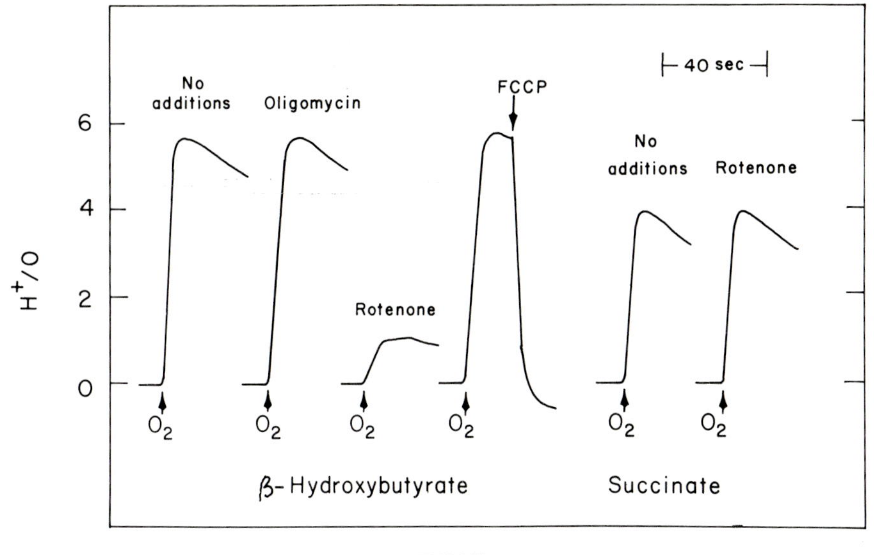

FIG. 6.17. Proton ejection by mitochondria. Rat liver mitochondria were pulsed with a known amount of oxygen either in the presence of β-hydroxybutyrate or succinate. The acidification of the external medium was followed in a pH meter. The presence or absence of inhibitors and uncouplers is indicated in each trace. (Mitchell and Moyle, 1967.)

tion IXA), the lower values of the membrane potential were attributed to the equalization of charge by the K^+ ions. Under normal conditions (absence of valinomycin), therefore, it may be expected that the membrane potential makes the major contribution to the protonmotive force, and the actual difference in pH is relatively small.

Two other important requirements of the chemiosmotic mechanism are (1) that electron transport and ATP hydrolysis be accompanied by an ejection of protons and (2) that the rate of ejection be consistent with the rate of ATP synthesis. These two requirements have been experimentally verified by Mitchell and others. When rat liver mitochondria are pulsed with oxygen in the presence of an oxidizable substrate, a rapid acidification of the medium is observed. This is followed by a slow rise in pH when the oxygen is exhausted (Fig. 6.17). From the known amount of oxygen added in the pulse and the H^+ ejected during the acidification phase, the stoichiometry of H^+/O has been determined to be 4 with succinate and 6 with NAD-linked substrates.* The initial rate of proton translocation is sufficiently fast to account for the rate of phosphorylation

*The stoichiometry of proton ejection during the oxidation of NADH or or succinate is still a matter of debate. Reynafarje and Lehninger have measured 4 H^+ per phosphorylation site with succinate as substrate. The higher values obtained by these authors have been explained as being caused by the presence of endogenous TCA cycle intermediates in mitochondria. For example, endogenous TCA cycle substrates can exchange with succinate, resulting in a simultaneous cotransport of protons. This would increase the apparent stoichiometry. Unfortunately, this controversy remains unresolved at present.

during state 3 respiration of rat liver mitochondria. The decay seen in the anaerobic phase has been interpreted to be caused by a leakiness of the membrane which allows protons to gradually equilibrate between the inner mitochondrial compartment and the exterior medium.

The release of protons appears to be an energy-dependent process requiring electron transfer and the existence of an intact membrane. It is prevented by electron transfer inhibitors and by uncouplers as well as detergents that disrupt the mitochondrial membrane (Fig. 6.17). Inhibitors of oxidative phosphorylation, however, have no effect. Mitchell has also shown that the addition of an uncoupler such as FCCP (another analogue related to *m*-CCCP) following the acidification phase increases the rate at which the pH gradient decays. This is in agreement with the postulated role of uncouplers which, according to the chemiosmotic model, make the membrane permeable to protons.

Mitchell and Moyle have also shown that mitochondria can eject protons when they are pulsed with ATP under anaerobic conditions. The stoichiometry of protons translocated to ATP hydrolyzed was measured as 2. This value came under some criticism because of a number of corrections that had to be made. Recently, however, a similar stoichiometry was found in submitochondrial particles where the corrections were not necessary. In contrast to the substrate-driven proton transport which is sensitive to uncouplers only, when ATP is used, the transport is inhibited by both uncouplers and inhibitors of oxidative phosphorylation.

Another interesting observation comes from studies with submitochondrial particles. Although such vesicles are capable of translocating protons in the presence of substrate or ATP, the direction of the proton flow is different. In oxygen or ATP pulse experiments of the type described above, the medium became more alkaline, indicating that there was an uptake instead of a release of protons. The energy-dependent uptake of protons by submitochondrial particles is predicted by the chemiosmotic mechanism in view of the inside-out orientation of their membranes.

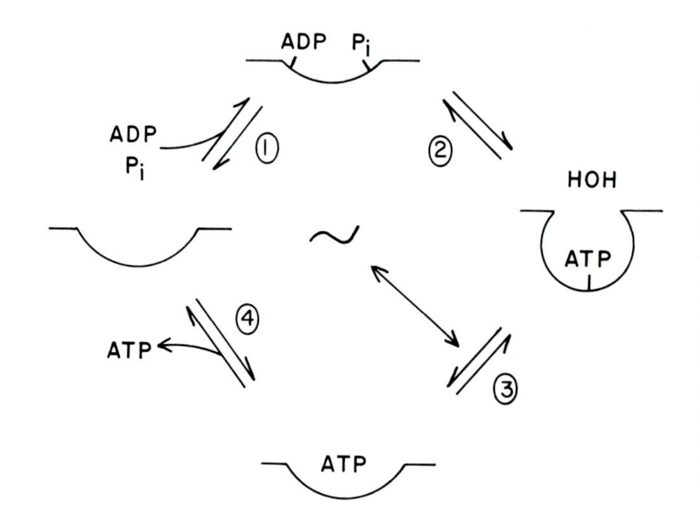

FIG. 6.18. Conformational hypothesis of ATP synthesis. Reaction 1: Binding of ADP and inorganic phosphate to the active site of the ATPase. 2: Esterification of ADP and phosphate. The ATP formed is tightly bound to the enzyme. 3: Energy-induced conformational change in the active site. 4: Release of ATP from the active site. Note that energy is only used for the conformational change that causes a weakening of binding affinity of the active site for ATP. (Boyer, 1974.)

C. Conformational Hypothesis

A conformational coupling mechanism of oxidative phosphorylation has been proposed by Boyer. In the original version of this hypothesis, the primary coupling event was postulated to be a conservation of the energy released during oxidation in a conformational state of a protein. In the energy-rich conformation, two reactive groups on the protein were thought to form a high-energy covalent compound that could be used for ATP synthesis in a series of reactions analogous to those of the chemical hypothesis.

A more recent interpretation is that the energy of oxidation is used to change the affinity of the active site of the ATP synthetase (ATPase) for ATP. In this model, ADP and P_i bind to a site on the enzyme where the dehydration reaction can proceed without any input of energy. The ATP formed is tightly bound to the enzyme and can only be released after an energy-induced conformational change occurs at the active site (Fig. 6.18). This novel mechanism was suggested by several experimental observations made in Boyer's laboratory. The first is the relative insensitivity of the P_i–$H_2{}^{18}O$ exchange reaction to uncouplers. According to the scheme of Fig. 6.18, this exchange occurs in the first two steps which result in the binding of the reactants to the active site and their dehydration to form ATP. The exchange, however, does not require that the ATP be released from the enzyme. Since ATP formation in the conformational hypothesis of Boyer occurs independently of an energy input, the exchange should not be inhibited by uncouplers. The other exchange reactions (ATP–P_i and ATP–$H_2{}^{18}O$), however, require that ATP be released from the active site and therefore, would be expected to be uncoupler sensitive.

Selected Readings

Ball, E. (1944) Energy relationships of the oxidative enzymes, *Ann. N.Y. Acad. Sci.* **45**:363.

Boyer, P. D. (1967) ^{18}O and related exchanges in enzymic formation and utilization of nucleoside triphosphates in *Current Topics in Bioenergetics* (R. Sanadi, ed.), Vol. 2, Academic Press, New York, pp. 99–149.

Boyer, P. D. (1974) Conformational coupling in biological energy transductions, in *Dynamics of Energy Transducing Membranes* (L. Ernster, R. W. Estabrook and E. C. Slater, eds.), Elsevier, Amsterdam, pp. 289–301.

Boyer, P. D., Cross, R. L., and Momsen, W. (1973) A new concept for energy coupling in oxidative phosphorylation based on a molecular explanation of the oxygen exchange reaction, *Proc. Natl. Acad. Sci. U.S.A.* **70**:2837.

Brierley, G. P., Murer, E., and Bachmann, E. (1964) Studies on ion transport. III. The accumulation of calcium and inorganic phosphate by heart mitochondria, *Arch. Biochem. Biophys.* **105**:89.

Chance, B., and Hollunger, G. (1960) Energy linked reduction of mitochondrial pyridine nucleotide, *Nature* **185**:666.

Chance, B., and Williams, G. R. (1955) A simple and rapid assay of oxidative phosphorylation, *Nature* **175**:1120.

Chance, B., and Williams, G. R. (1956) The respiratory chain and oxidative phosphorylation, *Adv. Enzymol.* **17**:65.

Greville, G. D. (1969) A scrutiny of Mitchell's chemiosmotic hypothesis of respiratory chain and photosynthetic phosphorylation, in *Current Topics in Bioenergetics* (R. Sanadi, ed.), Vol. 3, Academic Press, New York, pp. 1–78.

Hansen, M., and Smith, A. (1964) Studies on the mechanism of oxidative phosphorylation VII. Preparation of a submitochondrial particle (ETP_H) which is capable of fully coupled oxidative phosphorylation, *Biochim. Biophys. Acta* **81**:214.

Hanstein, W. G. (1976) Uncoupling of oxidative phosphorylation, *Biochim. Biophys. Acta* **456**:129.

Jacobs, E. E. (1960) Phosphorylation coupled to electron transport initiated by substituted phenyldiamines, *Biochem. Biophys. Res. Commun.* **3**:536.

Lardy, H., and Ferguson, S. M. (1969) Oxidative phosphorylation in mitochondria, *Annu. Rev. Biochem.* **38**:991.

Lardy, H. A., Johnson, D., and McMurray, W. C. (1958) Antibiotics as tools for metabolic studies. I. A survey of toxic antibiotics in respiratory, phosphorylative and glycolytic systems, *Arch. Biochem. Biophys.* **78**:587.

Lardy, H., Reed, P., and Lin, C. H. C. (1975) Antibiotic inhibitors of mitochondrial ATP synthesis, *Fed. Proc.* **34**:1707.

Lee, C. P., and Ernster, L. (1964) Studies of the energy transfer system of submitochondrial particles. I. Competition between oxidative phosphorylation and the energy linked nicotinamide adenine dinucleotide transhydrogenase reaction, *Eur. J. Biochem.* **3**:385.

Lee, C. P., Azzone, G. F., and Ernster, L. (1964) Evidence for energy coupling in non-phosphorylating electron transport particles from beef-heart mitochondria, *Nature* **201**:152.

Lehninger, A. L. (1964) *The Mitochondrion*, W. A. Benjamin Co., New York.

Lehninger, A. L., Carafoli, E., and Rossi, C. S. (1967) Energy-linked ion movements in mitochondrial systems, *Adv. Enzymol.* **29**:259.

Löw, H., and Vallin, I. (1963) Reduction of added DPN from the cytochrome *c* level in submitochondrial particles, *Biochem. Biophys. Res. Commun.* **9**:307.

Mitchell, P. (1961) Coupling of phosphorylation to electron and hydrogen transfer by a chemiosmotic type of mechanism, *Nature* **191**:105.

Mitchell, P. (1966) *Chemiosmotic Coupling in Oxidative and Photosynthetic Phosphorylation*, Glynn Research Ltd., Bodmin.

Mitchell, P. (1974) A chemiosmotic molecular mechanism for proton-translocating adenosine triphosphatases, *FEBS Lett.* **43**:189.

Mitchell, P., and Moyle, J. (1967) Respiration-driven proton translocation in rat liver mitochondria, *Biochem. J.* **105**:1147.

Mitchell, P., and Moyle, J. (1969) Estimation of membrane potential and pH differences across the cristae membrane of rat liver mitochondria, *Eur. J. Biochem.* **7**:471.

Nielsen, S. O., and Lehninger, A. L. (1955) Phosphorylation coupled to the oxidation of ferrocytochrome *c*, *J. Biol. Chem.* **215**:555.

Pullman, M. E., and Schatz, G. (1967) Mitochondrial oxidations and energy coupling, *Annu. Rev. Biochem.* **36**:539.

Racker, E. (1976) *A New Look at Mechanisms in Bioenergetics*, Academic Press, New York.

Reynafarje, B., and Lehninger, A. L. (1978) The K^+/site and H^+/site stoichiometry of mitochondrial electron transport, *J. Biol. Chem.* **253**:6331.

Rossi, C., and Lehninger, A. L. (1964) Stoichiometry of respiratory stimulated accumulation of Ca^{++} and phosphate, and oxidative phosphorylation in rat liver mitochondria, *J. Biol. Chem.* **239**:3971.

Schatz, G., and Racker, E. (1966) Partial resolution of the enzymes catalyzing oxidative phosphorylation. VII. Oxidative phosphorylation in the diphosphopyridine–cytochrome *b* segment of the respiratory chain: Assay and properties in submitochondrial particles, *J. Biol. Chem.* **241**:1429.

Slater, E. C. (1953) Mechanism of phosphorylation in the respiratory chain, *Nature* **172**:59.

Slater, E. C. (1955) Phosphorylation coupled with the oxidation of α-ketoglutarate by heart muscle sarcosomes. 3. Experiments with ferricytochrome *c* as hydrogen acceptor, *Biochem. J.* **59**:392.

Slater, E. C. (1966) Oxidative phosphorylation, in *Comprehensive Biochemistry* (M. Florkin and E. M. Stotz, eds.), Vol. 14, Elsevier, Amsterdam, pp. 327–387.

Slater, E. C. (1971) The coupling between energy-yielding and energy-utilizing reactions in mitochondria, *Q. Rev. Biophys.* **4**:35.

Wadkins, C. L., and Lehninger, A. L. (1963) Role of ATP–ADP exchange reaction in oxidative phosphorylation, *Fed. Proc.* **22**:1092.

7

The Mitochondrial Adenosine Triphosphatase

In the previous chapter we saw that the ATPase activity of coupled mitochondria is dramatically stimulated by uncouplers and more generally by conditions that interrupt the flow of energy from the electron transfer chain to the coupling device. These observations have for a long time been interpreted to indicate that the ATPase is involved in the terminal steps of oxidative phosphorylation and that under normal circumstances, i.e., in the coupled state, its primary function is that of an ATP synthetase or kinase. This is supported by the following evidence.

1. There is a good correlation between uncoupling and the appearance of ATPase activity in mitochondria.
2. Isolated components of the ATPase (coupling factors) stimulate oxidative phosphorylation in certain types of submitochondrial particles.
3. Inhibitors of oxidative phosphorylation are also potent inhibitors of the ATPase.
4. Antibodies against the ATPase inhibit oxidative phosphorylation in submitochondrial particles.
5. Purified preparations of the ATPase complex are capable of catalyzing partial reactions of oxidative phosphorylation (e.g., ATP–P_i exchange).
6. Oxidative phosphorylation can be reconstituted in simple systems consisting of a purified respiratory complex and of the ATPase.
7. Membranes that carry out the coupled synthesis of ATP (mitochondria, chloroplasts, bacterial membranes) contain ATPases that are remarkably similar in protein composition and structure.

Because of its central role in coupling, the ATPase of mitochondria has been extensively studied, and there is substantial knowledge of its structure and function. Some of this information will be reviewed in this chapter.

I. Purification and Properties of the ATPase Complex

The uncoupler-stimulated ATPase of mitochondria requires a divalent metal such as Mg^{2+} for optimal enzymatic activity. Another important property of the enzyme is its sensitivity to inhibitors of oxidative phosphorylation (Table 7.1). Oligomycin in particular has been used as a diagnostic tool to distinguish the mitochondrial from other cellular ATPases which are not inhibited by this antibiotic. Oligomycin has also been useful in assessing the intactness of the ATPase. Loss of sensitivity to oligomycin usually indicates partial or complete dissociation of the enzyme into its component parts. In the present discussion, preparations of the enzyme exhibiting oligomycin-sensitivity will be referred to as OS-ATPase or CF_0-F_1.

A number of procedures have been developed for the isolation of the OS-ATPase from mammalian, yeast, and other types of mitochondria. The lipophilic properties of the enzyme necessitate the use of surface-active reagents to release it from the membrane. The procedures used to purify the ATPase from beef heart and yeast mitochondria illustrate how it can be separated from the electron transfer complexes and are briefly described below.

A. Mammalian ATPase

Most of the methods used to purify the OS-ATPase depend on a solubilization step with bile acid followed by gel filtration or salt fractionation. The bovine ATPase has been solubilized either with deoxycholate or cholate. Ammonium sulfate precipitation of the soluble extract is usually sufficient to separate the ATPase from the electron transfer components. Rat liver OS-ATPase has been purified from a deoxycholate extract by chromatography on agarose gels. Most preparations of the ATPase differ in their phospholipid con-

TABLE 7.1. Inhibitors of Oxidative Phosphorylation and Mitochondrial ATPase

	Inhibition		
	Oxidative phosphorylation	OS-ATPase	F_1 ATPase
Class A			
Oligomycin	Yes	Yes	No
Rutamycin	Yes	Yes	No
Dicyclohexylcarbodiimide (DCCD)	Yes	Yes	No
Venturicidin	Yes	Yes	No
Tributyltin chloride	Yes	Yes	No
Mercurials	Yes	Yes	No
Class B			
Aurovertin	Yes	Yes	Yes
Dio-9	Yes	Yes	Yes
ADP	No	Yes	Yes
Natural inhibitor[a]	No	Yes	Yes

[a]This is a low-molecular-weight protein first isolated from beef-heart mitochondria by Pullman and Monroy.

tent and state of aggregation. The OS-ATPase of rat liver is fairly monodisperse. Preparations of the bovine ATPase are membranous since they are isolated relatively free of endogenous bile acid.

B. Yeast ATPase

The procedure used to isolate the yeast enzyme is very simple and is amenable to both large- and small-scale purifications. In essence two steps are used. The first is an extraction of yeast submitochondrial particles with a low concentration of the neutral detergent Triton X-100 which solubilizes the ATPase but leaves most of the other inner membrane proteins in the particulate fraction. The extract is centrifuged through a sucrose or glycerol gradient to separate the proteins according to their molecular size (Fig. 7.1). This centrifugation step yields a highly purified preparation essentially free of electron transfer enzymes. A similar procedure has been used to purify the ATPase from *N. crassa* except that deoxycholate instead of Triton X-100 is used as the solubilizing agent.

C. Enzymatic Properties of the Oligomycin-Sensitive ATPase

The isolated complex catalyzes the hydrolysis of ATP to ADP and inorganic phosphate. Other triphosphonucleotides are also hydrolyzed but at lower rates (Table 7.2). As will be seen in Chapter 8, some preparations of the OS-ATPase are capable of forming membranes that carry out ATP–P_i exchange.

The phospholipid content of the complex varies depending on the method of isolation. The enzymatic activity of preparations with a low endogenous phospholipid content is stimulated by externally added phospholipids. Although fatty acids also activate the enzyme, the resultant ATPase is not inhibited by oligomycin.

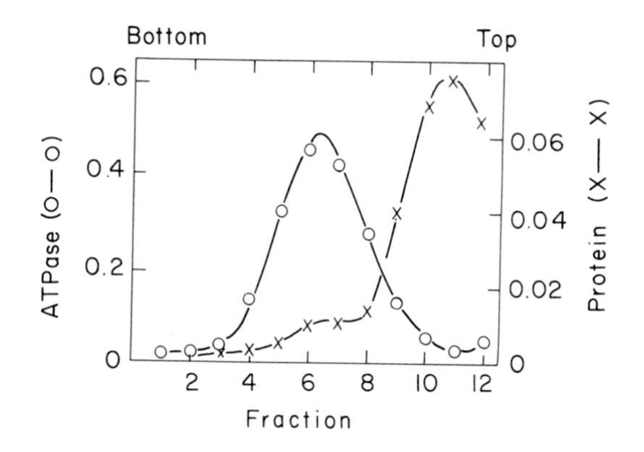

FIG. 7.1. Sedimentation of the yeast OS-ATPase on glycerol gradients. A Triton X-100 extract of yeast submitochondrial particles was centrifuged through a linear 5–15% glycerol gradient at 23,000 rpm for 17 hr. Thirteen fractions were collected and assayed for oligomycin-sensitive ATPase and protein.

TABLE 7.2. Hydrolysis of Triphosphonucleotides by Submitochondrial Particles, OS-ATPase, and F_1[a]

Enzyme	% of ATP			
	GTP	ITP	UTP	CTP
SMP	34	43	4	9
OS-ATPase	26	17	5	5
F_1	110	100	48	18

[a]The different preparations were derived from beef-heart mitochondria. Activities are expressed as percent of the rate of hydrolysis of ATP. (From Tzagoloff *et al.*, 1968a.)

D. Subunit Polypeptides of the Oligomycin-Sensitive ATPase

The OS-ATPases of beef heart, rat liver, *N. crassa*, and yeast mitochondria have been analyzed by SDS gel electrophoresis. In each case, similar patterns of proteins have been observed. Yeast OS-ATPase, which probably contains the fewest contaminants, is composed of at least ten different subunits with molecular weights ranging from 58,000 to 8,000 (Fig. 7.2). The functions of some of these proteins will be discussed later in this chapter.

II. Functional Components of the ATPase Complex

The OS-ATPase of bovine heart mitochondria has been resolved into three functional components, each essential for the expression of oligomycin sensitivity and other properties of the membrane-bound enzyme. The three components are: (1) F_1, an oligomycin-insensitive and water-soluble ATPase; (2) membrane factor, a unit consisting of four proteins that are insoluble in water

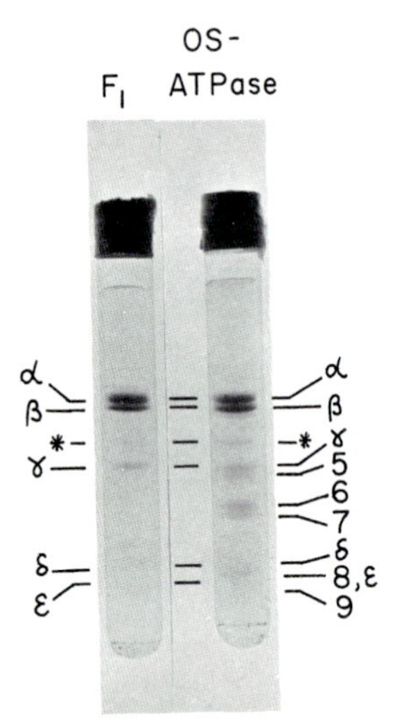

FIG. 7.2. Polyacrylamide gel electrophoresis of yeast F_1 and OS-ATPase in the presence of sodium dodecyl sulfate. The subunits are designated by the same nomenclature as in Table 7.7. The protein marked with an asterisk is only seen in some preparations of yeast F_1.

and contain the phospholipid- and oligomycin-binding sites of the complex; and (3) OSCP, a water soluble protein that binds to F_1 and to the membrane factor and probably serves as a physical link between the two. These components have also been isolated from the yeast ATPase and are a general feature of all mitochondrial ATPases.

A. F_1 ATPase

F_1 has been solubilized by mechanical disruption of mitochondria with glass beads or by extensive sonic irradiation of submitochondrial particles. The former method was first used by Pullman and his colleagues to purify F_1 from bovine heart mitochondria. Extraction of submitochondrial particles with chloroform has recently been introduced as an alternative means of solubilizing F_1. Chloroform apparently weakens the attachment of F_1 to the other proteins of the complex, allowing for a complete extraction of the enzyme in aqueous buffers.

Purified F_1 is a water-soluble polymer with a molecular weight of 360,000. It has a complex structure and consists of five different polypeptides. The subunits are firmly bound to each other but can be dissociated with sodium dodecyl sulfate or with strong chaotropes (guanidine hydrochloride, urea, etc.). Polyacrylamide gels of yeast F_1 dissociated in SDS show that the F_1 subunits are all present in the OS-ATPase (Fig. 7.2). Similar results have been obtained with the mammalian enzyme. The two largest polypeptides (α and β subunits) have molecular weights of 58,000 and 55,000, respectively, and make up about 85% of the total protein mass of F_1. The other polypeptides (γ, δ, and ϵ subunits) are present in lower concentrations. The stoichiometry of the subunits has been difficult to ascertain. A stoichiometry of 3α:3β:1γ:1δ:1ϵ has been proposed based on the relative stain intensities seen in SDS polyacrylamide gels. Most preparations of F_1 contain a sixth low-molecular-weight polypeptide (8,000 to 9,000 daltons) which suppresses the catalytic activity of the enzyme. This protein was purified by Pullman and Monroy from bovine mitochondria and has also been shown to be present in other types of mitochondria. In bovine mitochondria, the inhibitor masks as much as 80% of the potential ATPase activity. The inhibitor is considered to be part of the F_1 structure. Because of its weak attachment to F_1, however, most of the inhibitor is lost during the purification of the enzyme, thus explaining why only a small proportion of F_1 molecules contain bound inhibitor. The inhibitor probably has some regulatory function in the coupling activity; its precise function, however, has not been determined.

The molecular arrangement of the F_1 subunits cannot be discerned by conventional electron microscopy. Information on this point will probably be forthcoming from X-ray crystallographic data. Crystals suitable for X-ray analysis have recently been reported for the rat liver F_1. Preliminary diffraction patterns indicate a lattice structure with hexagonal cell dimensions.

Although all of the subunits of F_1 have been isolated in homogeneous form, they are functionally inactive (the ATPase inhibitor is an exception to this) and have not, therefore, been useful for functional studies. Other more indirect approaches, however, have helped to identify the subunits involved in the hydrolytic activity and binding of F_1 to the membrane factor. Several lines of evidence suggest that the hydrolytic site of F_1 is located on the β subunit. A number of reagents that react with tyrosine residues have been found to inhibit the ATPase activity of F_1. One such compound, 7-chloro-4-nitrobenzo-2-oxa-1,3 diazole (NBD-Cl), binds covalently to a tyrosine residue on the β subunit. Since NBD-Cl is a structural analogue of adenine, it has been proposed to react with an essential tyrosine at or near the hydrolytic site. Immunochemical studies also support a role for this subunit in the catalytic activity of the enzyme. Antibodies have been prepared against purified α and β subunits of yeast F_1. When tested against the native enzyme, only the antiserum directed against the β subunit inhibited the ATPase activity (Fig. 7.3).

The strongest evidence favoring a catalytic role of the β subunit comes from experiments on the reconstitution of the F_1 ATPase from a thermophilic bacterium. The bacterial ATPase (TF_1) has the same subunit composition as mitochondrial F_1. Kagawa and co-workers have isolated all five subunits of the enzyme and tested their ability to hydrolyze ATP either individually or in different combinations. These experiments indicated that although none of the subunits was active by itself, certain combinations were capable of hydrolyzing ATP. The enzymatically active combinations ($\alpha + \beta + \delta$ or $\beta + \gamma$) invariably had the β subunit (Table 7.3). The reconstitution of the catalytic activity was correlated with a physical association of the subunits into a complex with an electrophoretic mobility similar to native TF_1 (Fig. 7.4).

F_1 can bind to depleted membranes with concomitant restoration of oligomycin sensitivity. This reconstitution is very specific since other ATPases cannot be substituted for F_1, implying there may be one or more subunits of the enzyme involved in the binding. This is supported by studies with bacterial F_1 which can be purified with all five subunits or with the δ subunit missing. Although both types of preparations are equally active in catalyzing the hydrolysis of ATP, only the enzyme with the δ subunit is capable of rebinding to the

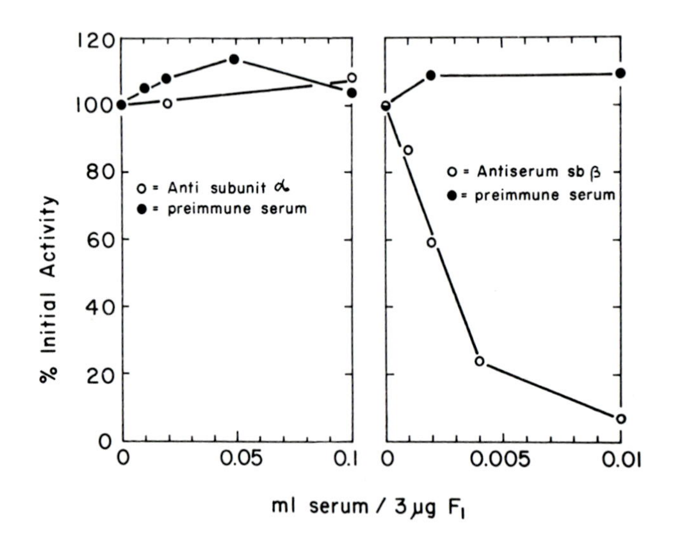

FIG. 7.3. Effect of antisera to yeast F_1 subunits on the ATPase activity. Yeast F_1 was preincubated with the indicated amounts of antisera to either the α or β subunits and assayed for ATPase.

TABLE 7.3. Reconstitution of ATPase from the Subunits of Bacterial F_1[a]

Subunits	ATPase (μmoles/min per mg protein)
α	0.00
β	0.04
γ	0.00
δ	0.00
ϵ	0.00
$\alpha + \beta$	0.00
$\alpha + \beta + \gamma$	1.90
$\alpha + \beta + \delta$	0.35
$\alpha + \beta + \gamma + \epsilon$	2.10
$\alpha + \beta + \gamma + \delta + \epsilon$	2.39
$\beta + \gamma$	2.38

[a]The purified subunits were mixed in buffer containing 5 mM Mg^{2+}, incubated overnight, and the ATPase activity assayed at 60°. (From Yoshida *et al.*, 1977.)

depleted bacterial membranes. In the bacterial system, therefore, the δ subunit appears to be essential for the interaction of F_1 with the rest of the complex.

F_1 and OS-ATPase catalyze the same reaction but are quite distinct in their physical and functional properties. Some of the differences are listed in Table 7.4. The OS-ATPase is a water-insoluble lipoprotein complex with a phospholipid requirement for enzymatic activity. F_1 is soluble in water and does not have a phospholipid dependence. The two preparations also differ in their nucleotide specificities and activation by divalent metals. For example, Ca^{2+} is effective in activating F_1 but not the OS-ATPase. Another important difference lies in the sensitivity of the two enzymes to inhibitors of oxidative phosphorylation. These can be divided into two groups based on their inhibitory effect on F_1 and the OS-ATPase (Table 7.1). Reagents such as Dio-9 and aurovertin inhibit both enzymes and probably act on a site present in F_1. Other compounds which include oligomycin (rutamycin) and DCCD inhibit only the complex and interact with sites located in the membrane factor.

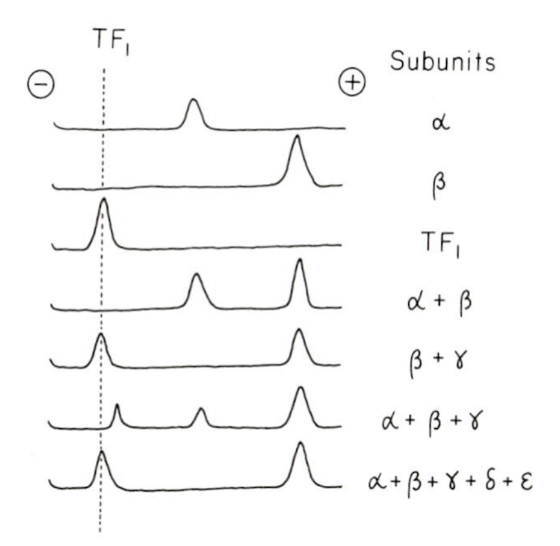

FIG. 7.4. Polyacrylamide gel electrophoresis of various combinations of TF_1 subunits. The subunits were mixed in the presence of 8 M urea, dialyzed, and separated on polyacrylamide in the absence of sodium dodecyl sulfate. The traces represent the distribution of protein stain along the length of the gels. The plus and minus signs refer to the anode and cathode, respectively. The migration of TF_1 is indicated by the dashed vertical line. (Taken from Yoshida *et al.*, 1977.)

TABLE 7.4. Properties of F_1 and of the OS-ATPase

Property	F_1	OS-ATPase
Phospholipid content	No	Yes
Water solubility	Yes	No
Cold lability	Yes	No
Membrane formation	No	Yes
Phospholipid requirement for activity	No	Yes
Inhibition by oligomycin, DCCD, etc.	No	Yes
Number of subunits	5	10
Molecular weights	360,000	520,000

B. Membrane Factor

Kagawa and Racker demonstrated that F_1 is part of a larger lipoprotein complex. They succeeded in isolating from bovine submitochondrial particles both an oligomycin-sensitive ATPase (CF_0–F_1) and a factor devoid of ATPase (CF_0) which altered the physical and catalytic properties of F_1. Among the more significant activities of CF_0 was its ability to confer oligomycin sensitivity on F_1 (Table 7.5). The procedure used for the isolation of CF_0–F_1 required a preliminary dissolution of the inner membrane with cholate followed by ammonium sulfate fractionation. In order to purify CF_0, the membranes were first treated with trypsin and urea to extract F_1. The resultant membranes (TU particles) were then fractionated with cholate and ammonium sulfate (Fig. 7.5). The CF_0 preparation obtained in this manner was insoluble in water and required phospholipid for the restoration of oligomycin sensitivity but not for the binding of F_1.

Studies of CF_0 and CF_0–F_1 led Kagawa and Racker to make a number of observations that clarified the role of the ATPase in the structure of the inner membrane. These were correlative studies establishing that the 90-Å spheres lining the matrix side of the inner membrane are equivalent to the F_1 component of the ATPase. Negatively stained inner membrane vesicles (SMP, ETP_H) have a full array of the 90-Å spheres projecting from the membrane surface. Kagawa and Racker showed that vesicles depleted of F_1 invariably lose the pro-

TABLE 7.5. Reconstitution of OS-ATPase from F_1, CF_0, and Phospholipids[a]

Additions to F_1		ATPase (μmoles/10 min)	
μg Phospholipids	μg CF_0	− Rutamycin	+ Rutamycin
0	0	1.16	1.16
0	100	0.17	
50	50	1.07	0.77
100	100	1.16	0.67
200	200	1.03	0.30

[a] CF_0 and soybean phospholipids were mixed with 2 μg F_1 in Tris buffer. After 5 min at 30°, the mixture was assayed for ATPase plus and minus rutamycin. (From Kagawa and Racker, 1966.)

jecting spheres (Fig. 7.6). Moreover, addition of F_1 to the stripped membranes led not only to a functional reconstitution of the oligomycin-sensitive ATPase but also of the original membrane morphology. The reappearance of the 90-Å spheres in the reconstituted membranes together with the fact that F_1 itself is a 90-Å spherical particle provided a strong case for the identity of F_1 with the projecting spheres seen on the inner membrane.

Similar reconstitutions were done with the isolated CF_0 and CF_0–F_1 complex. Both preparations are normally isolated with a low phospholipid content. When supplied with exogenous total mitochondrial or plant phospholipids (asolectin), CF_0 and CF_0–F_1 aggregate into vesicular membranes. The CF_0–F_1 vesicles are lined with 90-Å spheres and have a morphology similar to that of the native inner membrane. The CF_0 membranes are smooth but become lined with the projecting particles when reconstituted with F_1. Based on these results, Kagawa and Racker concluded that the oligomycin-sensitive ATPase of the mitochondrial inner membrane consists of at least two functionally and structurally different components. One is the hydrolytic unit F_1, and the other a hydrophobic set of proteins lodged in the lipid bilayer of the membrane. The membrane proteins of the CF_0 unit not only act to anchor F_1 to the membrane but also play an important role in oxidative phosphorylation.

The CF_0 unit of the complex has been further resolved into a soluble protein (OSCP) and a hydrophobic fraction referred to as the membrane factor. The membrane factor of the yeast OS-ATPase consists of four proteins with the following set of properties.

1. Binding of oligomycin and other inhibitors of oxidative phosphorylation.
2. Binding of phospholipids.
3. Binding of F_1 and OSCP.
4. Formation of membranes.
5. Conferral of oligomycin sensitivity on F_1.
6. Modulation of the catalytic and physical properties of F_1.

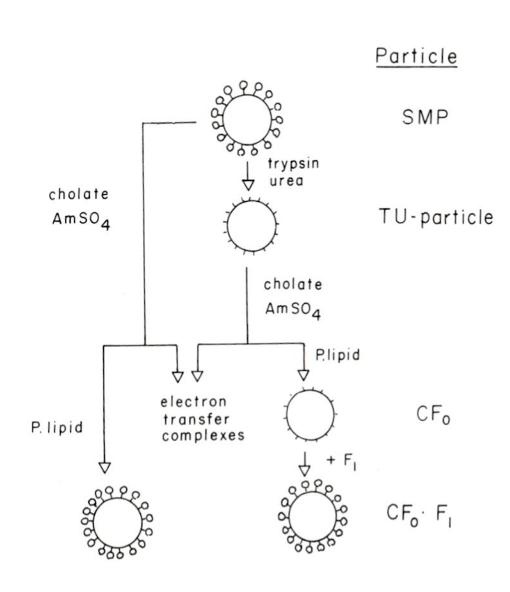

FIG. 7.5. Resolution of SMP into CF_0, F_1, and CF_0–F_1.

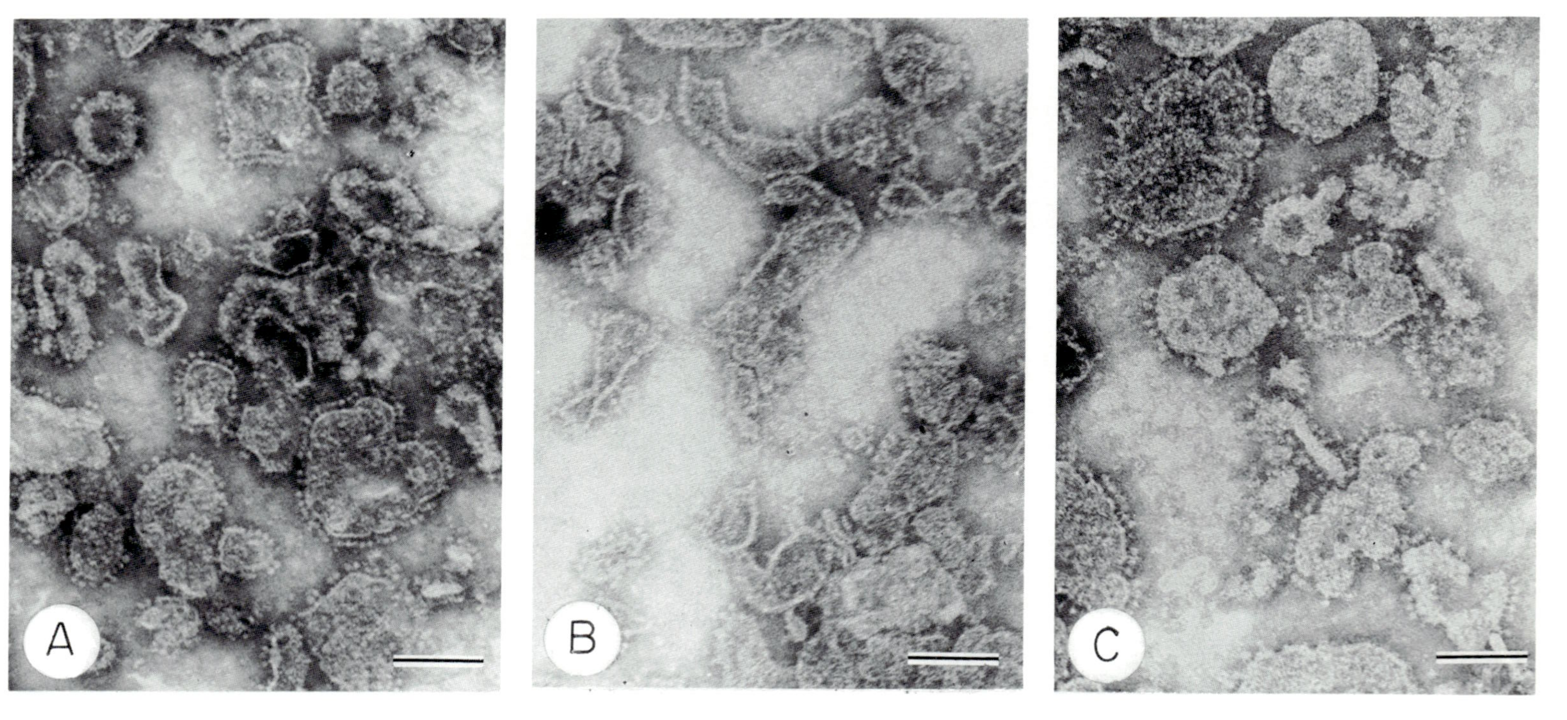

FIG. 7.6. Electron micrographs of bovine submitochondrial particles. A: ETP_H vesicles with projecting 90-Å particles. B: ETP_H(NaBr) vesicles prepared by extraction of ETP_H with 3.5 M NaBr. The vesicles are stripped of most of the projecting particles. C: Reconstituted vesicles. The ETP_H(NaBr) vesicles were mixed with F_1. The reconstituted vesicles have an almost complete array of projecting particles. All three preparations were stained with phosphotungstate. All scale bars represent 0.1 μm. (Courtesy of Dr. Junpei Asai.)

Three proteins thought to be subunits of the membrane factor have been purified from bovine heart mitochondria. F_6 has been isolated from submitochondrial particles depleted of F_1 and OSCP; F_6 is distinct from OSCP but, like the latter, promotes the binding of F_1 to the membrane factor. In bovine OS-ATPase, therefore, both OSCP and F_6 appear to be required for the attachment of F_1 to the hydrophobic proteins as well as reconstitution of oligomycin sensitivity.

Another component of the membrane factor is a low-molecular-weight proteolipid (8000 daltons). This protein has been purified from the ATPases of mitochondria, bacteria, and chloroplasts. Sebald has found substantial amino acid sequence homology among the various proteolipids (Fig. 7.7). The inhibitor DCCD has been shown to bind covalently to a glutamic acid residue of the mitochondrial proteolipid and to an aspartic acid in the case of bacterial proteolipid. Based on studies of oligomycin-resistant mutants of yeast, the proteolipid also seems to be involved in the binding of oligomycin (see Chapter 11, Section VIA). Although the function of the proteolipid in the catalytic activity of the ATPase has not been defined, one of the current ideas is that it forms a proton channel across the inner membrane. This is supported by model studies with phospholipid films whose permeability to H^+ is enhanced when the proteolipid is incorporated into the bilayer.

The third protein, factor B, has been extensively studied by Sanadi and his associates. Factor B stimulates energy-driven reactions in submitochondrial particles. Another characteristic of factor B is its inactivation by sulfhydryl-blocking reagents. Since the OS-ATPase is also inhibited by these reagents, it is possible that factor B may be involved in the hydrolytic activity of the complex.

Yeast	fMet-Gln-Leu-Val-Leu-Ala-Ala-Lys-Tyr-Ile-Gly-Ala-Gly-Ile-Ser-Thr-Ile-Gly-Leu-
N. crassa	Tyr-Ser-Ser-Glu-Ile-Ala-Gln-Ala-Met-Val-Glu-Val-Ser-Lys-Asn-Leu-Gly-Met-Gly-Ser-Ala-Ala-Ile-Gly-Leu-
Bovine	Asp-Ile-Asp-Thr-Ala-Ala-Lys-Phe-Ile-Gly-Ala-Gly-Ala-Ala-Thr-Val-Gly-Val-
Yeast	Leu-Gly-Ala-Gly-Ile-Gly-Ile-Ala-Ile-Val-Phe-Ala-Ala-Leu-Ile-Asn-Gly-Val-Ser-Arg-Asn-Pro-Ser-Ile-Lys-
N. crassa	Thr-Gly-Ala-Gly-Ile-Gly-Ile-Gly-Leu-Val-Phe-Ala-Ala-Leu-Leu-Asn-Gly-Val-Ala-Arg-Asn-Pro-Ala-Leu-Arg-
Bovine	Ala-Gly-Ser-Gly-Ala-Gly-Ile-Gly-Thr-Val-Phe-Gly-Ser-Leu-Ile-Ile-Gly-Tyr-Ala-Arg-Asn-Pro-Ser-Leu-Lys-
Yeast	Asp-Thr-Val-Phe-Pro-Met-Ala-Ile-Leu-Gly-Phe-Ala-Leu-Ser-Glu-Ala-Thr-Gly-Leu-Phe-Cys-Leu-Met-Val-Ser-
N. crassa	Gly-Gln-Leu-Phe-Ser-Tyr-Ala-Ile-Leu-Gly-Phe-Ala-Phe-Val-Glu-Ala-Ile-Gly-Leu-Phe-Asp-Leu-Met-Val-Ala-
Bovine	Gln-Gln-Leu-Phe-Ser-Tyr-Ala-Ile-Leu-Gly-Phe-Ala-Leu-Ser-Glu-Ala-Met-Gly-Leu-Phe-Cys-Leu-Met-Val-Ala-
Yeast	Phe-Leu-Leu-Leu-Phe-Gly-Val
N. crassa	Leu-Met-Ala-Lys-Phe-Thr
Bovine	Phe-Leu-Ile-Leu-Phe-Ala-Met

FIG. 7.7. Amino acid sequences of the ATPase proteolipid from yeast, *N. crassa*, and bovine mitochondria. (Sebald *et al.*, 1979.)

C. Oligomycin-Sensitivity-Conferring Protein

This basic protein has been purified from alkaline extracts of submitochondrial particles, CF_0, and the OS-ATPase. The purified protein is soluble in water and has been estimated to have a molecular weight of 18,000. The OSCP has already been indicated to be necessary for oligomycin sensitivity. Studies of the yeast ATPase complex indicate that it is also essential for the binding of F_1 to the membrane factor. In the experiment of Table 7.6, yeast submitochondrial particles were depleted of F_1 and OSCP by extraction with NaBr and ammonium hydroxide. The particles were then incubated sequentially with F_1 followed by OSCP or with the order reversed. The particles were isolated after each incubation and assayed for rutamycin-sensitive ATPase. The results of this experiment show that reconstitution of the OS-ATPase occurred only when the particles were first incubated with OSCP, suggesting that this component in conjunction with the hydrophobic proteins forms the F_1 binding site.

There are probably three proteins involved in the interaction of F_1 with the hydrophobic proteins of the complex—the δ subunit of F_1 and the F_6 and OSCP of the CF_0 unit. Since there is some evidence that OSCP can bind to F_1, it may act as the immediate link between F_1 and the rest of the complex. At present, the functions of the ATPase subunits are based largely on reconstitution experiments. Some of the proposed functions of the F_1 and CF_0 polypeptides are summarized in Table 7.7.

III. Ultrastructure of the ATPase

The structure of the OS-ATPase deduced by Kagawa and Racker from their studies of the CF_0-F_1 and CF_0 preparations have been confirmed by the

Table 7.6. Reconstitution of Oligomycin-Sensitive ATPase in Yeast Submitochondrial Particles[a]

		ATPase	
First incubation	Second incubation	− Rutamycin	+ Rutamycin
None	None	0.050	0.046
F_1	None	0.105	0.059
F_1	OSCP	0.105	0.076
OSCP	F_1	1.039	0.327
F_1 + OSCP	None	0.968	0.236

[a]Yeast submitochondrial particles were extracted with sodium bromide followed by ammonium hydroxide. The particles were incubated in the presence of the components indicated under the first incubation. The particles were then reisolated by centrifugation and incubated in the presence of the components indicated under the second incubation. After further centrifugation and washing, the particles were assayed for ATPase activity. The activities reported refer to μmoles of ATP hydrolyzed per min per mg particle protein. (From Tzagoloff, 1970.)

TABLE 7.7. Properties of the Subunit Polypeptides of the OS-ATPase Complex[a]

Subunit	M.W.	Function	Component
α	58,000	?	F_1
β	55,000	Hydrolytic site	F_1
	33,000[b]	?	F_1
γ	29,000	?	F_1
δ	14,000	Binding site for OSCP (?)	F_1
ϵ	8,000	?	F_1
7	18,000	Binding of F_1 to membrane factor	OSCP
5	28,000	?	Membrane factor
6	21,000	?	Membrane factor
8	12,000	F_6 (?) binding of F_1 to membrane factor	Membrane factor
9	8,000	Binding sites for DCCD and oligomycin	Membrane factor

[a]The composition is based on studies of the yeast OS-ATPase. The molecular weights of the subunits in the mammalian enzyme are slightly different.
[b]This protein is seen in preparations of yeast F_1.

ultrastructures of more dispersed preparations of the enzyme. F_1 is a round particle with a diameter of 90 Å. High-resolution electron microscopy reveals that it is made up of six domains arranged hexagonally to form a doughnut-shaped structure (Fig. 7.8). The six structural domains probably correspond to the α and β subunits.

The yeast OS-ATPase complex purified by the Triton procedure (see Section I) is monodisperse because of the presence of detergent. Electron micrographs of this preparation are shown in Fig. 7.8. Comparison of the electron micrographs of F_1 and the OS-ATPase indicates that the two enzymes have somewhat different ultrastructures. The complex is an oval particle approximately 90 Å wide and 150 Å long. The subunit arrangement of F_1 is not discerned in the complex. Electron micrographs of negatively stained rat liver OS-ATPase (Fig. 7.9) convincingly show a tripartite structure. The enzyme is seen to consist of a 90-Å spherical particle connected through a stalk to another globular unit of somewhat smaller dimensions. The smaller unit is probably an aggregate of the membrane factor proteins that are normally lodged in the lipid bilayer of the membrane.

Based on the dimensions of purified OSCP and on the observed changes of the mitochondrial inner membrane following removal of the protein, MacLennan and Asai have concluded that it is probably the stalk linking F_1 to the membrane factor. Electron micrographs of the inner membrane depleted of F_1 show the presence of naked stalks projecting from the membrane surface. The stalks become less evident, and the surface of the membrane acquires a smoother appearance after OSCP is extracted with alkali.

The above studies support the existence of three morphological domains in the ATPase. The best approximation of the enzyme's structure is that of a tripartite particle made of a globular membrane factor unit embedded in the phospholipid bilayer, a thin stalk corresponding to OSCP, and the 90-Å particle consisting of the F_1 subunits.

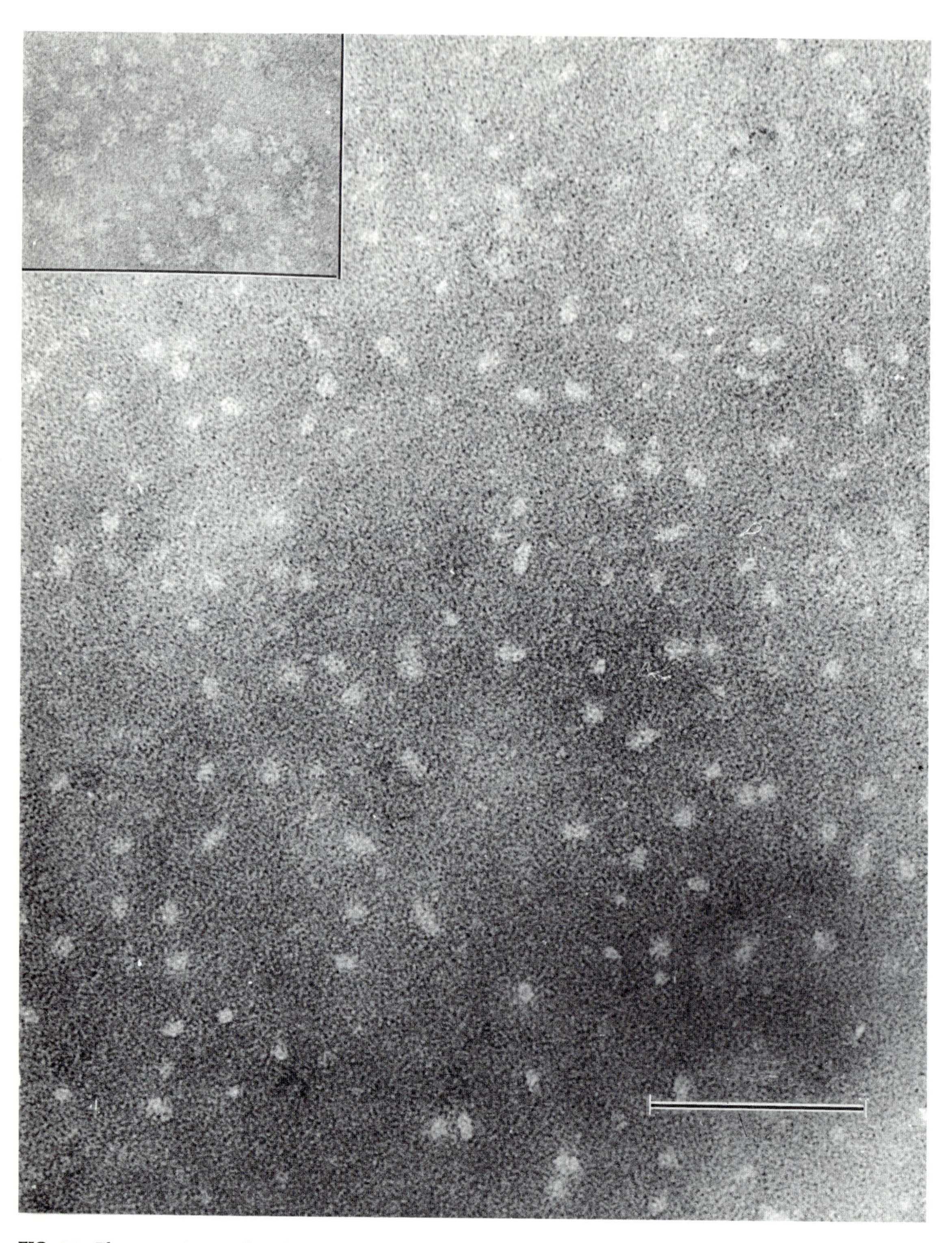

FIG. 7.8. Electron micrographs of OS-ATPase. The OS-ATPase was purified by extraction of yeast submitochondrial particles with Triton X-100 and centrifugation of the extract through a glycerol gradient. Yeast F_1 is shown in the inset. Both preparations were negatively stained with phosphotungstate. Scale bar represents 0.1 μm.

It is now generally viewed that the hydrolysis of ATP by mitochondria represents a reversal of the reactions entailed in ATP synthesis. The catalytic mechanism of the ATPase is therefore pertinent to the broader problem of the mechanism of oxidative phosphorylation. In the previous chapter, two possible mechanisms of the ATPase were considered in the context of the chemiosmotic and conformational models of energy coupling. Both mechanisms dispensed with the participation of high-energy phosphoryl intermediates.

The lack of experimental evidence for intermediates does not by itself constitute grounds for excluding chemical intermediates in ATP synthesis or hydrolysis. In fact, several observations are best rationalized in terms of two separate phosphoryl transfer reactions in the mechanism of the ATPase. For example, the difference in the enzymatic properties of F_1 and the OS-ATPase can be explained by assuming that the OS-ATPase catalyzes an internal phosphoryl transfer reaction that does not occur in F_1. This is shown in the following hypothetical scheme.

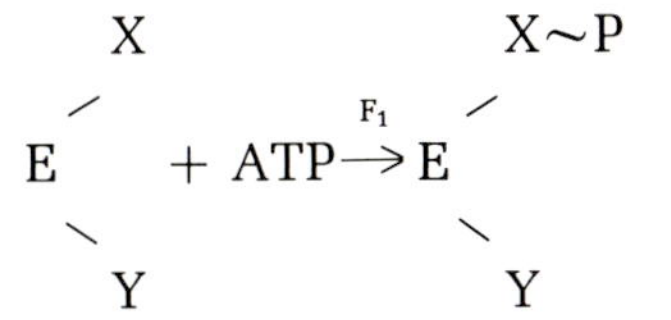

$$E\begin{matrix} \diagup X \\ \\ \diagdown Y \end{matrix} + ATP \xrightarrow{F_1} E\begin{matrix} \diagup X\sim P \\ \\ \diagdown Y \end{matrix} \qquad [7.1]$$

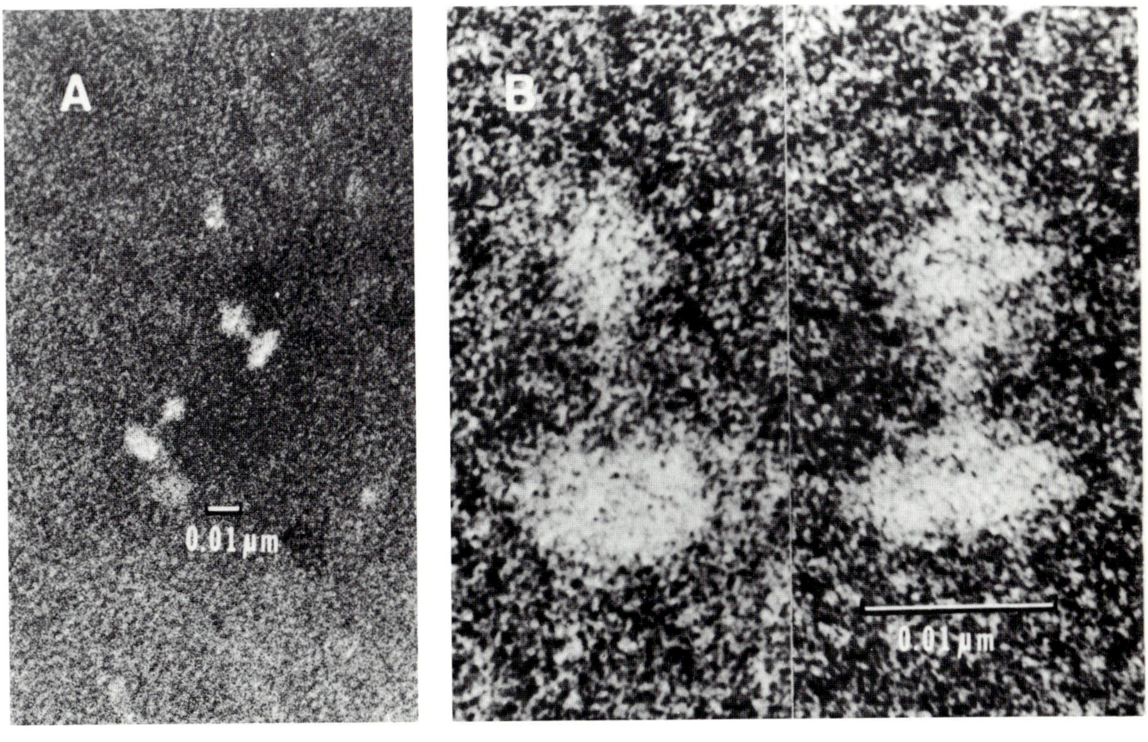

FIG. 7.9. Electron micrographs of rat liver OS-ATPase showing the tripartite structure of the enzyme. A dispersed preparation of the enzyme was negatively stained with phosphotungstate. (Courtesy of Dr. Peter Pedersen.)

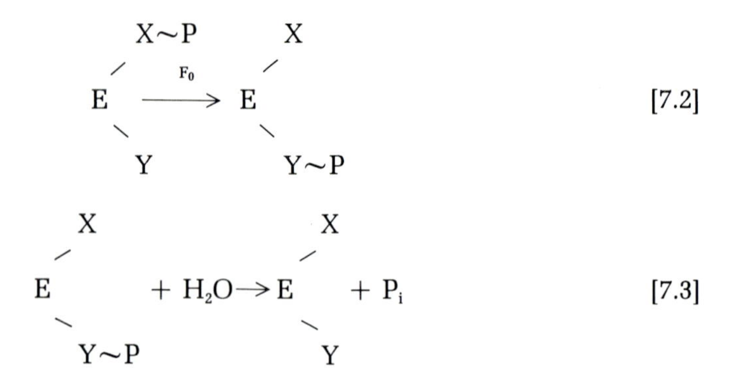

$$E\begin{matrix} X\sim P \\ Y \end{matrix} \xrightarrow{F_0} E\begin{matrix} X \\ Y\sim P \end{matrix} \qquad [7.2]$$

$$E\begin{matrix} X \\ Y\sim P \end{matrix} + H_2O \rightarrow E\begin{matrix} X \\ Y \end{matrix} + P_i \qquad [7.3]$$

In the first step, ATP phosphorylates a group, X, located on F_1. This is followed by an internal transfer of the phosphoryl from X to another accepting group Y present in the OS-ATPase. In the last reaction, Y~P is hydrolytically cleaved with the liberation of inorganic phosphate. According to this mechanism, reactions 7.2 and 7.3 are catalyzed by the membrane factor which is visualized to function as a phosphoryl transferase. This part of the catalytic mechanism would be dependent on phospholipids and sensitive to oligomycin, mercurials, and other class A inhibitors (Table 7.1). In the case of F_1, the hydrolysis of ATP may involve only the first reaction (7.1). The essential differences between the two enzymes are that the X~P intermediate can react with water only when F_1 is physically separated from the hydrophobic proteins and that in the complex there is an obligatory transfer of the phosphoryl group from X to Y.

V. Coupling Factor Activity of ATPase Components

The feasibility of resolving the components required for oxidative phosphorylation was first demonstrated unambiguously by Pullman and co-workers and by Linnane and Titchner who, in 1961, independently reported the isolation of soluble protein factors from bovine heart mitochondria capable of increasing the rate of ATP synthesis in submitochondrial particles. Even though such proteins were and continue to be called coupling factors, the choice of the term should not be interpreted to indicate that their mode of action is necessarily related directly to the coupling mechanism. In all instances, coupling factors have been found to stimulate ATP synthesis and ATP-driven but not substrate-driven reactions, suggesting that their effect is on the terminal phosphorylation process rather than on the coupling mechanism per se. As will be seen subsequently, the best-documented coupling factors have turned out to be constituents or subunits of the OS-ATPase, and for this reason the discussion of this topic has been left to this chapter.

The basic tactic used to resolve the mitochondrial phosphorylation system has been to isolate and purify soluble proteins or coupling factors that increase ATP synthesis and other energy-dependent reactions in submitochondrial particles. In such studies, it is necessary that the coupling factor be obtained in a reasonably pure state. This is important because in most instances the protein may have no intrinsic activity of its own and can only be assayed by its ability to restore phosphorylation in submitochondrial particles. The surest criterion of a new coupling factor therefore depends on the physical characterization of the purified protein. The second requirement is that there be a suitable submitochondrial particle for the assay of the coupling factor. Ideally, the test particle should be deficient specifically in the coupling factor for whose assay it is used. In practice this is seldom achieved, and usually one settles on particle preparations exhibiting the clearest response to the coupling factor. Finally, a practical assay has to be developed. This presents the least problem since coupling factors that stimulate ATP synthesis also increase the rate of other energy-coupled reactions such as ATP–P_i exchange and ATP-dependent pyridine nucleotide transhydrogenation or NAD^+ reduction by succinate. Any of these reactions can, therefore, be used in the assay of the coupling factor.

At present four proteins with coupling factor activity have been purified: F_1, F_6, OSCP, and factor B. Although a large number of other coupling factors have been reported in the literature, most of them have turned out to be modified forms or mixtures of two or more of the above four. It is of interest that each of four coupling factors is a part of the OS-ATPase, suggesting that the resolution of oxidative phosphorylation into soluble and particulate components occurs mainly at the level of the ATPase complex. In the following discussion, we shall review the effect of the purified coupling factors on various submitochondrial particles. Some of this information is summarized in Tables 7.8 and 7.9.

A. F_1

The best-studied and first coupling factor to be isolated is F_1 itself. Penefsky and his associates were able to show that submitochondrial particles (N particles) obtained after mechanical disruption of beef-heart mitochondria with glass beads had reduced P/O ratios which could be increased three- to fivefold by the addition of purified F_1. The extent of stimulation of phosphorylation was a function of the amount of F_1 added and had a saturable value. In addition to

TABLE 7.8. Coupling Factor Requirements in Various Submitochondrial Particles

Particle	Treatment	Requirement
ETP_H, SMP	Sonic irradiation of beef heart mitochondria in the presence of ATP, Mg^{2+}, Mn^{2+}	None
N particle	Mechanical disruption of beef heart mitochondria in a Nossal shaker	F_1
U particle	Extraction of SMP with 4 M urea	F_1
ETP_H (NaCl)	Extraction of ETP_H with 2 M NaCl	F_1 or F_1 subunits
A particle	Sonic irradiation of beef heart mitochondria in the presence of EDTA at pH 9.2	F_1 and OSCP
AE particle	Sonic irradiation of beef heart mitochondria in the presence of EDTA at pH 8.5	Factor B
STA particle	Extraction of A particles with 1% silicotungstate at pH 5.5	F_1, F_6, OSCP, Factor B

improving net phosphorylation, F_1 also stimulates the ATP–P_i exchange and ATP-dependent reactions such as pyridine nucleotide transhydrogenation and the reduction of NAD^+ by succinate.

Three types of submitochondrial particles have been described in which phosphorylation and ATP-driven reactions are stimulated by F_1: (1) particles that are depleted of most of their F_1 and have very low or no ATPase activity; (2) particles that have a high content of F_1 and have an active ATPase; and (3) particles containing F_1 that is enzymatically inactive because of either partial depolymerization or loss of some essential subunits of the enzyme.

Among the first type are particle preparations from which F_1 has been removed by extraction with urea or some other chaotropic agent (U particles). The restoration of phosphorylation by F_1 in these depleted membranes probably results primarily from the catalytic role of the enzyme in the terminal phosphorylation reactions.

The requirement of F_1 in particles with an already high endogenous content of F_1 (N and AE particles) has never been well understood. An explanation consistent with the available evidence is that besides its catalytic role, F_1 also confers structural integrity to the membrane. Thus, membranes that have enough F_1 to catalyze phosphorylation may be unable to do so because of a structural deficiency that leads to a dissipation of the nonphosphorylated intermediate. For example, in the chemiosmotic hypothesis, a partial deficiency of

TABLE 7.9. Properties of Coupling Factors of Oxidative Phosphorylation

Factor	Alternate name	Molecular weight	Intrinsic activity	Particle used for assay	Assay
F_1	—	360,000	ATPase	N particle A particle U particle	Stimulation of P/O ATP–P_i exchange NAD^+ reduction Transhydrogenation
F_6	F_c	8,000	None	STA particle	Stimulation of P/O ATP–P_i exchange NAD^+ reduction Transhydrogenation Binding of F_1
OSCP	—	18,000	None	A particle STA particle	Stimulation of P/O ATP–P_i exchange NAD^+ reduction Transhydrogenation Binding of F_1
Factor B	F_2	12,000–45,000	None	AE particle STA particle	Stimulation of P/O ATP–P_i exchange NAD^+ reduction Transhydrogenation Stimulation of ATP–P_i exchange in OS-ATPase

F_1 has been postulated to cause the membrane to be leaky to hydrogen ions and therefore incapable of maintaining a proton gradient of sufficient magnitude to facilitate the synthesis of ATP. This interpretation is supported by the observation that chemically modified F_1, even though enzymatically inactive, is nonetheless capable of restoring phosphorylation in N particles. This effect of F_1 may be related to the partial coupling of certain particles with low concentrations of oligomycin (see Chapter 6, Section VI).

The third type of submitrochondrial particle is obtained by extraction of ETP_H with 2 M NaCl [ETP_H(NaCl)]. The salt-extracted membranes have low ATPase activity but have been found in electron microscopic studies to contain a full complement of the 90-Å particles corresponding to F_1. The NaCl treatment appears to cause a partial depolymerization and loss of F_1 subunits. The F_1 molecules remaining on the membrane are enzymatically inactive because of an alteration in structure or absence of essential subunits. The ATPase activity and energy-coupled reactions of ETP_H(NaCl) can be reconstituted either with the active F_1 or the catalytically inactive subunits obtained by cold depolymerization of F_1. In this instance, therefore, subunits of F_1 can act as coupling factors.

The three types of reconstitutions discussed above are schematically illustrated in Fig. 7.10.

B. Coupling Factors That Are Related to F_1

Two coupling factors have been purified that are closely related to F_1. A preparation called factor A has been studied by Sanadi and his collaborators. Factor A contains all the subunits of F_1 but has a low basal ATPase activity which can be enhanced severalfold by heat treatment at 65°C. When the activated enzyme is cooled, the activity becomes masked, and this process can be repeated several times. Since the natural ATPase inhibitor is known to be

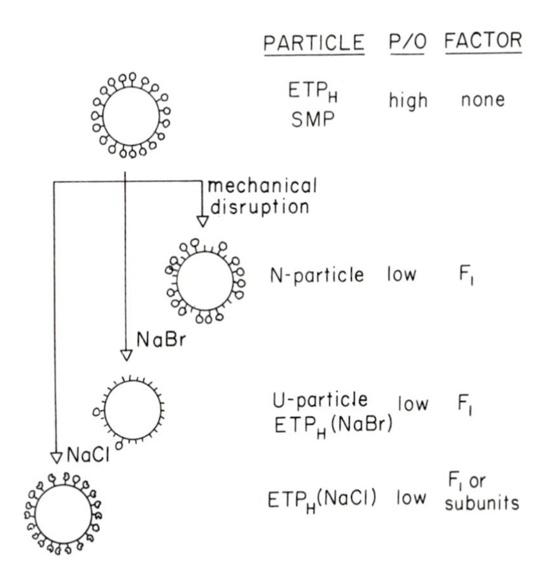

FIG. 7.10. Partial and complete resolution of F_1 from submitochondrial particles. The factor requirements for the reconstitution of oxidative phosphorylation in each particle are indicated in the right-hand column.

released from F_1 by heat, factor A is probably a complex of F_1 with the inhibitor. Alternatively, factor A may differ from F_1 in some aspect of its quaternary structure.

A still different form of F_1 has been isolated and studied by Van Dam. This factor, called $F_1 \cdot X$, has been shown to be a complex of F_1 and OSCP. Both $F_1 \cdot X$ and factor A can be substituted for F_1 in restoring phosphorylation and energy-coupled reactions in a large variety of submitochondrial particles.

C. Oligomycin-Sensitivity-Conferring Protein

This subunit of the OS-ATPase has also been shown to be effective in reconstituting oxidative phosphorylation and ATP-energized reactions in submitochondrial particles. The OSCP and F_1 cannot be substituted for each other and therefore act in an independent manner. The best assay for the coupling factor activity of OSCP utilizes submitochondrial particles that have been exposed to alkaline conditions (A particles). When such membrane preparations are tested for various coupled functions, neither F_1 nor OSCP alone exert any significant effect. Complete reconstitution, however, is achieved when both coupling factors are added together. This is shown in Table 7.10 for the ATP–P_i exchange reaction. The alkaline treatment required to extract OSCP also depolymerizes and removes the F_1 component from the particles. For this reason, it has not been feasible to obtain submitochondrial particles whose energy-coupled functions respond to OSCP alone.

Since OSCP has been found necessary for the reconstitution of OS-ATPase, its activity as a coupling factor is most likely related to a proper rebinding of F_1 to the depleted membranes.

D. F_6

This low-molecular-weight protein has recently been purified to homogeneity from the OS-ATPase. Two activities have been associated with F_6: (1) binding of F_1 to the membrane factor and (2) restoration of oxidative phosphorylation and ATP-dependent reactions in depleted particles. We have already seen that A particles prepared by exposure to alkaline pH acquire a require-

TABLE 7.10. Restoration of ATP–P_i Exchange in A Particles by F_1 and OSCP

Addition[a]	ATP–P_i exchange (nmoles/min per mg protein)
None	7
30 μg F_1	13
5 μg OSCP	8
30 μg F_1 + 5 μg OSCP	59

[a]Additions were made to 1 mg of A particles. (From MacLennan and Paulson, 1970.)

ment for F_1 and OSCP. Further extraction of A particles with silicotungstic acid yields STA particles with additional requirements for F_6 and factor B. The precise function of F_6 is not understood, but like OSCP, its stimulatory effect on coupled processes is probably related to the reconstitution of the membrane ATPase.

E. Factor B

Several different preparations of factor B have been reported. Their molecular weights range from 12,000 to 45,000. In other respects they are quite similar (all factor B preparations are inhibited by SH reagents) and are probably the same protein in different states of aggregation. Factor B is usually assayed by its ability to stimulate the ATP-driven reduction of NAD^+ by succinate. However, it also increases the efficiency of phosphorylation and other coupled reactions in a broad range of particles. A sole requirement for factor B has been found in AE particles prepared by sonic disruption of mitochondria at pH 8.5 in the presence of EDTA. In more extensively depleted particles (prepared at higher pH, i.e., A particles), factor B activity is measured in conjunction with F_1 and OSCP (A particles) or F_1, F_6, and OSCP (STA particles). The function of factor B in the ATPase is obscure, but similarities in the stimulation of coupled reactions by factor B and "oligomycin coupling" suggest that the protein may exert its effect by stabilizing the nonphosphorylated intermediate or, in the chemiosmotic interpretation, by decreasing the permeability of the membrane to protons.

F. Present Status of Coupling Factors

The study of coupling factors has been useful in several important respects. First, it has helped to identify proteins essential for oxidative phosphorylation. The four coupling factors for which such a role has been demonstrated are all subunit polypeptides of the ATPase complex. The fact that coupling factors are equally effective in restoring both ATP synthesis and its utilization in energy-dependent reactions further establishes the central role of the ATPase in all energy-coupled processes of the mitochondrial organelle. Second, coupling factors have provided important information about the subunit structure and organization of the mitochondrial ATPase.

In other respects, the coupling factor approach has been somewhat disappointing. The original anticipation that the dissection of the enzyme system involved in oxidative phosphorylation would help to reveal some of the more detailed aspects of the mechanism has not been borne out. This may be partially attributed to the fact that the overall reaction catalyzed by the ATPase is very simple, being either the formation or destruction of a pyrophosphate bond. Thus, there is a severe experimental limitation to the number of chemical phenomena that can be studied. The great deal of attention that is currently being devoted to various aspects of the ATPase will, it is hoped, point the way to clarifying, in a more specific way than has been possible up to now, how the coupling factors function in the phosphorylation mechanism.

Selected Readings

Abrams, A. (1976) Structure and function of membrane-bound ATPase in bacteria, in *The Enzymes of Biological Membranes* (A. Martonosi, ed.), Plenum Press, New York, pp. 57–73.

Amzel, L. M., and Pedersen, P. L. (1978) Adenosine triphosphatase from rat liver mitochondria. Crystallization and X-ray diffraction studies of the F_1 component of the enzyme, *J. Biol. Chem.* **253**:2067.

Andreoli, T. E., Lam, K. W., and Sanadi, D. R. (1965) Studies on oxidative phosphorylation X. A coupling enzyme which activates reverse electron transfer, *J. Biol. Chem.* **240**:2644.

Beechey, R. B., and Cattell, K. J. (1973) Mitochondrial coupling factors, in *Current Topics in Bioenergetics* (D. R. Sanadi and L. Packer, eds.), Academic Press, New York, pp. 306–357.

Cattell, K. J., Lindop, C. R., Knight, I. G., and Beechey, R. B. (1971) The identification of the site of action of N,N′-dicyclohexylcarbodiimide as a proteolipid in mitochondrial membranes, *Biochem. J.* **125**:169.

Criddle, R. S., Packer, L., and Shieh, P. (1977) Oligomycin-dependent ionophoric protein subunit of mitochondrial adenosine triphosphatase, *Proc. Natl. Acad. Sci. U.S.A.* **74**:4306.

Deters, D. W., Racker, E., Nelson, N., and Nelson, H. (1975) Partial resolution of the enzymes catalyzing photophosphorylation XV. Approaches to the active site of coupling factor 1, *J. Biol. Chem.* **250**:1041.

Joshi, S., Shaikh, F., and Sanadi, D. R. (1975) Restoration of P_i–ATP exchange in the oligomycin-sensitive ATPase: Effect of a coupling factor, *Biochem. Biophys. Res. Commun.* **65**:1371.

Kagawa, Y., and Racker, E. (1966) Partial resolution of the enzymes catalyzing oxidative phosphorylation IX. Reconstruction of oligomycin-sensitive adenosine triphosphatase, *J. Biol. Chem.* **241**:2467.

Kagawa, Y., and Racker, E. (1966) Partial resolution of the enzymes catalyzing oxidative phosphorylation X. Correlation of morphology and function in submitochondrial particles, *J. Biol. Chem.* **241**:2475.

Kanner, B. L., Serrano, R., Kandrach, M. A., and Racker, E. (1976) Preparation and characterization of homogeneous coupling factor 6 from bovine heart mitochondria, *Biochem. Biophys. Res. Commun.* **69**:1050.

Knowles, A. F., and Penefsky, H. S. (1972) The subunit structure of beef heart mitochondrial adenosine triphosphatase. Physical and chemical properties of isolated subunits, *J. Biol. Chem.* **247**:6624.

Lardy, H., Reed, P., and Lin, C. H. C. (1975) Antibiotic inhibitors of mitochondrial ATP synthesis, *Fed. Proc.* **34**:1707.

MacLennan, D. H., and Asai, J. (1968) Studies on the mitochondrial adenosine triphosphatase system V. Localization of the oligomycin-sensitivity conferring protein, *Biochem. Biophys. Res. Commun.* **33**:441.

MacLennan, D. H., and Paulson, C. W. (1970) Studies on the mitochondrial adenosine triphosphatase system VI. Coupling activity of F_1 subunits, *Can. J. Biochem.* **48**:1079.

MacLennan, D. H., and Tzagoloff, A. (1968) Studies on the mitochondrial adenosine triphosphatase system IV. Purification and characterization of the oligomycin-sensitivity conferring protein, *Biochemistry* **7**:1603.

MacLennan, D. H., Smoly, J. M., and Tzagoloff, A. (1968) Studies on the mitochondrial adenosine triphosphatase system I. Restoration of adenosine triphosphate dependent reactions in salt extracted submitochondrial particles, *J. Biol. Chem.* **243**:1589.

Norling, B., Glaser, E., and Ernster, L. (1978) Reconstitution of oligomycin- and dicyclohexylcarbodiimide-sensitive mitochondrial ATPase from isolated components, in *Frontiers of Biological Energetics* (P. L. Dutton, J. S. Leigh and A. Scarpa, eds.), Vol. I, Academic Press, New York, pp. 501–515.

Penefsky, H. S., Pullman, M. E., Datta, A., and Racker, E. (1960) Partial resolution of the enzymes catalyzing oxidative phosphorylation II. Participation of a soluble adenosine triphosphatase in oxidative phosphorylation, *J. Biol. Chem.* **235**:330.

Pullman, M. E., and Monroy, G. C. (1963) A naturally occurring inhibitor of mitochondrial adenosine triphosphatase, *J. Biol. Chem.* **238**:3762.

Pullman, M. E., Penefsky, H. S., Datta, A., and Racker, E. (1960) Partial resolution of the enzymes catalyzing oxidative phosphorylation I. Purification and properties of a soluble dinitrophenol stimulated adenosine triphosphatase, *J. Biol. Chem.* **235**:3322.

Sebald, W., Hoppe, J., and Wachter, E. (1979) Amino acid sequence of the ATPase proteolipid from mitochondria, chloroplasts and bacteria (wild type and mutants), in *Function and Molecular Aspects of Biomembrane Transport* (E. Quagliariello, E. Palmieri, S. Papa and M. Klingenberg, eds.), North-Holland, Amsterdam, pp. 63–74.

Senior, A. E., and Brooks, J. C. (1970) Studies on the mitochondrial oligomycin-insensitive ATPase I. An improved method of purification and the behavior of the enzyme in solutions of various depolymerization agents, *Arch. Biochem. Biophys.* **140**:257.

Shankaran, R., Sani, B. P., and Sanadi, D. R. (1975) Studies on the oxidative phosphorylation. Evidence for multiple forms of factor B activity, *Arch. Biochem. Biophys.* **168**:394.

Soper, J. W., Decker, G. L., and Pedersen, P. L. (1979) Mitochondrial ATPase complex. A dispersed, oligomycin-sensitive preparation from rat liver containing molecules with a tripartite structural arrangement, *J. Biol. Chem.* **254**:11170.

Stigall, D. L., Galante, Y. M., and Hatefi, Y. (1978) Preparation and properties of an ATP–P_i exchange complex (complex V) from bovine heart mitochondria, *J. Biol. Chem.* **253**:956.

Tzagoloff, A. (1970) Assembly of the mitochondrial membrane system III. Function and synthesis of the oligomycin-sensitivity conferring protein of yeast mitochondria, *J. Biol. Chem.* **245**:1545.

Tzagoloff, A., and Meagher, P. (1971) Assembly of the mitochondrial membrane system V. Properties of a dispersed preparation of the rutamycin-sensitive adenosine triphosphatase of yeast mitochondria, *J. Biol. Chem.* **246**:7328.

Tzagoloff, A., Byington, K. H., and MacLennan, D. H. (1968a) Studies on the mitochondrial adenosine triphosphatase system II. The isolation and characterization of an oligomycin-sensitive adenosine triphosphatase from bovine heart mitochondria, *J. Biol. Chem.* **243**:2405.

Tzagoloff, A., MacLennan, D. H., and Byington, K. H. (1968b) Studies on the mitochondrial adenosine triphosphatase system III. Isolation from the oligomycin sensitive adenosine triphosphatase complex of the factors which bind F_1 and determine oligomycin-sensitivity of bound F_1, *Biochemistry* **7**:1596.

Van der Stadt, R. J., Kraaipoel, R. J., and Van Dam, K. (1972) $F_1 \cdot X$, a complex between F_1 and OSCP, *Biochim. Biophys. Acta* **267**:25.

Yoshida, M., Sone, N., Hirata, H., and Kagawa, Y. (1977) Reconstitution of adenosine triphosphatase of thermophilic bacterium from purified individual subunits, *J. Biol. Chem.* **252**:3480.

8

Resolution and Reconstitution of Electron Transport and Oxidative Phosphorylation

The resolution of complex biological systems and subsequent reconstitution of part or the whole of the original function from the separated components have been powerful tools in biochemistry and molecular biology. This approach has also been used in mitochondrial studies, and much of our current knowledge of electron transport and energy-coupling mechanisms stems from attempts at dissecting and reconstituting the multienzyme system that catalyzes these processes. This chapter will review some of the older and more recent experiments on the reconstitution of the respiratory chain and of the ATP synthetase.

I. Electron Transport Chain

The first successful resolution of the respiratory chain was reported by Keilin and King. These authors were able to selectively inactivate the succinate oxidase of submitochondrial particles by alkali treatment. The oxidase activity could be restored by the addition of a soluble flavoprotein that had no oxidase activity on its own but catalyzed the reduction of artificial dyes by succinate. This enzyme was later purified and shown to be the dehydrogenase of the succinate–coenzyme Q reductase complex. A similar reconstitution has been achieved with the NADH oxidase. Under appropriate conditions submitochondrial particles can be depleted of the NADH dehydrogenase. Such particles are rendered fully active when supplied with the extracted NADH dehydrogenase. These early studies demonstrated the feasibility of removing and reattaching individual members of the electron transfer chain to crude preparations of the mitochondrial inner membrane. Because both primary dehydrogenases are water-soluble proteins, they were the first carriers to be purified and studied in detail.

Taking advantage of the newly discovered property of bile acids and detergents to solubilize membrane proteins, Hatefi and co-workers resolved the mitochondrial respiratory chain into four enzyme complexes. The compositions and catalytic activities of the purified enzymes (see Chapter 4, Section IVB) suggested that together they could account for the complete oxidation of succinate and NADH by molecular oxygen. This was confirmed by a series of reconstitution experiments in which various combinations of the enzymes were shown to restore part of the overall pathways for the oxidation of the two substrates. Before describing these important experiments, it is necessary to recall that the final preparations of the complexes contain 10–30% phospholipid and bile acids (cholate and/or deoxycholate) approximately equal to the amount of protein. Concentrated solutions of the purified complexes are consequently dispersed (optically clear) because of the detergent effect of the mixed bile acid–phospholipid micelles.

The spans of the electron transport chain formed from different combinations of the complexes are listed in Table 8.1. These include succinate- and NADH–cytochrome *c* reductase; succinate and NADH oxidase; and $CoQH_2$ oxidase. The reconstitutions of the oxidase activities required the addition of cytochrome *c*, since this low-molecular-weight carrier is not present in any of the purified complexes. There was no requirement for coenzyme Q, which, because of its lipophilic properties, cofractionates with each of the four complexes. The restoration of succinate or NADH oxidase was highly efficient as evidenced by the high final specific activity. Furthermore, the reconstituted activities mimicked the native respiratory chain in their responses to classical inhibitors of electron transport (Table 8.2).

In defining the conditions necessary for the reconstitution of an integrated electron transport activity, it was noted that the component complexes had to be mixed at high protein concentration prior to dilution in buffer for the enzymatic assays. The efficiency of reconstitution was drastically decreased when the enzymes were diluted separately, even though the individual enzymes in the final mixture were fully active (Table 8.3). This crucial observation indicated that reconstitution involved an association of the components into a physical unit, the association being in some manner dependent on the particular conditions used to attain maximal activity.

The explanation for the need to premix the complexes at high concentrations for optimal reconstitution stemmed from the discovery that the purified

TABLE 8.1. Segments of the Respiratory Chain Reconstituted from the Purified Complexes

Complexes[a]	Reconstituted activity
I + III	NADH–cytochrome *c* reductase
II + III	Succinate–cytochrome *c* reductase
III + IV + cytochrome *c*	Coenzyme QH_2 oxidase
I + III + IV + cytochrome *c*	NADH oxidase
II + III + IV + cytochrome *c*	Succinate oxidase
I + II + III + IV + cytochrome *c*	NADH and succinate oxidase

[a]Complex I, NADH–coenzyme Q reductase; complex II, succinate–coenzyme Q reductase; complex III, coenzyme QH_2–cytochrome *c* reductase; complex IV, cytochrome oxidase.

TABLE 8.2. Reconstitution of NADH and Succinate Oxidase from the Purified Complexes and Cytochrome *c*

Additions	Specific activity	
	Succinate	NADH
I + cytochrome *c*	0	0
II + cytochrome *c*	0	0
III + cytochrome *c*	0	0
IV + cytochrome *c*	0	<1
I + II + III + cytochrome *c*	3.1	2.1
I + II + III + IV + cytochrome *c*	28.7	14.5
I + II + III + IV + cytochrome *c* + cyanide	0	0
I + II + III + IV + cytochrome *c* + antimycin	0	0
I + II + III + IV + cytochrome *c* + amobarbital	25.7	0
I + II + III + IV + cytochrome *c* + thenoyltrifluoroacetone	0.7	n.d.

[a]The components indicated were preincubated at high protein concentration, diluted to 0.1 mg/ml in buffered sucrose, and assayed immediately. The inhibitors were added to the assay medium. The specific activity refers to μmoles of substrate oxidized/min per mg protein. (From Hatefi *et al.*, 1962.)

enzymes are capable of spontaneously aggregating into membranes provided there is sufficient phospholipid and proper steps are taken to remove or decrease the concentration of endogenous bile acids (see Chapter 5, Section III). In fact, the dilution used in the reconstitution assay was found to cause a rapid aggregation of each complex or of mixtures of complexes into vesicular membranes (Fig. 8.1). This effect of dilution can be attributed to the more favorable partitioning of bile acids in the water phase when their concentrations are reduced below the critical micellar concentration.

The tendency of the complexes to form membranes suggested a physical basis for the reconstitution of an integrated electron chain, namely, that the component enzymes must be present within the same membrane continuum. The sole condition required for the formation of such hybrid membranes was that the complexes be diluted from a common dispersed mixture.

TABLE 8.3. Effect of Enzyme Concentration in the Premixture on the Reconstitution of Succinate–Cytochrome *c* Reductase[a]

Complex II (mg protein/ml)	Complex III (mg protein/ml)	Specific activity of succinate–cytochrome *c* reductase
0.001	0.001	0.9
0.010	0.010	2.6
0.10	0.10	13.2
6.0	6.0	39.0

[a]The two enzymes were diluted to the indicated concentrations, mixed, and assayed for succinate–cytochrome *c* reductase. Specific activity refers to μmoles of cytochrome *c* reduced/min per mg protein. (From Hatefi *et al.*, 1962.)

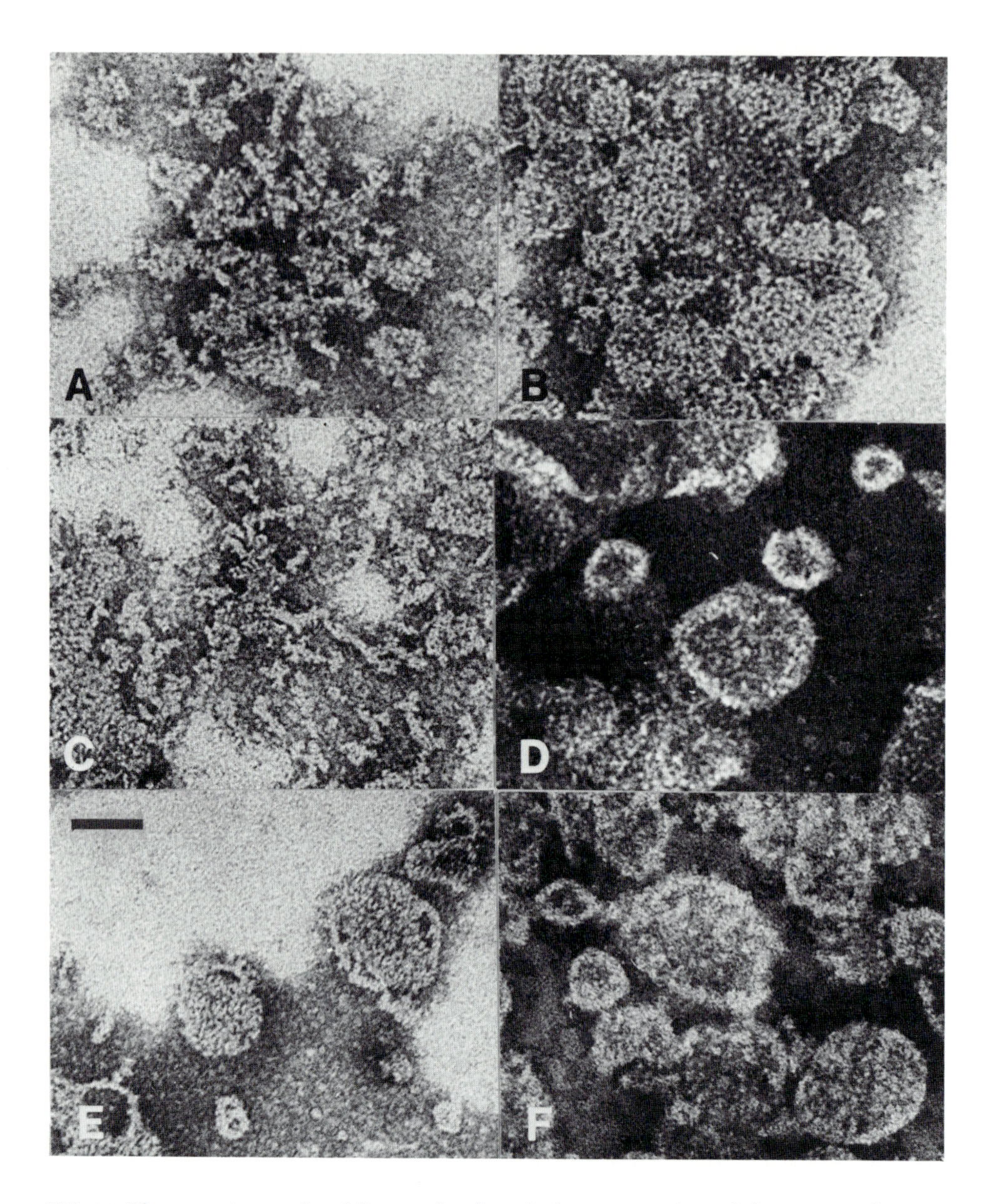

FIG. 8.1. Electron micrographs of dispersed and particulate preparations of electron transfer complexes. Dispersed preparations (A–C): NADH–cytochrome *c* reductase (A), cytochrome oxidase (B), and a mixture of the two (C) at a protein concentration of 5 mg/ml were negatively stained with phosphotungstate. Particulate preparations (D–F): NADH–cytochrome *c* reductase (D), cytochrome oxidase (E), and a mixture of the two (F) were diluted to 0.25 mg/ml before staining with phosphotungstate. The experiment shows the presence of membrane vesicles in the diluted but not the concentrated samples of the complexes. The scale bar represents 0.1 μm.

In the experiment of Fig. 8.2, NADH–cytochrome *c* reductase (this preparation is composed of complexes I and III) and cytochrome oxidase were separately diluted, mixed, and then centrifuged on a sucrose gradient. When this order was followed, two distinct bands were formed on the gradient, one consisting of NADH–cytochrome *c* reductase and the other of cytochrome oxidase. The difference in the density of the two membranes is caused by the lower phospholipid content of cytochrome oxidase. Since hybrid membranes were not detected, it may be concluded that there is little or no exchange of the enzymes between membranes. In the same experiment, the component enzymes were mixed at high protein concentration before dilution. The membranes formed under these conditions banded with a density intermediate between those of the individual enzymes. The membranes were fully active in catalyzing NADH oxidase, indicating that they contained all three complexes.

Several other interesting findings emerged from the reconstitution studies. In mitochondria, the carriers of the electron transport chain are present in a

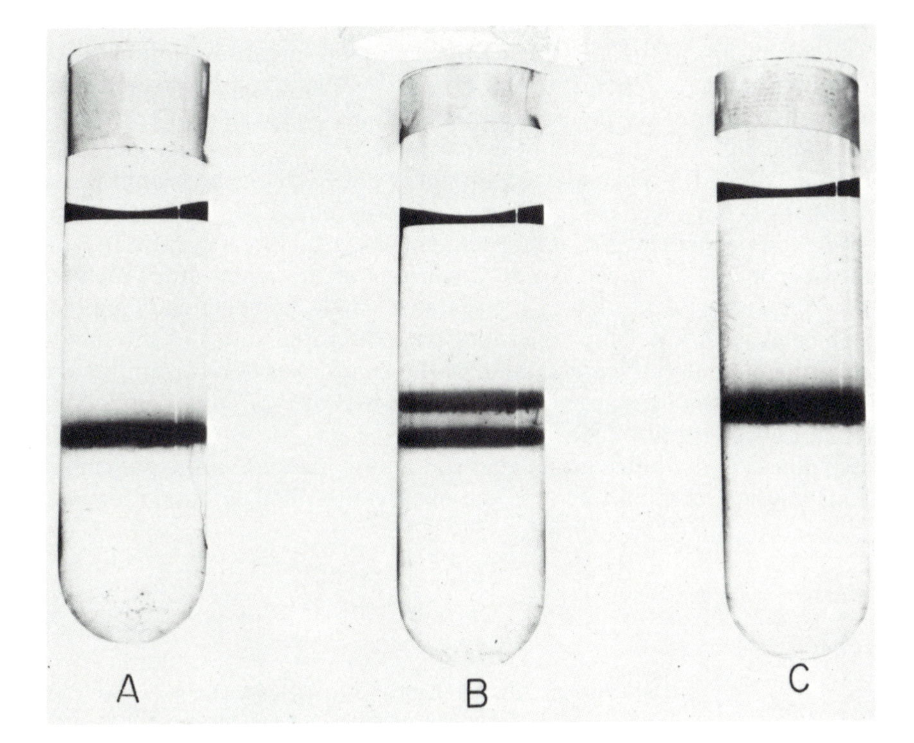

FIG. 8.2. Density gradient centrifugation of electron transfer complexes. NADH–cytochrome *c* reductase and cytochrome oxidase were treated as follows. A: A concentrated solution of cytochrome oxidase was diluted to a protein concentration of 0.25 mg/ml. B: Cytochrome oxidase and NADH–cytochrome *c* reductase were separately diluted to a protein concentration of 0.25 mg/ml and mixed. C: Cytochrome oxidase and NADH–cytochrome *c* reductase were mixed at a high protein concentration and diluted to a final concentration of 0.25 mg/ml. The samples were applied to a linear 0.5–2 M sucrose gradient and centrifuged at 39,000 rpm for 2 hr. The lower band in tube B was identified as containing only cytochrome oxidase, and the upper band NADH–cytochrome *c* reductase membranes.

TABLE 8.4. Concentration of Cytochromes a, a_3, and b in Membranes Reconstituted with Different Ratios of Complexes I + III/Complex IV[a]

Membrane	Cytochrome content (nmoles/mg protein)		
	$a + a_3$	b	$(a + a_3)/b$
A	1.2	1.4	0.86
B	3.3	0.80	4.10
C	5.4	0.35	15.35

[a]The concentrations of the cytochromes were estimated from the spectra of Fig. 8.3. The following weight ratios of NADH–cytochrome c reductase (Complexes I + III) and cytochrome oxidase (Complex IV) were used in the premixture: A, 1:5; B, 1:1; C, 5:1. (From Tzagoloff *et al.*, 1967.)

fixed stoichiometry. In the case of the reconstituted membranes, however, the stoichiometry could be varied at will simply by changing the molecular proportions of the complexes in the starting mixture. These results are shown in Table 8.4 and Fig. 8.3. The ratio of cytochrome $a + a_3$ to cytochrome b in the membranes isolated from dispersed mixtures of NADH–cytochrome c reductase and cytochrome oxidase was approximately equal to that of their input. That the ratios reflected the actual compositions of the reconstituted membranes was confirmed by the banding patterns of the membranes on sucrose gradients (Fig. 8.4). Each mixture generated a single homogeneous population of membranes with a density consistent with the input ratios of the enzymes. These results indicated the association of the complexes to be a random event, the final membranes being a mosaic of individual enzyme units embedded in a common phospholipid bilayer.

Two pieces of evidence argue against the organization of the enzymes into an actual physical complex. First is the observation that the final composition

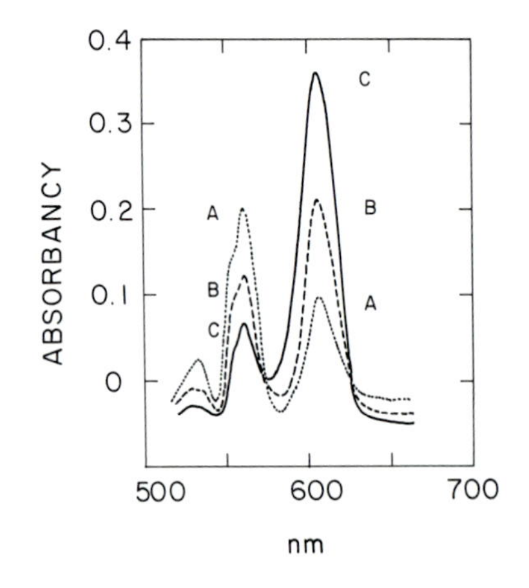

FIG. 8.3. Ratios of NADH–cytochrome c reductase and cytochrome oxidase in membranes reconstituted from mixtures containing different proportions of the two enzymes. The membranes were prepared by mixing NADH–cytochrome c reductase and cytochrome oxidase in the following proportions: A, 1:0.2; B, 1:1; C, 0.2:1. The mixtures were diluted to a final protein concentration of 0.25 mg/ml and centrifuged. The membrane pellets were suspended in buffered sucrose, and their spectra recorded. Cytochromes $a + a_3$ have absorption maxima at 605 nm, cytochrome b at 562 nm, and cytochromes $c + c_1$ (shoulder) at 550 nm.

of the membrane is determined solely by the ratio of the enzymes in the initial mixture. Second, the efficiency of electron transport in the reconstituted vesicles was not affected by their compositions. Rather, the specific activity depended only on the concentration of the rate-limiting enzyme in the membrane. These findings suggested that the mobile carriers coenzyme Q and cytochrome *c* can promote an efficient transfer of electrons between complexes as long as they are present in the same membrane.

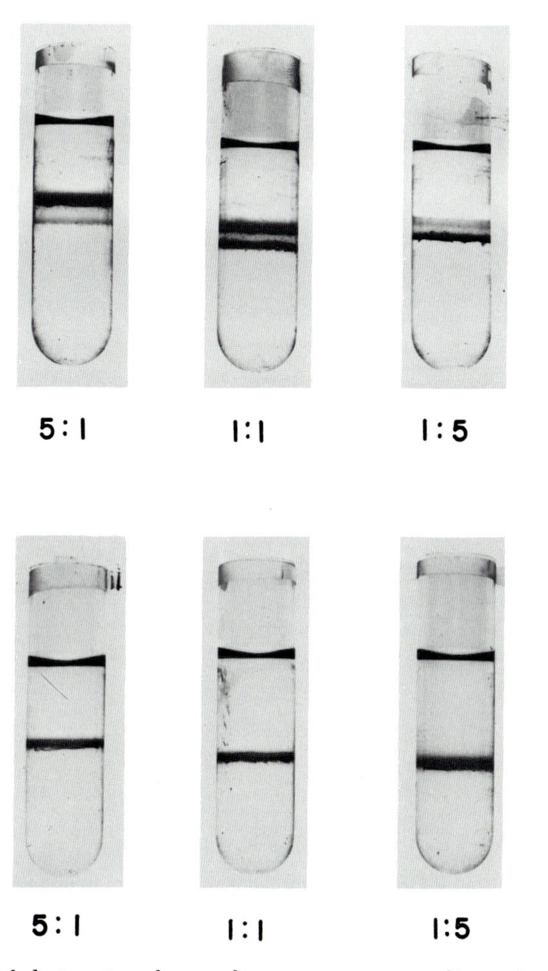

FIG. 8.4. Banding of electron transfer membranes on sucrose gradients. Concentrated solutions of NADH–cytochrome *c* reductase and cytochrome oxidase were either diluted separately and mixed in the indicated proportions (upper tubes) or diluted after mixing (bottom tubes). The membranes were isolated by centrifugation from the diluted mixtures and applied to linear sucrose gradients. The conditions of centrifugation were the same as in Fig. 8.2. The top band seen in the upper tubes corresponds to NADH–cytochrome *c* reductase membranes, and the lower band to cytochrome oxidase membranes. The three gradients shown in the bottom part of the figure have a single band consisting of hybrid membranes.

To study the role of cytochrome *c* in the reconstitution, different amounts of the carrier were added to a mixture of NADH–cytochrome *c* reductase and cytochrome oxidase either before or after the hybrid membranes had been allowed to form. The enzymes were also diluted separately in the presence of cytochrome *c*. The most efficient reconstitution was seen when cytochrome *c* was present in the initial mixture of the complexes (Fig. 8.5). Although addition of cytochrome *c* subsequent to membrane formation also resulted in efficient reconstitution, the amounts required to achieve high specific activities were considerably higher. As might be anticipated, virtually no oxidase activity was measured in mixtures of the two different membranes, even though each contained an amount of cytochrome *c* in excess of that in the hybrid membranes. Cytochrome *c*, therefore, can be incorporated into preformed membranes in a manner that permits effective coupling of the coenzyme QH_2–cytochrome *c* reductase and cytochrome oxidase reactions. The equilibration of cytochrome *c* between two different membranes, like that of the complexes, appears to be negligible.

II. Reconstitution of the Electron Transfer Complexes

The reconstitution of the respiratory complexes from their subunit polypeptides has proven very difficult; consequently, this approach has not been very rewarding in elucidating the functions of the proteins. The NADH- and succinate–coenzyme Q reductases have been fractionated into their respective primary dehydrogenases by methods similar to those employed earlier for the selective extraction of the flavoproteins directly from submitochondrial particles. In both instances, the coenzyme Q reductase activity has been restored by adding the purified flavoproteins back into the lipoprotein fraction of each complex. Attempts to obtain other reconstitutively active subunits from these complexes have not been successful.

The reconstitution of cytochrome oxidase and coenzyme QH_2–cytochrome *c* reductase has also been marked by a singular lack of success. One of the

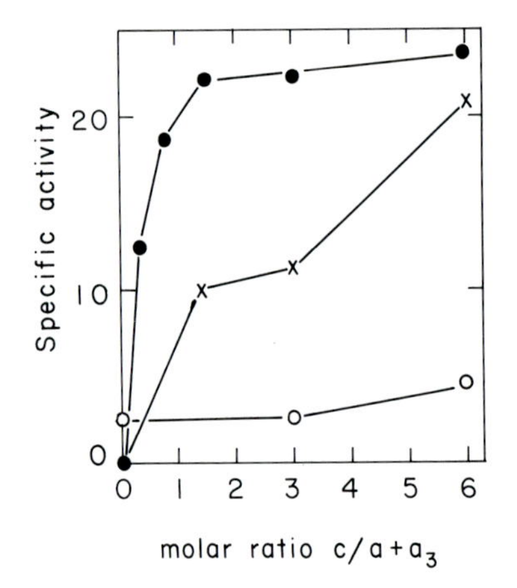

FIG. 8.5. Effect of cytochrome *c* on the reconstitution of NADH oxidase. Concentrated solutions of NADH-cytochrome *c* reductase and cytochrome oxidase were mixed with the indicated amounts of cytochrome *c* under the following conditions. Cytochrome *c* was added to the concentrated mixture of the complexes prior to dilution (●—●); cytochrome *c* was added to the reconstituted membranes formed after dilution (X—X); NADH-cytochrome *c* reductase and cytochrome oxidase were diluted separately in the presence of cytochrome *c* and then mixed (o—o). The samples were assayed for NADH oxidase polarographically. Specific activity refers to μmoles NADH oxidized/min per mg protein.

principal reasons for the failure to reconstitute these enzymes is the unusually strong interactions that hold the subunits together. As a result, conditions necessary to dissociate the complexes also cause a loss of the tertiary structure of the proteins. Until now, it has not been possible to reform the native structure of the subunits, thus reducing any chances of reconstituting the more complex structure of the enzymes themselves.

III. Resolution of Oxidative Phosphorylation in Submitochondrial Particles

The resolution of the enzymatic machinery responsible for ATP synthesis has followed a difficult and tortuous path over the past 30 years. After the initial description of the phenomenon, it was soon realized that the structural intactness of mitochondria was essential for a high efficiency of oxidative phosphorylation. In a sense, the dependence of an enzymatic process on the preservation of a biological structure was a new concept to enzymologists whose experience up to that time had been confined almost exclusively to the isolation and characterization of soluble enzymes.

In the 1950s, several laboratories reported conditions for the fragmentation of mitochondria into smaller membrane vesicles which retained the capacity for phosphorylation. These studies culminated in preparations of inner membrane vesicles (ETP_H) with P/O ratios nearly identical to those measured in intact mitochondria (Chapters 6 and 7). This was an important achievement, for it meant that any further attempts at resolution had to be directed toward enzymes localized in the inner membranes.

A second milestone in the resolution of oxidative phosphorylation occurred in 1960 when Pullman and co-workers reported the purification of the coupling factor F_1 which was capable of restoring oxidative phosphorylation in certain types of submitochondrial particles. As pointed out in Chapter 7, the discovery of F_1 was extremely important for the future development of the field. First, it demonstrated the practicality of resolution and reconstitution as a means of probing the phosphorylation mechanism. A great many efforts were subsequently devoted to the isolation of other coupling factors. Second, it provided substance to the idea of a terminal reversible ATP synthetase, the ATPase activity being an expression of the reverse reaction. Finally, the purified F_1 led to the identification of the 90-Å particles lining the inner membrane.

The reconstitution of the ATPase from F_1 and the hydrophobic components and the relationship of the subunit polypeptides to the various coupling factors have already been discussed in Chapter 7. These studies have provided a fair amount of information about the functions of the ATPase proteins. It is important to bear in mind that the roles of these subunits are defined largely in operational terms that probably have little bearing on their real catalytic functions. In this respect, the progress in our understanding of the coupling mechanism made via the reconstitution approach has been disappointing. It would seem that with the armory of pure ATPase constituents now available, new efforts aimed at examining the more specific roles of the proteins in energy coupling are warranted. Despite the fact that the experimental momentum in the coupling factor field has diminished in recent years, reconstitution may still be the most viable approach at unraveling the details of how ATP is synthesized.

IV. Reconstitution of a Coupled ATPase Membrane

The reconstitution of the ATPase complex described in Chapter 7 was aimed at restoring an oligomycin-sensitive hydrolytic activity. It was implied by a large body of data, however, that ATP hydrolysis is an index of uncoupling or proton leakiness according to the chemiosmotic hypothesis of energy coupling. Under normal circumstances, when mitochondria synthesize ATP, the ATPase works in the forward mode, i.e., as an ATP synthetase. Kagawa and Racker, therefore, explored conditions that might lead to a reconstitution of ATP synthetase membranes from the same set of components previously used to reconstitute oligomycin-sensitive ATPase.

Since the ATPase cannot generate net ATP in the absence of the respiratory chain, these authors used the ATP–P_i exchange reaction as an expression of the coupled state of the membrane. Based on the predictions of the chemiosmotic hypothesis, ATP hydrolysis should also be coupled to a stoichiometric release or uptake (depending on the orientation of the enzyme in the membrane) of protons. It should be mentioned that in their earlier studies, the reconstituted CF_0–F_1 membranes were unable to catalyze either ATP–P_i exchange or proton translocation. Essentially, these membranes operated in the uncoupled mode.

Using the chemiosmotic model as a guide, Kagawa and Racker reasoned that to reconstitute a coupled membrane it was essential to effect a unique orientation of the ATPase with respect to the two sides of the membrane. Membranes having a randomly oriented ATPase would translocate protons bidirectionally, thereby preventing the generation of a net proton gradient needed for coupling (Fig. 8.6). A second important property of coupled membranes was that they be reasonably proton impermeable.

To achieve the desired orientation of the enzyme, the reconstitutions were done with separated components rather than the whole complex. Kagawa and Racker first reconstituted membranes from phospholipids and the hydrophobic membrane proteins. The vesicles containing only the membrane factor unit were then reconstituted with purified OSCP and F_1. This sequence insured the attachment of F_1 exclusively to the outer surface of the membrane vesicles. In

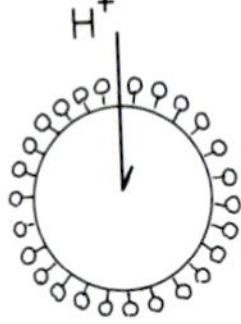

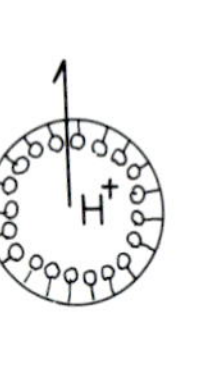

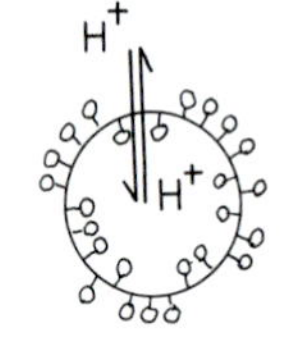

FIG. 8.6. Directionality of proton translocation in vesicles with different orientation of the ATPase.

other words, even if the primary membranes had some percentage of the membrane factor proteins improperly oriented (OSCP and F_1 binding sites on the inside), these would be reconstitutively inactive because of the inability of F_1 and OSCP to penetrate into the vesicle lumen. The second requirement, that of proton impermeability, was achieved by a cholate dialysis procedure that was subsequently found to have general applicability for a wide variety of reconstitutions. In this procedure, the proteins are mixed with phospholipids and cholate. The concentration of cholate is of prime importance, since it has to be sufficiently high to keep the proteins dispersed and yet be below the critical micellar concentration. By choosing appropriate concentrations of cholate, membrane formation could be induced by a simple dialysis of the starting mixture. The final concentration of the bile acid was below the level found to be inhibitory to energy-dependent processes of mitochondria.

The ATPase vesicles reconstituted by the cholate dialysis method catalyzed ATP–P_i exchange and were able to effect a unidirectional transfer of protons. The ATP–P_i exchange was sensitive to uncouplers and oligomycin (Table 8.5). The experiment of Fig. 8.7 shows that during ATP hydrolysis the suspending medium becomes more alkaline as a result of a net movement of protons to the interior of the vesicles. This direction of proton movement is consistent with an orientation of the ATPase that is the same as that in submitochondrial particles but converse to that of mitochondria. It should be recalled that in mitochondria the proton flux is outward. The traces of Fig. 8.7 also illustrate the uncoupler and oligomycin sensitivity of proton translocation. In contrast to oligomycin which completely inhibits proton uptake, the uncoupler FCCP actually causes a release of the internalized protons.

Kagawa and Racker also examined the specificity of the phospholipid requirement for the reconstitution of the coupled membrane. These interesting experiments revealed that the highest ATP–P_i exchange activity was attained when the membranes were reconstituted with a mixture of phospholipids that included the acidic phospholipid cardiolipin (Table 8.6). This is to be contrasted with the oligomycin-sensitive ATPase which could be maximally activated by

TABLE 8.5. Reconstitution of ATP–P_i Exchange in ATPase Membrane Vesicles[a]

Additions	ATP–$^{32}P_i$ exchange
None	8.7
F_6	7.8
OSCP	7.9
F_6, OSCP	3.7
F_1	151.1
F_1, F_6	181.5
F_1, OSCP	224.0
F_1, F_6, OSCP	406.5

[a]Membranes were reconstituted from the hydrophobic protein fraction by the cholate dialysis procedure. To 0.1 mg of membrane protein were added F_1 (40 μg), OSCP (15 μg) and F_6 (11 μg) as indicated in the table. The ATP–P_i exchange refers to nmoles $^{32}P_i$ incorporated into ATP/10 min per mg protein. (From Kagawa and Racker, 1971.)

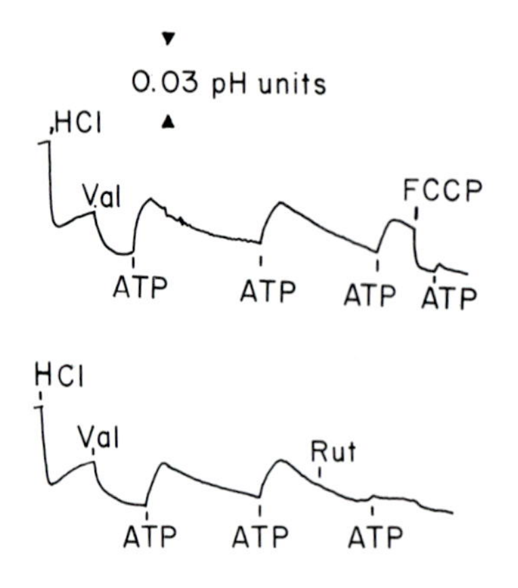

FIG. 8.7. Accumulation of protons by reconstituted CF_0–F_1 vesicles. Vesicles were formed by the cholate dialysis method in the presence of soybean phospholipids and suspended in a medium containing 150 mM KCl and 2 mM glycylglycine buffer, pH 6.25. A small amount of HCl (24.2 nmoles) was added initially to determine the buffering capacity of the medium. Acidification is seen as a downward trace. The addition of ATP causes the external medium to become more alkaline because of proton uptake by the vesicles. Both an uncoupler (FCCP) and an inhibitor (rutamycin) of oxidative phosphorylation abolish the response of the vesicles to further additions of ATP. (From Kagawa *et al.*, 1973.)

single phospholipids. Not all phospholipid combinations studied yielded vesicles with identical structures. An important correlation, however, was established between the reconstitution of single-layered closed vesicles and the generation of the coupled state.

V. Reconstitution of Coupled Electron Transport Membranes

The reconstitution of ATPase membranes capable of ATP–P_i exchange and proton translocation was soon followed by similar reconstitutions of coupled membranes consisting of purified electron transport complexes. In these experiments, the conditions used to form the membranes were patterned on the original cholate dialysis method of Kagawa and Racker. Again, the criteria for a coupled electron transport membrane were based on the principles of the chemiosmotic hypothesis. These were: (1) respiration-dependent proton translocation; (2) respiration-supported transport of a cation such as K^+; and (3) stimulation of respiration by uncoupling agents. All three criteria were met for membrane vesicles consisting of either NADH–coenzyme Q reductase, coenzyme QH_2-cytochrome *c* reductase, or cytochrome oxidase.

TABLE 8.6. Effect of Phospholipid Composition on ATP-P_i Exchange[a]

Phospholipids added to hydrophobic proteins	ATP–P_i exchange (nmoles/min per mg protein)
Cardiolipin	0.0
PC + cardiolipin	0.0
PE + cardiolipin	0.9
PC + PE + cardiolipin	58.9

[a]The hydrophobic proteins were mixed with the indicated phospholipids (20 μmoles of each phospholipid per mg protein) in the presence of cholate. The mixture was dialyzed, reconstituted with F_1 and OSCP, and assayed for ATP–P_i exchange. (From Kagawa *et al.*, 1973.)

TABLE 8.7. Properties of Reconstituted Coenzyme QH_2–Cytochrome *c* Reductase Vesicles[a]

Additions	Reduction of cytochrome *c* (μmoles/min per mg)	R.C.
None	3.1	—
m-CCCP, 10 μM	34.1	11.0
Valinomycin, 50 ng + nigericin, 50 ng	31.7	10.2

[a]The reconstituted membranes were assayed for the reduction of cytochrome *c* by reduced coenzyme Q_2. (From Leung and Hinkle, 1975.)

The reconstitution of coupled electron transport depended on the removal of cholate from a starting solution of the purified complex, cholate, and an artificial or natural mixture of phospholipids. Some of the salient properties of coenzyme QH_2–cytochrome *c* reductase vesicles are shown in Tables 8.7 and 8.8 and Figs. 8.8 and 8.9. The reduction of cytochrome *c* by coenzyme QH_2 is enhanced some tenfold in the presence of the uncoupler *m*-CCCP or a combination of valinomycin and nigericin whose ionophoric activities destroy the proton gradient across the membrane (Chapter 9, Section IXA). The stimulation of respiration by these uncouplers has been equated to the phenomenon of respiratory control in mitochondria and indicates that the energy of oxidation of coenzyme QH_2 is conserved in a high-energy state or proton gradient. It is of interest that the reconstituted membranes are more tightly coupled than preparations of submitochondrial particles with high phosphorylation efficiency. For example, the respiratory activities of ETP_H are known to be unaffected by uncoupling agents.

The oxidation of coenzyme QH_2 is accompanied by an expulsion of protons into the suspending medium (Fig. 8.8). The number of protons transferred is stoichiometric with the amount of substrate oxidized. In this reaction, ferricyanide was used as the final oxidant, and valinomycin was added to allow potas-

TABLE 8.8. Effect of Phospholipid Content of Coenzyme QH_2–Cytochrome *c* Membranes on the Respiratory Control Ratio[a]

Phospholipids	Cytochrome *c* reduction (μmoles/min per mg protein)		
	Alone	+*m*-CCCP	R.C.
PC:PE, 1:1	2.1	9.0	4.3
PC:PE, 1:2	7.1	9.0	1.3
PC:PE, 2:1	2.8	11.9	4.3
PC:PE:cardiolipin, 1:1:1	7.8	25.9	3.3
Total soybean phospholipids	5.6	30.6	5.5

[a]Purified coenzyme QH_2–cytochrome *c* reductase was mixed with phospholipids in the indicated proportions. The membranes, isolated after removal of cholate, were assayed for the reduction of cytochrome *c* by reduced coenzyme Q_2 in the absence and presence of *m*-CCCP. (From Leung and Hinkle, 1975.)

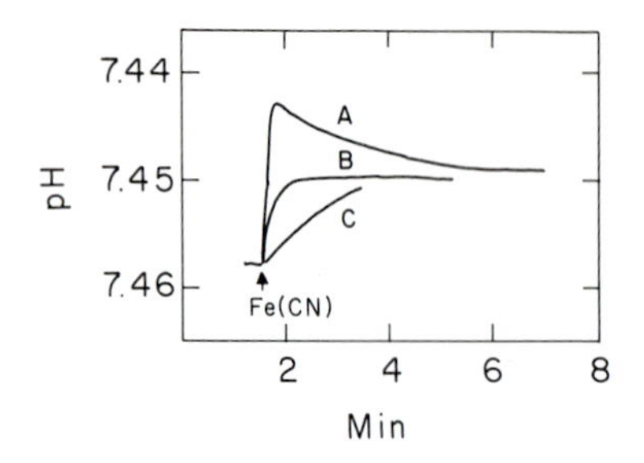

Fig. 8.8. Proton translocation by reconstituted coenzyme QH_2-cytochrome *c* reductase vesicles. Vesicles of the purified enzyme formed by the cholate dialysis procedure were suspended in 150 mM KCl, 5 mM $MgCl_2$, 1 μg valinomycin, 70 μM reduced coenzyme Q_2, and 12 μM cytochrome *c*. Potassium ferricyanide (2 nmoles) was added as the electron acceptor to initiate the reaction. Trace A: no other additions. Trace B: same as A except for the addition of *m*-CCCP. C: same as A except for the addition of antimycin A. (From Leung and Hinkle, 1975.)

sium ion counterflow to compensate for the outward movement of protons (see scheme of Fig. 8.10).

As seen in Fig. 8.8, when the membranes are pulsed with ferricyanide, there is an immediate acidification of the medium followed by a slow decay to a new plateau value. In the presence of *m*-CCCP, the number of protons produced in the medium is equivalent to the final equilibrium value seen in the absence of the uncoupler. The uncoupler-insensitive acidification has been interpreted to be caused by the oxidation of coenzyme QH_2 per se, whereas the sensitive component is a measure of proton translocation. The measured ratio of protons to electrons transported in this system is one that agrees with the stoichiometry predicted by the chemiosmotic model. The scheme of Fig. 8.10 postulates the existence of a proton carrier (Z) that must be an intrinsic constituent of the enzyme complex. That the proton translocation is coupled to K^+ uptake is shown in Fig. 8.9 where the concentration of the external cation was monitored with a K^+ electrode. The addition of ferricyanide to a suspension of coenzyme QH_2-cytochrome *c* reductase vesicles caused an immediate uncoupler-sensitive uptake of potassium with a $K^+/2e^-$ of 2.

The results described above for the second segment of the respiratory chain have been essentially duplicated with the NADH-coenzyme Q reductase and cytochrome oxidase complexes. Each enzyme, therefore, can be used to generate membranes capable of coupling the oxidation of their respective substrates to proton translocation and ion uptake.

VI. Reconstitution of Oxidative Phosphorylation

Reconstitutions involving the formation of a coupled ATPase or respiratory membrane represent the two separate halves of the oxidative phosphorylation

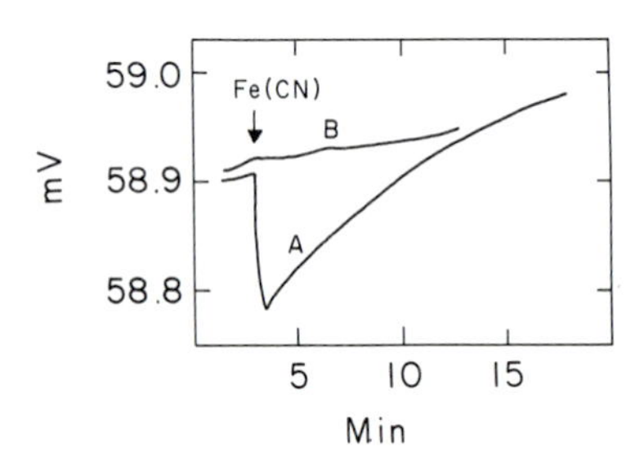

FIG. 8.9. Potassium translocation by reconstituted coenzyme QH_2-cytochrome *c* reductase vesicles. The vesicles were suspended in 100 mM choline chloride, 140 μM reduced coenzyme Q_2, 12 μM cytochrome *c*, and 1 μg/ml valinomycin buffered at pH 7.5. The reaction was started by the addition of potassium ferricyanide as the ultimate electron acceptor. Trace A: no other additions. Trace B: in the presence of 10 μM *m*-CCCP. A downward trace indicates uptake of K^+ from the medium. (From Leung and Hinkle, 1975.)

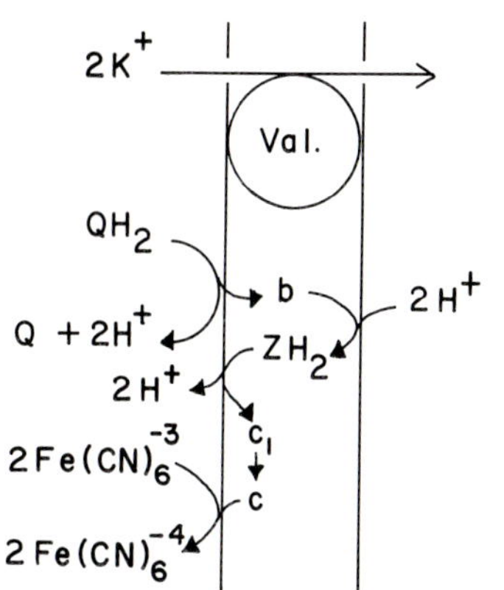

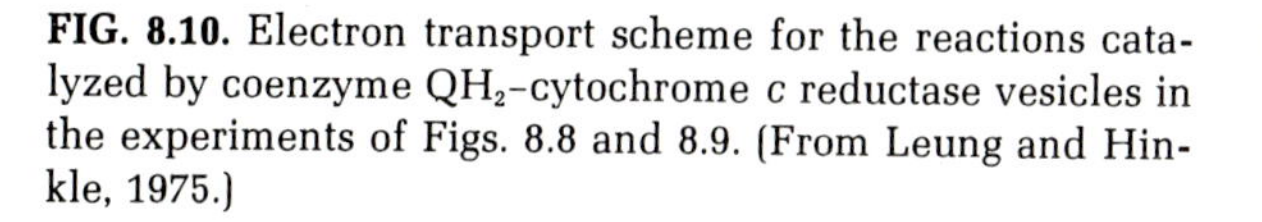
FIG. 8.10. Electron transport scheme for the reactions catalyzed by coenzyme QH_2–cytochrome *c* reductase vesicles in the experiments of Figs. 8.8 and 8.9. (From Leung and Hinkle, 1975.)

The membranes formed in this manner were able to couple the oxidation of a cytochrome oxidase substrate (ascorbate plus PMS) to ATP synthesis with P/O as high as 0.6 or 60% of the efficiency of mitochondria. As shown in Fig. 8.12, the best phosphorylation was found in the range of 50–100 μg of cytochrome oxidase protein per milligram of hydrophobic proteins. There were also optimal concentrations of cytochrome *c* and phospholipids. In confirmation of the earlier results on the coupling of the ATPase, the reconstitution of oxidative phosphorylation was influenced by the nature of the phospholipids, the most effective being mixtures of phosphatidylcholine, phosphatidylethanolamine, and cardiolipin.

Similar reconstitutions of oxidative phosphorylation have been reported for the first and second segments of the respiratory chain. Although all of the initial reconstitutions were done by the cholate dialysis procedure, Racker and system—the respiratory complex functions to couple the oxidation event to the production of a stable form of energy, whereas the ATPase uses the trapped energy to esterify ADP and P_i.

A physical synthesis of the two halves of the process was achieved by Racker and collaborators for each of the three segments of the respiratory chain. These workers showed that membranes reconstituted from an electron transport complex and the ATPase carry out a net synthesis of ATP. This will be illustrated for the cytochrome oxidase complex. The procedure used to obtain phosphorylating membranes was based on the prior results with the separate enzymes. In order to properly orient the two systems, purified cytochrome oxidase, cytochrome *c*, and the hydrophobic proteins of the ATPase complex were first mixed with cholate and phospholipids. The membranes induced by dialysis should therefore have cytochrome *c* on the inner side of the vesicles.* The resultant membranes were further reconstituted with OSCP and F_1 to complete the ATP synthetase. According to this protocol, the relative orientation of cytochrome *c* (inside) and F_1 (outside) should be the same as in submitochondrial particles (Fig. 8.11).

*External cytochrome *c* was removed by centrifuging the membranes through a sucrose gradient.

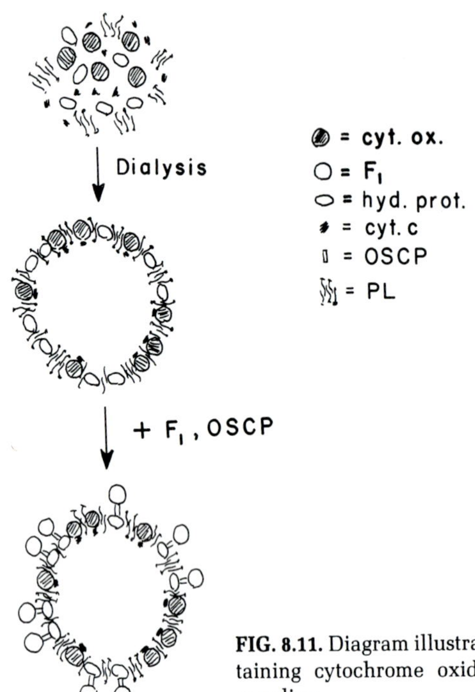

FIG. 8.11. Diagram illustrating the reconstitution of a hybrid membrane containing cytochrome oxidase and CF_0–F_1 in the proper orientation for coupling.

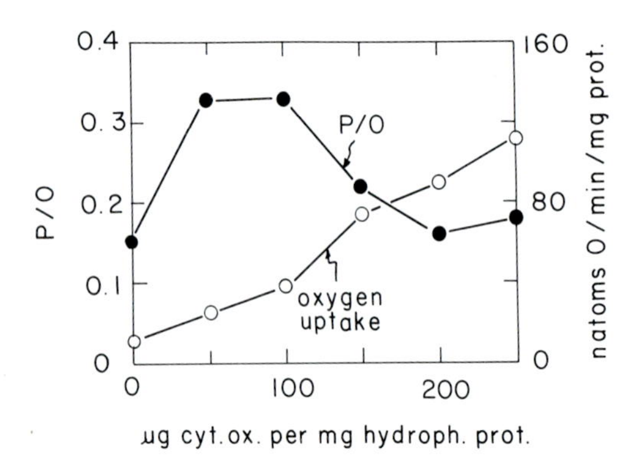

FIG. 8.12. Oxidative phosphorylation catalyzed by hybrid cytochrome oxidase and ATPase vesicles. The vesicles were reconstituted from cytochrome oxidase and purified ATPase components by the cholate dialysis procedure with the indicated proportions of the hydrophobic protein fraction. (From Racker and Kandrach, 1973.)

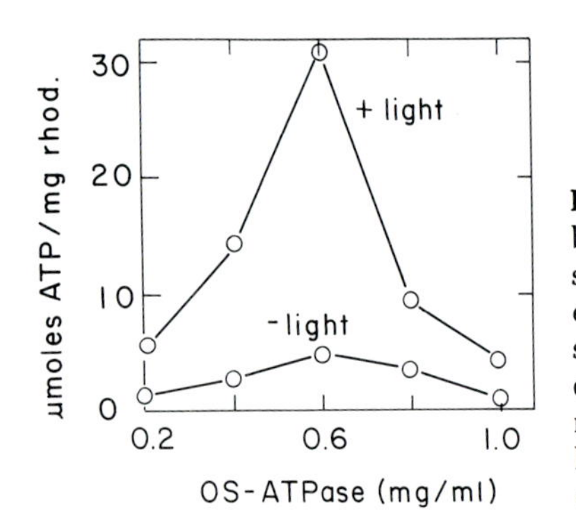

FIG. 8.13. Reconstitution of coupled phosphorylation by bacteriorhodopsin membranes. Bacteriorhodopsin was sonicated in the presence of a mixture of phosphatidylethanolamine, phosphatidylcholine, and phosphatidylserine. To the clarified suspension was added the indicated amounts of bovine OS-ATPase. The resultant membranes were assayed for their ability to catalyze light-dependent esterification of ADP and P_i. (From Eytan *et al.*, 1976.)

co-workers subsequently devised a simpler method which involves sonication of a mixture of the ATPase and respiratory complexes with preformed liposomes (sealed phospholipid vesicles). Under these conditions, the enzyme complexes are incorporated into the liposomes in the proper orientation yielding fully coupled membranes.

One of the most dramatic experiments reported by Racker and Stoeckenius, and one that demonstrates the full potential of this approach, involved the reconstitution of photophosphorylation from bacteriorhodopsin and the mitochondrial ATPase. Purified bacteriorhodopsin, when incorporated into phospholipid vesicles, can carry out a light-dependent expulsion of protons into the medium. Racker and Stoeckenius showed that the proton gradient generated by the bacterial pigment can be used by the bovine mitochondrial ATPase to synthesize ATP (Fig. 8.13). The reconstitution of this hybrid photophosphorylation system required only the presence of bacteriorhodopsin and the ATPase in the same membrane. This simple experiment provides the strongest evidence to date in support of the chemiosmotic hypothesis, for it shows that the primary energy conservation event in membrane-associated energy metabolism (oxidative phosphorylation, photosynthesis, photophosphorylation) is the generation of a proton gradient. This energy is subsequently converted to a chemically stable pyrophosphate bond by a universal ATPase device sharing many common structural and catalytic features among totally unrelated organelles.

Selected Readings

Eytan, G. D., Matheson, M. J., and Racker, E. (1976) Incorporation of mitochondrial membrane proteins into liposomes containing acidic phospholipids, *J. Biol. Chem.* **251**:6831.

Fowler, L. R., and Richardson, S. H. (1963) Studies on the electron transfer system. L. On the mechanism of reconstitution of the mitochondrial electron transfer system, *J. Biol. Chem.* **238**:446.

Hatefi, Y., Haavik, A. G., Fowler, L. R., and Griffiths, D. E. (1962) Studies on the electron transfer system. XLII. Reconstitution of the electron transfer system, *J. Biol. Chem.* **237**:2661.

Hinkle, P. C. (1973) Electron transfer across membranes and energy coupling, *Fed. Proc.* **32**:1988.

Kagawa, Y. (1972) Reconstitution of oxidative phosphorylation, *Biochim. Biophys. Acta* **265**:297.

Kagawa, Y., and Racker, E. (1971) Partial resolution of the enzymes catalyzing oxidative phosphorylation. XXV. Reconstitution of vesicles catalyzing $^{32}P_i$–adenosine triphosphate exchange, *J. Biol. Chem.* **246**:5477.

Kagawa, Y., Kandrach, A., and Racker, E. (1973) Partial resolution of the enzymes catalyzing oxidative phosphorylation, *J. Biol. Chem.* **248**:676.

King, T. E. (1963) Reconstitution of the respiratory chain enzyme systems. XII. Some observations on the reconstitution of the succinate oxidase system from heart muscle, *J. Biol. Chem.* **238**:4037.

Leung, K. H., and Hinkle, P. C. (1975) Reconstitution of ion transport and respiratory control in vesicles formed from reduced coenzyme Q–cytochrome *c* reductase and phospholipids, *J. Biol. Chem.* **250**:8467.

Racker, E., and Kandrach, A. (1973) Partial resolution of the enzymes catalyzing oxidative phosphorylation. XXXIX. Reconstitution of the third segment of oxidative phosphorylation, *J. Biol. Chem.* **248**:5841.

Racker, E., and Stoeckenius, W. (1974) Reconstitution of purple membrane vesicles catalyzing light-driven proton uptake and adenosine triphosphate formation, *J. Biol. Chem.* **249**:662.

Ragan, C. I., and Hinkle, P. C. (1975) Ion transport and respiratory control in vesicles formed from the reduced nicotinamide adenine dinucleotide coenzyme Q reductase and phospholipids, *J. Biol. Chem.* **250**:8472.

Ragan, C. I., and Racker, E. (1973) Partial resolution of the enzymes catalyzing oxidative phosphorylation. XXVIII. The reconstitution of the first segment of energy conservation, *J. Biol. Chem.* **248:**2563.

Tzagoloff, A., MacLennan, D. H., McConnell, D. G., and Green, D. E. (1967) Studies on the electron transfer system. LXVIII. Formation of membranes as the basis of the reconstitution of the mitochondrial electron transfer chain, *J. Biol. Chem.* **242:**2051.

9

Mitochondrial Transport Systems

The specialized role of mitochondria in intermediary metabolism requires that only certain substrates, cofactors, and metals be accessible to their interior compartments. Of the substrates that must be capable of entering the matrix space, the most important are O_2, H_2O, ADP, phosphate, pyruvate, and fatty acids. At the same time, products of mitochondrial oxidations and phosphorylation must have a means of exiting from the organelle. These include CO_2 and ATP. Virtually all mitochondria, irrespective of their source, have been shown to be either freely permeable or to have specific transport systems that accommodate an efficient passage of these essential metabolites across the permeability barriers separating the matrix space from the surrounding cytoplasm.

The ability of mitochondria to take up other substrates, e.g., intermediates of the TCA cycle, depends on the tissue of origin and its special metabolic needs. Being concerned exclusively with the synthesis of ATP, blowfly muscle mitochondria lack most of the transport systems found in liver and other types of mitochondria that have a greater enzymatic diversity and perform other essential functions for the cell.

There are several reasons for the growing interest in mitochondrial transport. First of all, such studies are of fundamental importance in understanding the mechanisms by which solutes are transported across biological membranes in general. Mitochondria utilize a broad spectrum of different types of transport systems, all of which are extremely active. Since homogeneous preparations of mitochondria are easy to obtain, they are a convenient experimental material for transport studies. Second, the shuttle of certain intermediates between the cytoplasm and mitochondria influences and regulates the overall metabolic balance of the cell. Both of these aspects will be discussed in this chapter.

I. Some General Considerations of How Molecules Cross Biological Membranes

Most membranes are impermeable to polar compounds (sugars, acids), to cations (K^+, Mg^{2+}), and to anions (Cl^-, $H_2PO_4^-$). On the other hand, low-molecular-weight compounds with lipophilic properties such as alcohols and hydrocarbons are freely permeable. This property is thought to arise from the structure of biological membranes in which the phospholipid bilayer acts as a barrier that prevents the diffusion of hydrophilic but not of hydrophobic substances. This rule, however, does not apply to water which is capable of diffusing through all known membranes. In general, the flux of freely permeable molecules across a membrane shows the characteristics of a simple diffusion process with a rate that is concentration and temperature dependent. The net flux is always down the concentration gradient and proceeds until the concentrations on both sides of the membrane are equal.

Neutral and charged substrates can also penetrate across membranes by mechanisms that utilize specific *porter* or *carrier* proteins. Such systems are referred to as *facilitated diffusion* if they operate independently of energy and as *active transport* when they are coupled to some energy-yielding reaction. Finally, substrates may be chemically modified in the process of being transported. This type of transport is known as *group translocation*. Some of the properties of each of these three modes of transport are reviewed below.

A. Facilitated Diffusion

In free and facilitated diffusion, no energy is required for the movement of the permeant species. This means that in both cases net transfer of the permeant solute will occur only in the presence of a concentration gradient and will stop when the gradient is abolished. Facilitated diffusion is characterized by the following set of properties.

1. The rate of flux is usually higher than would be predicted for free diffusion based on the chemical structure of the solute.
2. In free diffusion, the rate is directly proportional to the concentration of the solute. This is not true of facilitated diffusion which shows saturation kinetics that can be described by the Michaelis–Menten equation (Fig. 9.1).
3. Transport by facilitated diffusion is usually inhibited by structural analogues of the solute. The inhibition is kinetically similar to competitive inhibition of an enzymatic process. Many facilitated diffusion systems are also blocked by specific inhibitors.

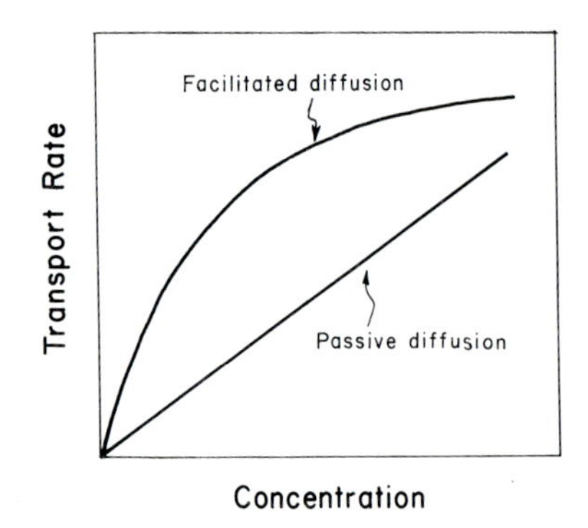

FIG. 9.1. Kinetics of transport as a function of substrate concentration in passive and facilitated diffusion.

Each of the above properties is indicative of a catalytic process and has led to the concept of a special class of membrane proteins acting as carriers of various ions and cellular metabolites. Such carriers are thought to diffuse or to undergo conformational changes, allowing otherwise impermeable solutes to cross the hydrophobic barrier of the membrane. The list of compounds now known to be transported by facilitated diffusion is very long and includes sugars, amino acids, and TCA cycle intermediates as well as different cations and anions.

B. Active Transport

In order to maintain high concentrations of substrates needed for optimal metabolic activity, cells and organelles must be capable of carrying out transport against a concentration gradient. Since the transfer of a solute to a phase of higher concentration is accompanied by an increase in free energy, active transport systems must be coupled to some other, energy-yielding process. The energy needed to support the accumulation of a solute against an electrochemical gradient is described by the equation,

$$\Delta G^\circ = RT \ln \frac{[S]_{in}}{[S]_{out}} + nF\Delta\psi \qquad [9.1]$$

where R is the gas constant, T is the absolute temperature, $[S]_{in}$ and $[S]_{out}$ are the concentrations of the solute in the two phases (the direction of the transfer being from the outside to the inside), n is the charge on the solute, F is the faraday constant, and $\Delta\psi$ is the electrical potential difference between the two phases or the membrane potential. For uncharged molecules, the last term drops out, and the change in free energy is determined solely by the difference in concentration of the solute. At physiological temperatures, the free energy change or the energy required to transfer 1 mole of an electrically neutral solute to a phase where it is at a tenfold higher concentration is approximately 1.35 kcal.

When the solute is charged, the second term of the equation comes into play, and the free energy change will be determined not only by the difference in concentration but also by the number of charges transported and the membrane potential. Although it is possible, using the above equation, to calculate the energy expenditure involved in maintaining any given gradient of a particular substrate or ion, such estimates do not take into account possible backflows of the solute because of a leakiness in the membrane. The actual energy input or work done during transport, therefore, usually exceeds the theoretical value, and the extent of the underestimation will depend on the rate of back diffusion.

The movement of a solute against a concentration or electrochemical gradient by itself is not a sufficiently rigorous criterion for an active transport mechanism. Some solutes can be transported against a gradient without any net input of energy if they are linked to secondary fluxes of other species. Another consideration is the chemical activity of the solute in the two phases. If the accumulated substrate or ion is removed from solution by being bound or converted to some other chemical form, it will not contribute to the osmotic force, and the gradient may be only an apparent one. The best criteria for active transport are (1) that the solute be transferred to a compartment where its chemical activity is demonstrably higher and (2) that the driving force does not come from secondary compensating fluxes involving the movement of other solutes down their electrochemical gradient. A consequence of these conditions is that the transport must be coupled to some other chemical event that provides the energy required for the osmotic work.

Active transport systems share many properties in common with facilitated diffusion. They are saturated at high substrate concentrations, and the rate of transport is lowered in the presence of analogues of the substrate or compounds acting as specific inhibitors. Active transport is also blocked by inhibitors of intermediary metabolism that affect the cellular production of ATP. This is not true of facilitated diffusion and is an important test to distinguish the two modes of transport.

Current models of active transport generally incorporate some carrier or porter protein capable of specifically binding the substrate. The actual mechanism of translocation, however, is still unresolved. For example, it is not clear whether there is a diffusion of the carrier through the thickness of the membrane, whether the substrate binding site on the carrier alternates between the two sides as a result of a conformational change in the protein, or whether the substrate penetrates through a channel in the carrier (Fig. 9.2). Also lacking is a molecular explanation of how energy (usually in the form of ATP) is used to

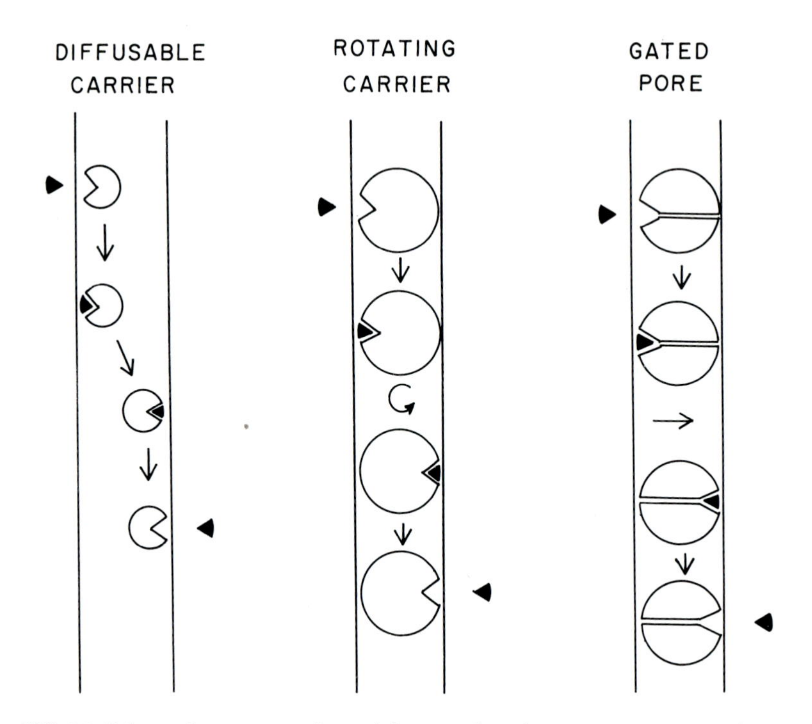

FIG. 9.2. Schematic representations of three modes of transport across membranes.

preferentially transfer substrates to the phase of higher concentration, although, here, a reasonable assumption has been that the binding of ATP decreases the affinity of the carrier for the substrate. It is probably true that the basic translocation mechanisms in facilitated diffusion and active transport are similar, since in some active transport systems the carriers can effect a transfer of the substrate even in the absence of an energy source provided transport is down the concentration gradient.

C. Group Translocation

In facilitated diffusion and in active transport, there is no chemical modification of the substrate during any stage of the transfer. The binding of the substrate to its carrier occurs through electrostatic, hydrophobic, or a combination of both kinds of interactions. In group translocation, however, the substrate is enzymatically changed to a new compound and is either released as such or converted back to its original form following translocation.

The accumulation of substrates by group translocation is not necessarily energy dependent. The transport of fatty acyl substrates into mitochondria by the carnitine shuttle mechanism (see Chapter 3, Section IIA) is an example of group translocation that does not depend on any energy input. Other systems, such as glucose transport in bacteria, require ATP which is used in the phosphorylation of the sugar to glucose-6-phosphate.

II. Methodology for Studying Mitochondrial Transport

Three methods have been used to study the rate and extent of transport of substrates and ions by mitochondria: (1) direct estimation of the concentration of the substrates in the internal mitochondrial space, (2) measurements of mitochondrial swelling under conditions of massive accumulation of the substrate, and (3) measurements of the reduction of endogenous NAD^+ by externally added substrates.

A. Direct Measurements of Substrate Concentration

This method is suitable for studying the uptake of substrates and ions as well as their release from preloaded mitochondria. Most often, mitochondria are separated from the assay medium by quick filtration or by sedimentation through a layer of sucrose or silicone. The intramitochondrial concentration of the permeant can be estimated from the known volume of the impermeable space. If the transport or steady-state concentration of the substrate is affected by the metabolic state of the mitochondria, the washing procedure is usually combined with fixation in acid. This is done by centrifuging the mitochondria through an intermediary layer of silicone into an acid solution. If the internalized substrate is metabolized, it may be necessary to include a suitable inhibitor in the uptake medium.

Although the direct method is the most versatile in that uptake, release, and exchange of solutes can be studied, because of the time required in the physical manipulations, it is not always suitable for kinetic measurements, especially when initial rates of transport are desired.

B. Mitochondrial Swelling

Mitochondria behave as perfect osmometers in response to external osmotic pressure. The net transport of a solute into the matrix space is accompanied by an influx of water. When the amount of a solute taken up is large, mitochondria undergo large-amplitude swelling which can be detected by a decrease in their light-scattering properties. The extent of mitochondrial swelling or contraction is conveniently measured in a spectrophotometer at a wavelength where there is little absorption of light by cytochromes or other chromophoric groups. This technique has been extensively used by Chappell and his co-workers to study the transport of dicarboxylic and tricarboxylic acid substrates. It is equally suitable for measuring the transport of other anions and cations.

C. Oxidation of Substrates

The oxidation of certain substrates of the TCA cycle and β-oxidation are linked to the reduction of NAD^+ to NADH. The transport of these substrates into the matrix space can be studied by measuring the reduction of endogenous pyridine nucleotides spectrophotometrically or fluorimetrically. This method has the advantage of permitting rapid kinetic measurements to be made at concentrations of substrate approaching physiological conditions. A drawback of the method, however, is that in those cases where the reduction of NAD^+ is limited by the activity of the dehydrogenase, the rates of transport will be underestimated.

III. Passive Diffusion of Gases and Water

The inner mitochondrial membrane is permeable to O_2, CO_2, and H_2O. All existing evidence is consistent with the notion that the diffusion of these compounds through the lipid barrier occurs independently of any carrier systems.

A. Water

The extrusion of water from actively respiring mitochondria is an extremely rapid process. Under phosphorylating conditions, heart mitochondria will oxidize pyruvate at a rate of 0.07 μmoles/min per mg protein. Since the complete oxidation of 1 mole of pyruvate yields 17 moles of water, the minimal rate of water efflux must be of the order of 1 μmole/min per mg protein. Water fluxes several thousandfold greater have been estimated during swelling and contraction cycles of mitochondria. It is, therefore, quite apparent that the inner membrane is freely permeable to water, and the potential capacity for water efflux is orders of magnitude greater than that required for metabolism.

B. Oxygen

Oxygen is the ultimate acceptor of the electrons and hydrogens released from the various intermediates of the TCA cycle and the β-oxidation pathway. The oxygen-utilizing step is catalyzed by cytochrome oxidase and is thought to occur on the matrix side of the inner membrane. That the diffusion of oxygen into the matrix compartment is not a rate-limiting step in the oxidation–reduction reactions of the respiratory chain can be inferred from the following considerations. The highest oxidation rates in isolated mitochondria are seen with NAD-linked substrates such as pyruvate, malate, or glutamate. The rates of oxygen consumption in heart mitochondria in state 3 respiration with these substrates range from 0.2–0.4 μmoles oxygen/min per mg protein. The diffusion rate of oxygen, however, must be considerably higher, since the measured rates of oxygen consumption when cytochrome oxidase is assayed in the presence of ascorbate–PMS (see Chapter 5, Section I) can achieve values as high as 1–2 μmoles oxygen/min per mg protein. Assuming that this corresponds to the maximal diffusion of oxygen through the inner membrane, it is still five times higher than required for optimal rates of oxidation of NAD-linked substrates.

C. Carbon Dioxide

The inner membrane appears to be as permeable to carbon dioxide as it is to oxygen. Since mitochondria contain carbonic anhydrase, it has been suggested that CO_2 may be extruded as HCO_3^-. This, however, is unlikely in view of the inability of externally added HCO_3^- to penetrate into mitochondria to any significant extent. At present, it must be assumed that carbon dioxide crosses the mitochondrial membrane by passive diffusion.

IV. Mechanisms of Substrate Transport in Mitochondria

Mitochondrial substrates have different charge properties. They can be neutral (or zwitterionic), anions of weak acids, or cations of weak bases. These solutes cross the mitochondrial inner membrane either by passive diffusion or by carrier-mediated processes. The latter may or may not require an energy source. Before proceeding to a discussion of specific substrates, it may be useful to first consider some general features of the different mechanisms that have been implicated in mitochondrial transport of substrates.

A. Passive Diffusion

Anionic or cationic solutes can diffuse through the mitochondrial membrane in the absence of an energy source and in a non-carrier-mediated fashion. The transport can be either electroneutral if the permeant diffuses as the uncharged acid or electrogenic if it is ionized. In mechanism A of Fig. 9.3., the anion diffuses through the membrane in the un-ionized form. This mechanism has been invoked for the transport of acetate and other monocarboxylic acids. In the presence of a permeant cation such as ammonium ion, acetate and most other monocarboxylates cause a rapid swelling of mitochondria. Ammonium permeates the membrane as free ammonia and serves to neutralize the proton released in the interior as a result of the ionization of the acid. Mechanism B shows another means for the diffusion of anions. Here the transport is electrogenic since the permeant is charged. The characteristic of anions transported by this mechanism is that mitochondrial swelling requires the presence of both a permeant cation and an uncoupling agent. The uncoupler is visualized as facilitating an inward transport of protons that are needed to neutralize the internal production of hydroxide ions. Nitrate and thiocyanate have been shown to be transported by this mechanism.

B. Carrier-Mediated Transport

Mitochondrial substrates have been shown to be transported on carriers acting either as uniports or as antiports that exchange external for internal anions. Even though most carrier-mediated transport systems function by facil-

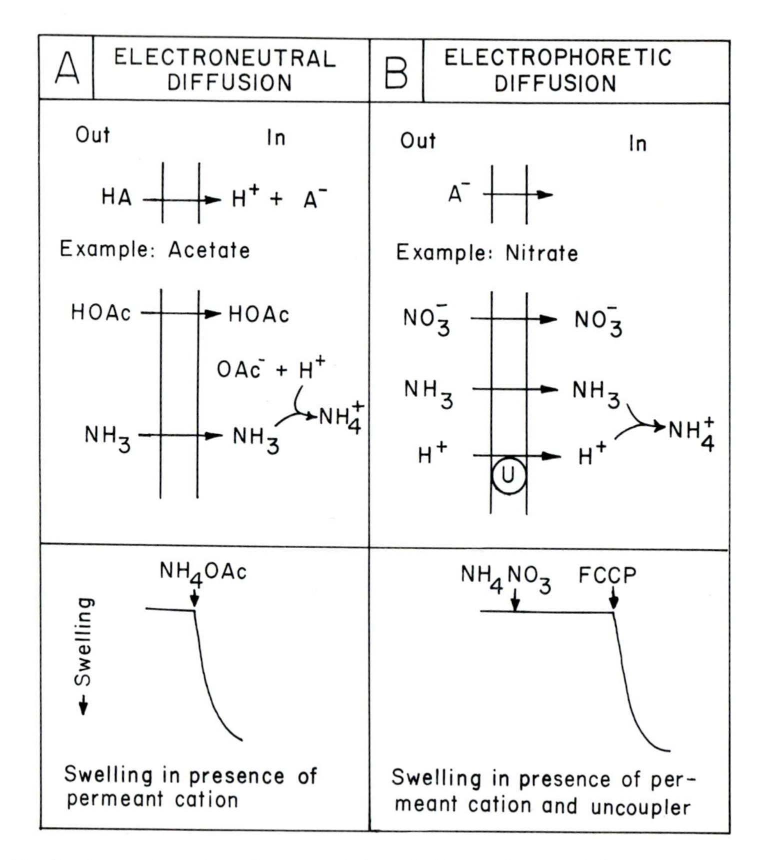

FIG. 9.3. Mechanisms of passive electroneutral and electrophoretic diffusion. In electrophoretic diffusion, the uncoupler (U) serves to neutralize the internal negative charge by promoting the uptake of protons.

itated diffusion, they can be influenced in different ways by the energy state of the mitochondrion.

A number of mechanisms of carrier-mediated transport of substrates have been proposed. These can be distinguished by the conditions required to elicit mitochondrial swelling (Fig. 9.4). Some neutral or zwitterionic solutes are transported by an electroneutral process as shown in mechanism C. Such substrates induce mitochondrial swelling in the absence of any other electrolytes. Other anionic substrates may be exchanged for hydroxide ions, in which case swelling occurs only in the presence of a permeant cation. The electroneutral anion/hydroxide exchange (mechanism D) is formally analogous to an electroneutral diffusion (mechanism A of Fig. 9.3) except that it utilizes a carrier that permits the substrate to penetrate as the ionized species. In mechanism E, the carrier catalyzes an electroneutral exchange of two anions. Substrates exchanged by such antiports do not cause mitochondria to swell unless they are coupled to some other transport system that promotes a net uptake of the external anion. In the example shown, malate by itself has no effect, but in combination with phosphate, it induces rapid swelling of mitochondria through an exchange of malate with phosphate. In this system, phosphate is catalytically cycled

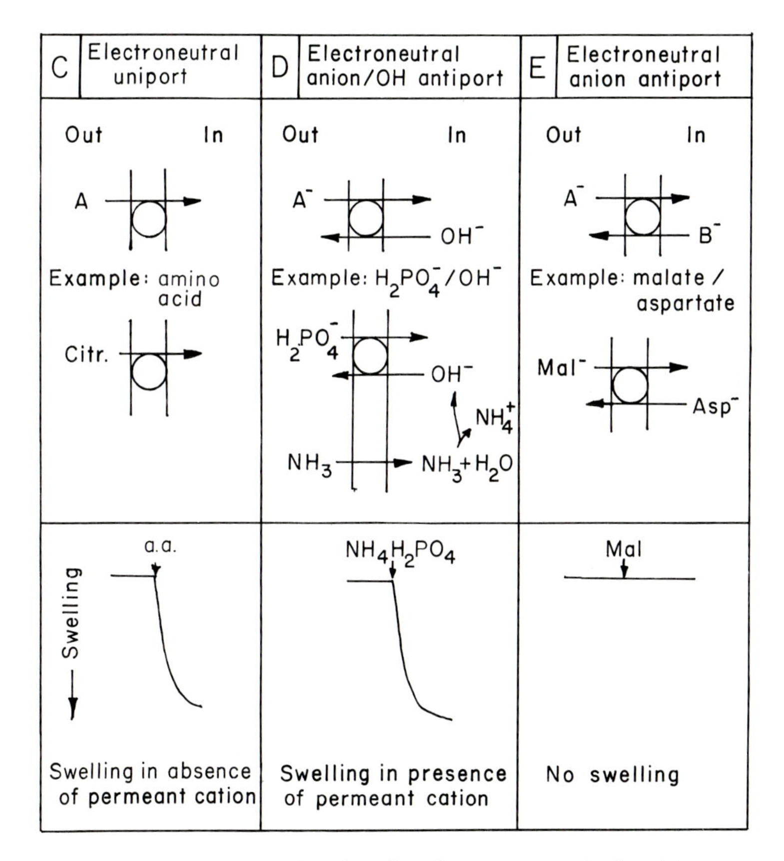

FIG. 9.4. Carrier-mediated modes of transport in mitochondria.

between the external and internal compartments by the phosphate/OH^- antiport.

C. Energy-Coupled Transport

During the oxidation of NADH or succinate by the electron transfer chain, there is an expulsion of protons into the medium and an accumulation of hydroxide ions in the matrix compartment. This creates what Mitchell calls a protonmotive force consisting of two energetic components—a pH differential and a membrane potential caused by the separation of protons and hydroxide ions (see Chapter 6, Section VII.B). Whether or not the anisotropic discharge of protons and hydroxide is a primary or secondary event in the conservation of oxidizable energy (as held by antagonists of the chemiosmotic hypothesis) is not important in the present context. Suffice it to say that there is good experimental evidence that the energy of both the pH gradient and the membrane potential can play an important role in regulating and promoting the transport of substrates and ions.

Since the membrane potential generated by electron transport is negative

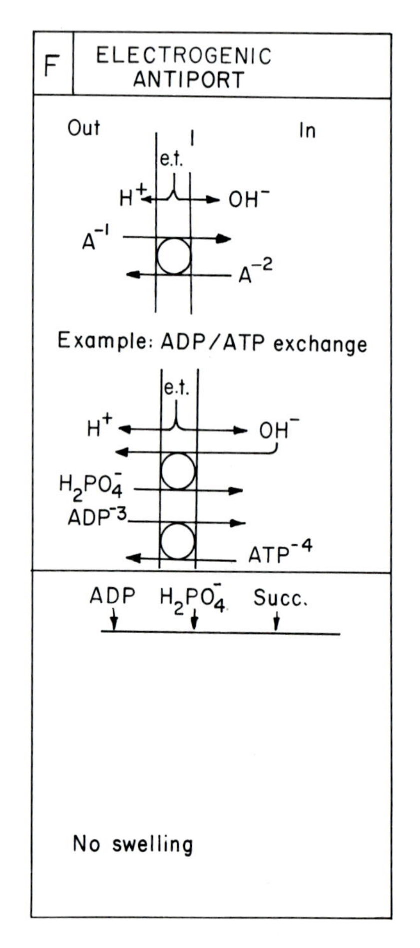

FIG. 9.5. Electrogenic antiport mechanism for the exchange of ADP and ATP.

on the inside, it is capable of supporting an electrophoretic exchange of more negatively charged internal for less negatively charged external substrates. This is shown in Fig. 9.5 (mechanism F) for the preferential exchange of external ADP for internal ATP by the adenine nucleotide carrier when mitochondria are energized with substrate.

V. Transport of Specific Substrates

A. Anions

1. Monocarboxylic Acids (Mechanism A)

Acetate, propionate, pyruvate, hydroxybutyrate, and acetoacetate are permeant anions that penetrate mitochondria by non-carrier-mediated diffusion. The dissociation constants of monocarboxylates are sufficiently low to permit a significant rate of transport of the undissociated acid at physiological pH. It is generally considered that monocarboxylic acids freely diffuse through the inner membrane as the un-ionized acids in the absence of any compensatory efflux of counterions. This is supported by studies on the rates of uptake of different monocarboxylates which are found to increase with increasing pK of the acid.

2. Pyruvate (Mechanism D)

Although pyruvate is considered to be a permeant anion, there is good evidence that its uptake is also mediated by a specific carrier transport system. In rat liver mitochondria, pyruvate uptake has been shown to follow saturation kinetics and to be inhibited by α-cyano-4-hydroxycinnamic acid. Whether this transport route is essential for mitochondrial oxidation of pyruvate is not known at present.

3. Phosphate (Mechanism D)

Phosphate is transported into mitochondria by a carrier-dependent process thought to be an $H_2PO_4^-/OH^-$ antiport. When placed in a medium containing high concentrations of ammonium phosphate (0.1–0.2 M), mitochondria swell because of an osmotically induced uptake of water accompanying the inward flux of NH_4^+ and $H_2PO_4^-$. The hydroxide exchange mechanism is based on the reasoning that at pH 7, the usual conditions of the swelling experiments, there is virtually no undissociated phosphoric acid present. Consequently, phosphate uptake is presumed to be compensated by hydroxide ion efflux to maintain charge neutrality. Mitchell has pointed out, however, that in theory the transported species could equally well be phosphoric acid, since the concentrations of the reactants and the proton activity at the active site of the carrier would not necessarily have to be those of the external medium. Accordingly, Mitchell favors a simpler uniport of the undissociated acid as shown in Fig. 9.6. Monofluoro- but not difluorophosphate has been shown to be transported into mitochondria. This suggests that the carrier binds the divalent form of phosphate

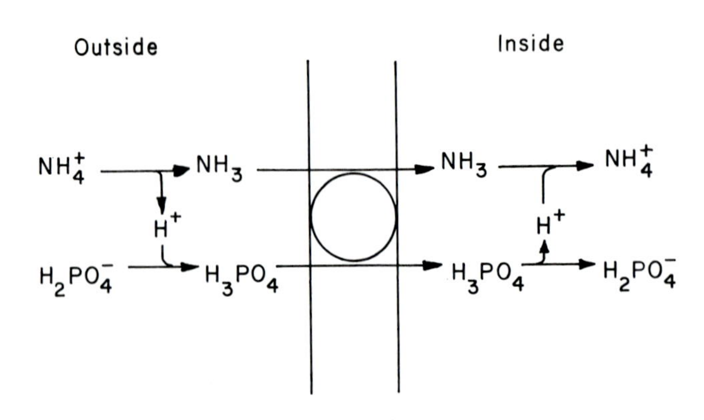

FIG. 9.6. Uniport mechanism for the transport of phosphate as the undissociated acid.

and either exchanges it for two hydroxide ions (antiport mechanism) or cotransports two protons. In the context of the present discussion, we shall assume phosphate to be transported by a hydroxide antiporter. Although a hydroxide-compensated exchange is mechanistically different from proton cotransport, formally the two are equivalent.

The transport of phosphate is inhibited by mersalyl, *N*-ethyl maleimide, and other sulfhydryl-blocking reagents, indicating that a sulfhydryl group is probably involved in the binding of the substrate. The use of mercurials and inhibitors that bind covalently (*N*-ethyl maleimide) is now being explored in attempts to purify the carrier protein. To date, however, this goal has not been achieved.

4. Adenine Nucleotides (Mechanism F)

During state 3 respiration, mitochondria synthesize ATP at a rate of 0.6–1.2 μmoles/min per mg protein. Almost all of this ATP is destined to be exported to the cytoplasm, and consequently, mitochondria must be able to replenish their endogenous adenine nucleotide pool with an equivalent amount of ADP. The adenine nucleotide exchange has been extensively studied in the laboratories of Klingenberg and of Vignais and is probably the best-understood transport system of mitochondria.

Klingenberg demonstrated that for every mole of adenine nucleotide transported into mitochondria, a mole of endogenous adenine nucleotide is exported, thus establishing the exchange nature of the process. It was found quite early that the adenine nucleotide exchange is specifically inhibited by atractyloside, a plant steroidal glycoside (Fig. 9.7). Since ADP relieves the inhibition by atractyloside, it was evident that a carrier is involved. The atractyloside-binding protein has recently been purified and shown to have properties expected of the carrier.

Adenine nucleotide transport has been studied by (1) following the exchange of external radioactive adenine nucleotides with a cold internal pool—the extent of exchange is determined by measuring the specific activity of the endogenous nucleotides; (2) preloading the endogenous pool with a

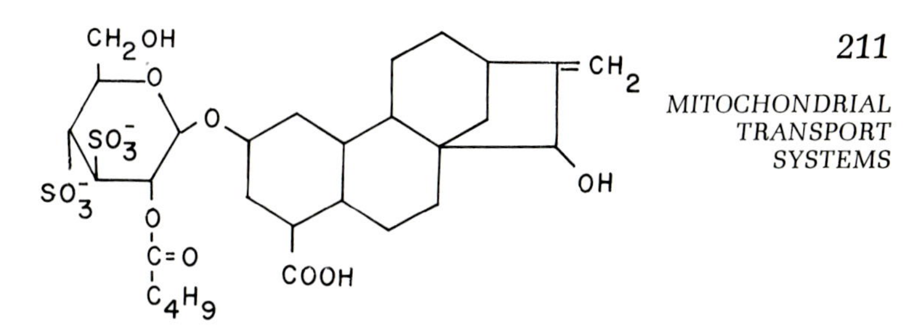

FIG. 9.7. Chemical structure of atractyloside.

radioactive nucleotide and determining the exchange with external cold nucleotides; and (3) preloading the internal pool with a nucleotide labeled with one isotope and following its exchange with an external nucleotide labeled with another isotope. In each method, a 1:1 exchange of internal and external nucleotides has been demonstrated. Some of the salient properties of the transport system are summarized in Table 9.1. The exchange is rapid and fully capable of importing ADP at a sufficiently fast rate to maintain oxidative phosphorylation at maximum capacity. The specificity is restricted to an exchange of ADP and ATP only. Adenosine monophosphate is virtually completely excluded, and so are the other ribose and deoxyribose nucleotides. The transfer is inhibited by atractyloside, carboxyatractyloside, and bongkrekic acid. The number of binding sites for ADP and the inhibitors is approximately the same and is in the range of the cytochrome oxidase content of mitochondria. The dissociation constant of ADP is tenfold higher than atractyloside and 100-fold higher than carboxyatractyloside. The latter inhibitor has a dissociation constant of 0.02 μM and behaves essentially as a noncompetitive inhibitor. The high affinity for carboxyatractyloside has turned out to be very useful in purification studies, since it remains bound to the carrier even under denaturing conditions.

Even though mitochondria can transport both external ADP and ATP,

TABLE 9.1. Properties of the Adenine Nucleotide Carrier System[a]

Exchange activity (at 18°)	0.2 μmoles/min per mg protein in rat liver mitochondria; 0.6 μmoles/min per mg in beef-heart mitochondria
Turnover number (18°)	About 500 min^{-1} based on the number of carboxyatractyloside binding sites
Specificity	Equally active with ADP and ATP; no activity with AMP, GDP, CDP, UDP, and IDP
Inhibitors	Atractyloside, carboxyatractyloside, and bongkrekic acid
Kinetic constants	Dissociation constant for ADP, 2 μM; atractyloside, 0.2 μM; carboxyatractyloside, 0.02 μM
Activation energy	30 kcal/mole
Number of binding sites	0.1–0.3 nmoles/mg protein for ADP and ATP in rat liver mitochondria, and 0.6–1.0 in beef-heart mitochondria; 0.15–0.35 nmoles/mg protein for atractyloside in rat-liver mitochondria, and 1.2–1.8 nmoles/mg protein in beef-heart mitochondria

[a]Data from Klingenberg (1970).

there is a strong bias in favor of an exchange of external ADP for internal ATP under energized or phosphorylating conditions. In the experiment shown in Fig. 9.8, the transport of external ATP was measured in energized and uncoupled mitochondria. At low concentrations of ATP, the inward-directed exchange of ATP is seen to be 3–4 times faster when mitochondria are uncoupled with FCCP. Klingenberg has shown that when both ADP and ATP are externally added to mitochondria under energized conditions, the rate of exchange of ADP is ten times higher than that of ATP. The opposite is true when the outward exchange of endogenous nucleotides is measured: here there is a ten times greater preference for the exchange of ATP. In the presence of an uncoupler, this effect is abolished, and the exchange of the two nucleotides is nearly identical.

The ability of phosphorylating mitochondria to generate a high external ATP/ADP ratio (about 30) through their transport system is of physiological significance. Klingenberg has suggested that the explanation lies in the electrogenic nature of the exchange and that the carrier itself is not affected by the metabolic state of the mitochondrion. Since ATP carries one negative charge more than ADP, the exchange of external ADP for internal ATP creates a membrane potential positive on the inside. The energy needed to drive the electrogenic exchange is thought to be derived from electron transport which generates a membrane potential of opposite polarity (see chemiosmotic hypothesis, Chapter 6, Section VII.B). In uncoupled mitochondria, the membrane is permeable to protons, and the electrogenic exchange of ATP for ADP is compensated by an efflux of protons.

The two inhibitors, atractyloside and bongkrekic acid, have provided some clues about the mechanism of adenine nucleotide transport. Some pertinent

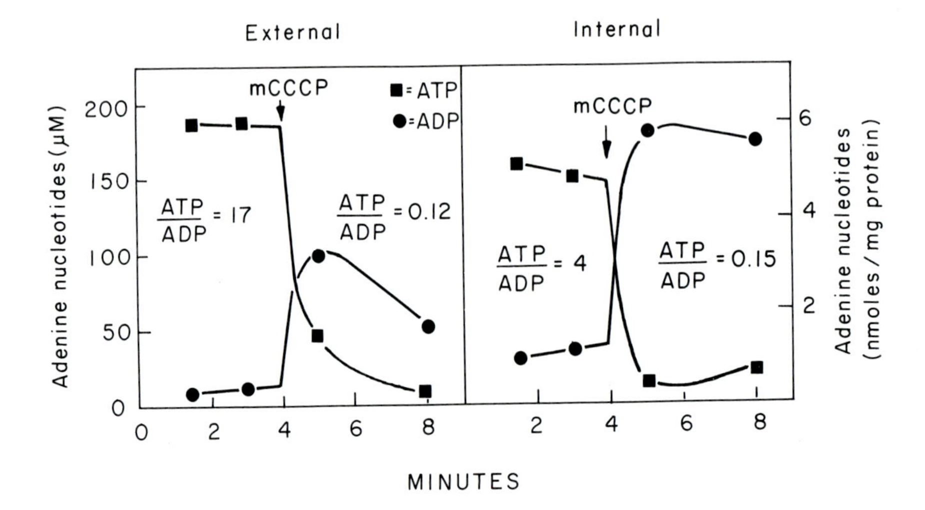

FIG. 9.8. External and internal ratios of ATP/ADP under phosphorylating and nonphosphorylating conditions. Mitochondria were incubated in the presence of substrate, P_i, and ^{14}C-ADP. The uncoupler *m*-CCCP was added at 4 min. In the absence of uncoupler, the external ratio is higher than the internal ratio because of the preferential exchange of external ADP for internal ATP. Under uncoupled conditions, the two ratios are nearly identical, indicating an unbiased exchange of the two adenine nucleotides by the carrier. (Taken from Heldt *et al.*, 1972.)

observations made with these inhibitors are listed in Table 9.2. Bongkrekic acid is a tricarboxylic acid which acts as a noncompetitive inhibitor of adenine nucleotide exchange. Its inhibitory activity is pH dependent and is most pronounced at acidic values where partial protonation of the acid groups occurs. Another interesting property of bongkrekic acid is the observed time lag in inhibition. These findings have been interpreted to indicate that bongkrekic acid must penetrate the inner membrane to the matrix side before it can react with the carrier. The binding of bongkrekic acid on the inner side is also supported by the fact that in sonically prepared submitochondrial particles, which have an inverted orientation, neither the time lag nor the pH effect is observed. In contrast, atractyloside and carboxyatractyloside inhibit mitochondrial adenine nucleotide transport without any lag and at pH values where the two sulfate groups are fully ionized. These inhibitors have no effect on transport in submitochondrial particles and most probably react with the carrier on the outer side of the inner membrane. The last point to be mentioned is that bongkrekic acid appears to increase the affinity of the carrier for ADP.

Based on these observations and other kinetic data, Klingenberg has proposed that atractyloside binds to the carrier on the outer surface of the inner membrane, thus tying up carrier molecules and preventing net transfer of adenine nucleotides. Bongkrekic acid, on the other hand, by virtue of its lipophilic properties at acid pH, penetrates the membrane and reacts with the carrier on the inner side. According to the mechanism, carrier molecules situated on the outside must transport ADP to the inner side of the membrane in order to react with bongkrekic acid, and this accounts for the apparent increase in the affinity of the carrier for substrate in the presence of the inhibitor. The model also implies that there is a structural difference in the active site of the carrier depending on its orientation with respect to the sides of the membrane. Klingenberg has pointed out that such conformational changes in the carrier are more consistent with a gated pore model of transport, since it invokes a different geometry of the active site on each side of the membrane. In the diffusable and reorienting carrier models, the active site is postulated to be the same on both sides (Fig. 9.2). This interpretation is obviously based on rather circumstantial evidence, and alternative explanations are equally tenable at present. For example, the two sides of the membrane may have different lipid environments which could influence and account for the differences in the binding of the two inhibitors.

The purification of the adenine nucleotide carrier has been achieved by

TABLE 9.2. Properties of Atractyloside and Bongkrekic Acid

Atractyloside	Bongkrekic acid
Competitive inhibitor	Noncompetitive inhibitor
No time lag for inhibition	Inhibits with a time lag
No pH dependence	More effective at acidic pH values
Does not modify affinity for ADP	Causes an apparent increase in the affinity for ADP
Does not inhibit adenine nucleotide exchange in submitochondrial particles	Inhibits exchange in submitochondrial particles. Time lag not seen.

labeling mitochondria with radioactive carboxyatractylate. Most of the label is associated with a 30,000-dalton protein that is soluble in high concentrations of Triton X-100 and can be purified by chromatography on hydroxylapatite and sieving gels. This component is estimated to make up 5% of the total protein of the inner membrane. The radioactive carboxyatractyloside remains bound to the protein throughout the purification procedure but can be displaced by ADP. It is interesting that the inhibitor appears to protect the carrier against denaturation. If the carrier is isolated in the absence of inhibitor, the final protein is modified and fails to bind either carboxyatractyloside or ADP. The availability of a purified carrier should permit more detailed studies of the molecular mechanism of adenine nucleotide transport.

5. Dicarboxylates (Mechanism E)

Both phosphate and adenine nucleotide transport systems are present in all types of mitochondria. This is understandable in view of the absolute requirement of these substrates for oxidative phosphorylation. The transport of anionic substrates and amino acids is restricted to certain types of mitochondria. Table 9.3 shows a partial list of substrates and the distribution of their transport systems in liver, heart, and insect muscle mitochondria. Whereas liver mitochondria are capable of transporting a large number of different substrates, this is not true of muscle mitochondria such as those of blowfly which lack all the known anion transport systems and do not oxidize externally added intermediates of the TCA cycle. Heart mitochondria appear to have an intermediate spectrum of transport systems.

At neutral pH, the concentration of undissociated di- and tricarboxylic acids is too low for any significant permeation. The uptake of these substrates is therefore dependent on carrier-mediated transport systems. The transport of anionic substrates has been studied by following the reduction of intramitochondrial pyridine nucleotides, by the swelling technique in massive uptake experiments, and also by direct measurements of the incorporation into mitochondria of the radioactive form of the substrates.

The dicarboxylate carrier exchanges L-malate, succinate, malonate, or oxalacetate for inorganic phosphate. Alternatively, a dicarboxylic acid such as

TABLE 9.3. Transport Systems of Different Mitochondria[a]

Carrier	Liver	Heart	Blowfly muscle
Phosphate	+	+	+
Adenine nucleotide	+	+	+
Dicarboxylate	+	+	−
Tricarboxylate	+	−	−
α-Ketoglutarate	+	−	−
Glutamate	+	+	−
Calcium	+	+	−

[a]Taken from Lehninger (1971).

malate can be exchanged for another dicarboxylic acid, e.g., succinate. Net transport, however, can occur only when phosphate is the exchanging ion. The efficiency of transport is greatest with malate and succinate. Oxalacetate is transported, but only poorly and in the absence of other competing dicarboxylic acids.

Transport of dicarboxylates is specifically inhibited by *n*-butylmalonate and phenylsuccinate. Sulfhydryl reagents also block the transport but at higher concentrations than required for the inhibition of the phosphate carrier. The fact that butylmalonate inhibits the transport of the different substrates suggests that a single carrier is involved. Direct evidence, however, is still lacking and will have to await purification of the carrier.

6. Tricarboxylates (Mechanism E)

The tricarboxylate carrier acts as a citrate/malate antiport. This carrier can also transport isocitrate and phosphoenolpyruvate. In both cases, the exchange is inhibited by *n*-ethylcitrate. The exchange of citrate for malate can be demonstrated with radioactive substrates; however, in the presence of citrate alone, there is no net uptake of the anion, and consequently, mitochondrial swelling does not occur. The oxidation of citrate and mitochondrial swelling are induced by the addition of malate and phosphate. Under these conditions, the transport of citrate is coupled to the exchange of malate with phosphate, thus permitting a net uptake of the anion (Fig. 9.9).

7. α-Ketoglutarate (Mechanism E)

α-Ketoglutarate is transported in exchange for malate. This carrier system is insensitive to *n*-butylmalonate and is not activated by phosphate, indicating that it is distinct from the dicarboxylate carrier responsible for the transport of malate and succinate.

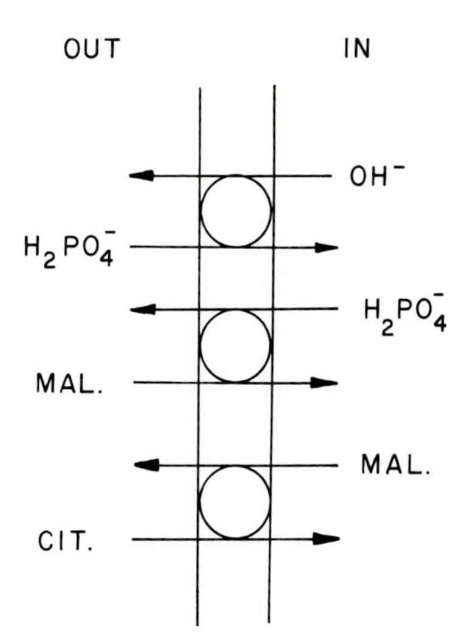

FIG. 9.9. Coupling between the citrate/malate, malate/phosphate, and phosphate/OH^- antiports.

8. Regulation of Anion Transport

All of the anion antiports have the general properties of facilitated diffusion and catalyze the uptake of substrates in a non-energy-linked fashion provided the transport is down the concentration gradient. It is known, however, that mitochondria can retain high internal concentrations of anionic substrates even when the external concentrations are low. The ability of mitochondria to accumulate anions against a concentration gradient indicates that the carrier systems are in some way controlled by the energy state of the mitochondrion.

To understand the role of energy in anion transport, let us first consider two important features of the transport mechanisms. From the above descriptions of the di- and tricarboxylate carriers, it is evident that net uptake of TCA cycle intermediates is either directly or indirectly coupled to the transport of inorganic phosphate. The central role of phosphate is illustrated in the scheme of Fig. 9.9 which shows how the influx of citrate occurs at the expense of an extrusion of a dicarboxylic acid such as malate which in turn is coupled to phosphate transport. The second important aspect of all anion antiports is that they catalyze electroneutral exchanges; i.e., there is no membrane potential formed as the result of the exchange. In the case of the citrate/malate antiport which on the surface appears to violate this rule, it has been experimentally verified that there is a cotransport of citrate with a proton. This exchange is therefore also electroneutral.

A predictable consequence of the coupling of phosphate uptake to proton (or OH^-) movement and the dependence of other electroneutral anion exchange systems on phosphate transport is that both the transport and distribution of anions between the external and internal phases of mitochondria should be affected by the pH of the medium. The anion distribution at equilibrium for an ideal two-phase system is described by the equation:

$$\log \frac{[A_i]}{[A_e]} = \log f + n\Delta pH \qquad [9.2]$$

where A_i and A_e are the internal and external concentrations of the anion, f is the ratio of the external to internal activities of the anion, n is the charge on the anion, and ΔpH is the difference in pH between the two phases. The relationship is seen to be a linear function in which the slope is equal to the number of charges on the anion.

Palmieri and co-workers have looked at the distribution of various anions in liver mitochondria as a function of the pH in the suspending medium. These studies indicated that the distribution of each anion was proportional to the external pH or the pH gradient across the membrane (Fig. 9.10A). More significant was the finding that the slopes of the lines relating the logs of anion and proton distributions correlated well with the expected values of the charge based on the known dissociation constants of the carboxylic acids at pH 7. The slopes were determined to be 0.8 for acetate, 1.5 for malate, and 1.6 for phosphate (Fig. 9.10B). A reasonable conclusion is that anionic fluxes are responsive primarily to the extrusion of protons that accompanies the oxidation–reduction reactions of the respiratory chain. Part of the energy of the proton gradient established by mitochondria under energized conditions may therefore be used for the maintenance of an energetically unfavorable gradient of anionic substrates. It is not ruled out, however, that anion transport may be subject to other controls as well. As defined by Mitchell, the total protonmotive force is a sum of the pH gradient and of the membrane potential. Conceivably, cation movements that change the relative contribution of the two components (increase in membrane potential and decrease in pH gradient) also exert a regulatory effect on anion uptake.

B. Amino Acids

1. Glutamate (Mechanisms D and F)

Two transport pathways have been found for the entry of glutamate into mitochondria. The first operates as a glutamate/OH^- antiport. This carrier has a relatively low affinity for glutamate and is specifically inhibited by the antibiotic avenaceolide which bears some structural similarities to glutamic acid. A proteolipid capable of binding glutamate and increasing the permeability of phospholipid films to glutamate has been purified from pig-heart mitochondria. Since the reconstituted system is sensitive to avenaceolide, the protein has been postulated to be the glutamate carrier.

The second means of importing glutamate is through an exchange with intramitochondrial aspartate. This exchange has been shown to be electrogenic in that there is approximately one proton taken up with each molecule of glutamate. The glutamate/aspartate antiport is similar to the adenine nucleotide antiport. For example, the directionality is determined by the energy state of mitochondria. Coupled mitochondria do not take up any aspartate even though they can actively oxidize glutamate and extrude aspartate. In the energized state, therefore, the exchange is predominantly in favor of glutamate uptake

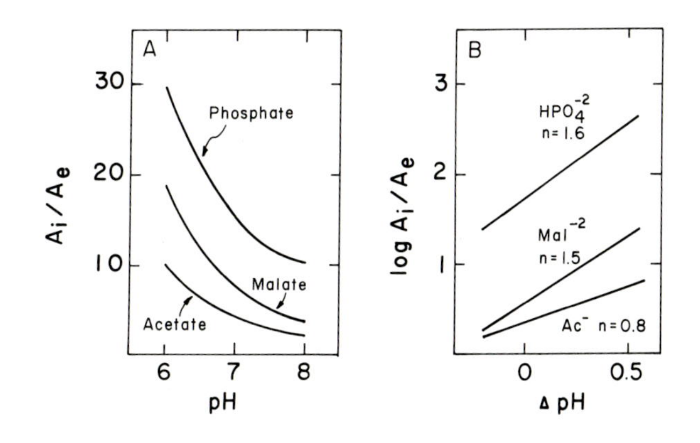

FIG. 9.10. Distribution of anions between the external medium and internal compartment of mitochondria as a function of pH. Mitochondria were incubated in the presence of radioactive phosphate, malate, or acetate and inhibitors of oxidative phosphorylation and respiration. The external and internal concentrations of the anions were measured after 1 min (left). The results of the experiment are replotted to show that the log of the anion distribution between the two compartments is proportional to ΔpH, with slopes increasing as a function of the charge on the anion. (From Palmieri *et al.*, 1970.)

and aspartate extrusion. The oxidation of glutamate by uncoupled mitochondria is accompanied by an accumulation of internal aspartate which does not appear to be exchanged.

2. Glutamine (Mechanism F)

Chappell and co-workers have recently reported that kidney mitochondria carry out an uptake of glutamine in exchange for glutamate. During state 4 respiration (absence of ADP), kidney mitochondria quantitatively convert glutamine to glutamate. This reaction is catalyzed by a glutaminase present in the matrix. The deamination of glutamine is blocked by uncouplers and inhibitors of the respiratory chain. When ADP is added to induce state 3 respiration, part of the glutamate formed is metabolized to aspartate. To account for these observations Chappell has proposed that glutamine is first exchanged for glutamate by an electrogenic antiport (hence the requirement for electron transport). The glutamate formed is then exchanged for aspartate by the glutamate carrier.

This transport system is not present in liver and heart mitochondria and probably plays an important role in maintaining the proper acid–base balance in kidney through the production of ammonia.

3. Ornithine

Liver mitochondria house two enzymes of the urea cycle. Both carbamyl phosphate synthetase and ornithine transcarbamylase are located in the matrix compartment. Liver mitochondria must be capable of importing ornithine and exporting citrulline, the substrate and product of the transcarbamylase.

Although ornithine was thought to enter the matrix space by an electrogenic uniport mechanism, more recent evidence indicates that the ornithine carrier effects an electroneutral cation/H^+ exchange. The charge neutrality is preserved through the generation of a proton in the matrix during subsequent reactions leading to the formation of citrulline. As already mentioned, the efflux of citrulline is electroneutral.

4. Neutral Amino Acids

The transport of neutral (zwitterionic) amino acids is mediated by a carrier that appears to have a broad range of specificity. At high concentrations, citrulline, alanine, valine, proline, etc. cause mitochondrial swelling independent of the presence of a permeant anion or cation. The transport of these amino acids is mutually competitive, suggestive of a single porter molecule.

C. Fatty Acids

Long-chain fatty acids are transported into mitochondria by the carnitine shuttle described in Chapter 3. Fatty acids and fatty acyl-CoAs are not permeable and must be converted to fatty acyl carnitine derivatives to permeate across the inner membrane. This mechanism meets the criteria of transport by group translocation.

VI. Substrate Transport and Coordination of Mitochondrial and Cytoplasmic Metabolism

The primary function of mitochondria is to supply the cell with the ATP made during oxidative phosphorylation. This is accomplished by the adenine nucleotide carrier that preferentially exchanges internal ATP for external ADP. The adenine nucleotide carrier is present in all mitochondria and along with the phosphate carrier is one of the most widely distributed transport systems in eucaryotes. Another function of mitochondria is to regulate the level of reduced pyridine nucleotides in the cytoplasm. Because of the impermeability of the inner membrane to pyridine nucleotides, their oxidation or reduction by the terminal respiratory pathway occurs indirectly by means of special substrate shuttle mechanisms. The transfer of reducing equivalents is probably a property of all mitochondria, although the particular shuttle(s) used may depend on the tissue. Finally, liver mitochondria possess some of the enzymes necessary for glucose and urea synthesis. The delivery of intermediates of gluconeogenesis and ureogenesis to and from mitochondria is also dependent on some of the transport systems described in this chapter.

A. Shuttles That Oxidize External NADH

1. Malate-Aspartate Shuttle

This shuttle achieves a transfer of reducing equivalents from cytoplasmic NADH to mitochondrial NAD^+ and involves two separate transport systems: the glutamate/aspartate and the malate/α-ketoglutarate antiports (Fig. 9.11). The overall mechanism can be broken down into the following exchanges: (1)

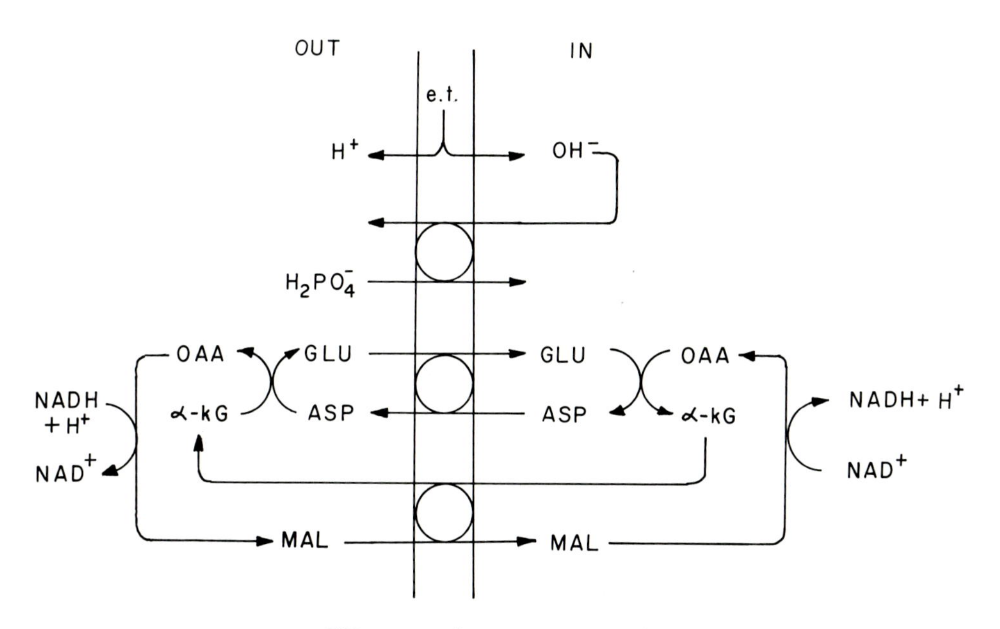

FIG. 9.11. Malate-aspartate shuttle.

an electrogenic exchange of external glutamate for aspartate driven by the membrane potential generated from electron transport; (2) an electroneutral exchange of external malate for internal α-ketoglutarate; (3) a transamination of glutamate and oxalacetate yielding aspartate and α-ketoglutarate (This reaction is catalyzed by aspartate transaminase located in the matrix. The reverse transamination occurs in the cytoplasm and is catalyzed by a distinct transaminase); and (4) an oxidation of intramitochondrial malate to oxalacetate and a reduction of cytoplasmic oxalacetate to malate. These reactions are catalyzed by two different NAD-dependent malate dehydrogenases. The combined reactions result in a net oxidation of cytoplasmic NADH without actual transfer of the pyridine nucleotide to the mitochondrial matrix. The oxidation of cytoplasmic NADH is stoichiometrically linked to the reduction of an equivalent amount of mitochondrial NAD^+ which can be further oxidized by the respiratory chain. Since mitochondrial NADH oxidation is coupled to the synthesis of three molecules of ATP, the shuttle provides an efficient mechanism for conserving the energy of oxidation of cytoplasmic NADH.

The malate–aspartate shuttle has been demonstrated to work in isolated mitochondria, and there is good evidence that it also operates *in vivo* in heart and liver tissue. For example, the oxidation of ethanol to acetaldehyde which requires the presence of cytoplasmic NAD^+ is stimulated in heart and liver cells by addition of malate, glutamate, or α-ketoglutarate. Each of these substrates alone or in combination should increase the level of cytoplasmic NAD^+ if the shuttle exists.

2. Malate–Citrate Shuttle

An alternative shuttle for the oxidation of cytoplasmic NADH is shown in Fig. 9.12. This shuttle utilizes the citrate/malate antiport only. Cytoplasmic citrate is converted to oxalacetate and acetyl CoA by an ATP-dependent citrate lyase. The oxalacetate is then reduced to malate, consuming reducing equivalents from NADH. The cycle is completed by the oxidation of intramitochon-

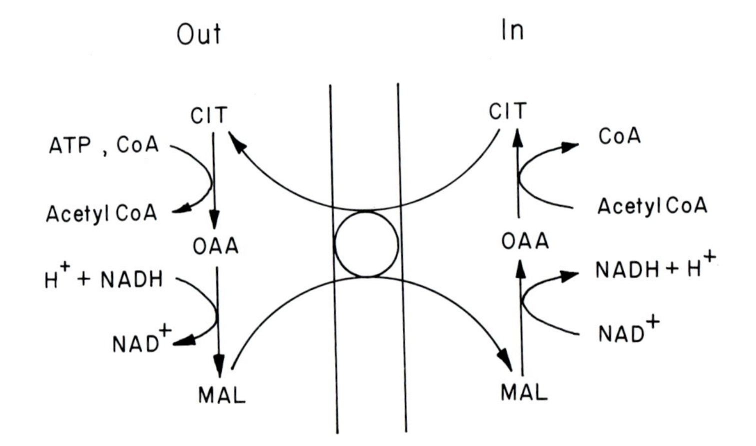

FIG. 9.12. Malate–citrate shuttle.

drial malate to oxalacetate and condensation of oxalacetate with acetyl-CoA to form citrate. In addition to transferring reducing equivalents, the shuttle also transfers intramitochondrial acetyl-CoA to the cytoplasm. Experiments with perfused liver indicate that the malate–citrate shuttle contributes significantly to NADH oxidation during glycolysis. Thus, hydroxycitrate, a specific inhibitor of the lyase, causes a threefold increase in the ratio of lactate to pyruvate. Lowenstein has proposed that this shuttle may be important during periods of high fatty acid synthesis in liver.

3. Fatty Acyl-CoA Shuttle

Liver and heart mitochondria contain a fatty acyl-CoA elongase in the outer membrane. This enzyme condenses a fatty acyl-CoA with acetyl-CoA to form a new fatty acyl-CoA two carbons longer. The reaction depends on NADH. The fatty acyl-CoA can be transported to the matrix by the carnitine shuttle where it may be oxidized by the β-oxidation system to regenerate the original fatty acyl-CoA and acetyl-CoA. Whether this shuttle is functional is difficult to evaluate at present. A requirement of the fatty acyl-CoA shuttle is that the fatty acyl-CoAs produced during β-oxidation exist as free diffusable intermediates. There is some controversy at present on this point, since most of the evidence suggests that no such intermediates are formed during β-oxidation. Studies with liver mitochondria, however, have revealed very sizable stimulation of external NADH oxidation in the presence of catalytic amounts of a long-chain fatty acid and CoA.

4. α-Glycerophosphate Shuttle

Insect flight muscle mitochondria contain high levels of α-glycerophosphate dehydrogenase. This enzyme differs from the cytoplasmic glycerophosphate dehydrogenase in that the hydrogen acceptor is flavin rather than NAD^+ (see Chapter 4, Section VI). The mitochondrial dehydrogenase is located in the inner membrane, but its orientation does not require the substrate to penetrate into the matrix. Bücher and co-workers showed that in flight muscle most of the NADH produced during glycolysis can be oxidized by a shuttle mechanism involving the reduction of dihydroxyacetone phosphate to α-glycerophosphate by the cytoplasmic dehydrogenase and a symmetrical reoxidation of α-glycerophosphate by the mitochondrial dehydrogenase (Fig. 9.13). There is some evidence that this shuttle may also operate in liver and muscle of hyperthyroid rats, but its relative importance under normal metabolic conditions is not known.

B. Transport Systems That Function in Gluconeogenesis and Ureogenesis

The discovery of mitochondrial anion transport systems has stimulated experiments aimed at assessing their role in the metabolism of liver and other types of tissues. Much of this work has been done by using specific inhibitors of mitochondrial carriers in perfusates and cell suspensions and determining

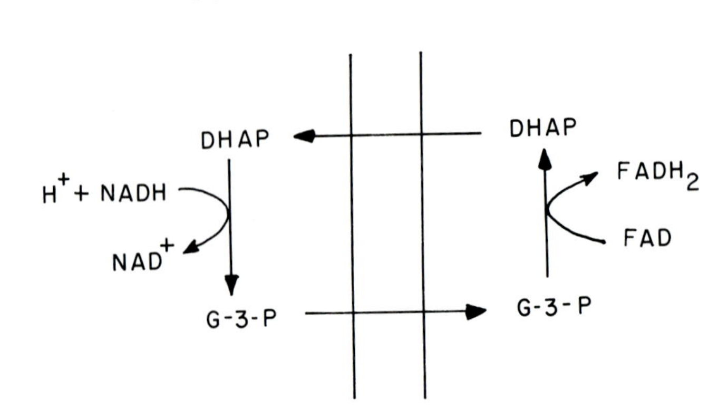

FIG. 9.13. α-Glycerophosphate shuttle. α-Glycerol phosphate (G-3-P) formed in the cytoplasm is freely permeable to the outer membrane and is oxidized to dihydroxyacetone phosphate (DHAP) by the FAD-dependent α-glycerol phosphate dehydrogenase located in the inner membrane.

how they affect the levels of various intermediary metabolites. There is now good evidence that the transport systems studied in isolated mitochondria also function in whole cells and contribute to the regulation of glucose and urea synthesis in liver.

1. Gluconeogenesis

In gluconeogenesis, the endproducts of glycolysis (pyruvate and lactate) are reutilized to make glucose. This happens mainly in the liver and under normal circumstances is not a very active process. However, following vigorous activity, the lactate produced in muscle tissue is transported to the liver where it is converted to glucose. Similarly, in the absence of dietary glucose, certain amino acids released from protein degradation are converted to pyruvate which can serve as a precursor of glucose. Under both conditions, gluconeogenesis is markedly stimulated.

Gluconeogenesis involves a reversal of all of the catalytic steps of glycolysis with two exceptions. In glycolysis, the dephosphorylation of phosphoenolpyruvate to pyruvate is catalyzed by pyruvate kinase. This reaction is essentially irreversible, and the conversion of pyruvate to phosphoenolpyruvate in gluconeogenesis occurs by an alternate pathway. Pyruvate is first carboxylated to oxalacetate which is then reduced to malate. The two reactions are catalyzed in turn by pyruvate carboxylase and malate dehydrogenase, both enzymes being located in the mitochondrial matrix. The malate can be transported to the cytoplasm where it is oxidized back to oxalacetate by the cytoplasmic malate dehydrogenase. Finally, a cytoplasmic GTP-dependent phosphoenolpyruvate kinase decarboxylates oxalacetate to phosphoenolpyruvate. In some mammals (guinea pig) and in birds, phosphoenolpyruvate kinase in equally distributed in mitochondria and the cytoplasm, and in this case phosphoenolpyruvate can be synthesized in both compartments of the cell. The rest of gluconeogenesis occurs in the cytoplasm and need not concern us here.

When *n*-butylmalonate is used in perfused rat liver to block mitochondrial transport of malate on the dicarboxylate carrier, there is a severe inhibition of glucose formation. Since the inhibition is observed when either pyruvate or lactate is used as the precursor, malate must be extruded from mitochondria in both cases. The transport schemes shown in Fig. 9.14 have been proposed to account for these observations. With pyruvate as substrate, the NADH used for the reductive steps of gluconeogenesis is derived from the oxidation of malate to oxalacetate in the cytoplasm, and the outward movement of malate must predominate. With lactate, however, the situation is more complicated, since NADH is produced in the oxidation of lactate to pyruvate. Under these conditions, the extrusion of aspartate is more important, and the dicarboxylate carrier functions mainly as a malate/α-ketoglutarate antiport. According to the mechanism shown in Fig. 9.15, no net flux of malate is required in the lactate system, and aspartate is the main substrate exported from mitochondria. The two schemes are also supported by the finding that inhibitors of aspartate transaminase such as aminooxyacetate and difluorooxalacetate prevent glucose synthesis from lactate but not from pyruvate.

2. Ureogenesis

The enzymes of the urea cycle are distributed between the mitochondrial matrix and the cytoplasm. We have seen that there are special carriers that mediate the entry of ornithine into and extrusion of citrulline from mitochondria. Other anion transport systems, however, have also been implicated in the regulation of ureogenesis.

The urea cycle starts with the synthesis of carbamyl phosphate from bicarbonate and ammonia in the mitochondrion. The immediate source of ammonia

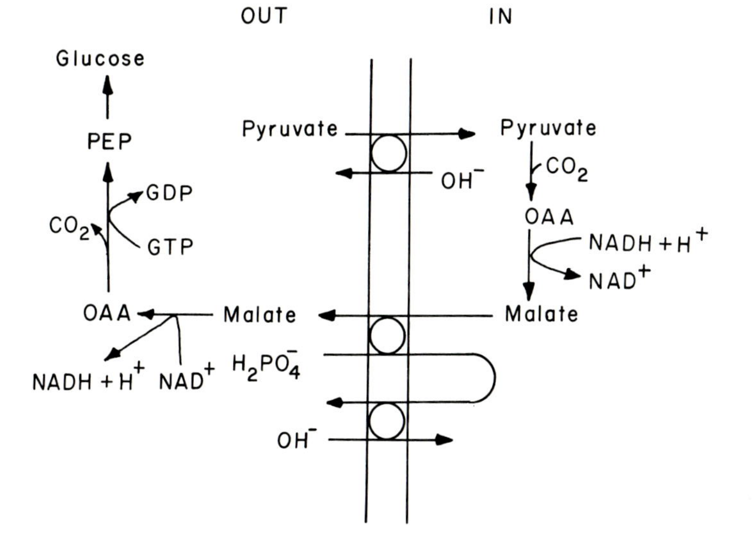

FIG. 9.14. Mitochondrial carrier systems involved in glucose synthesis from pyruvate. (Based on Williamson, 1976.)

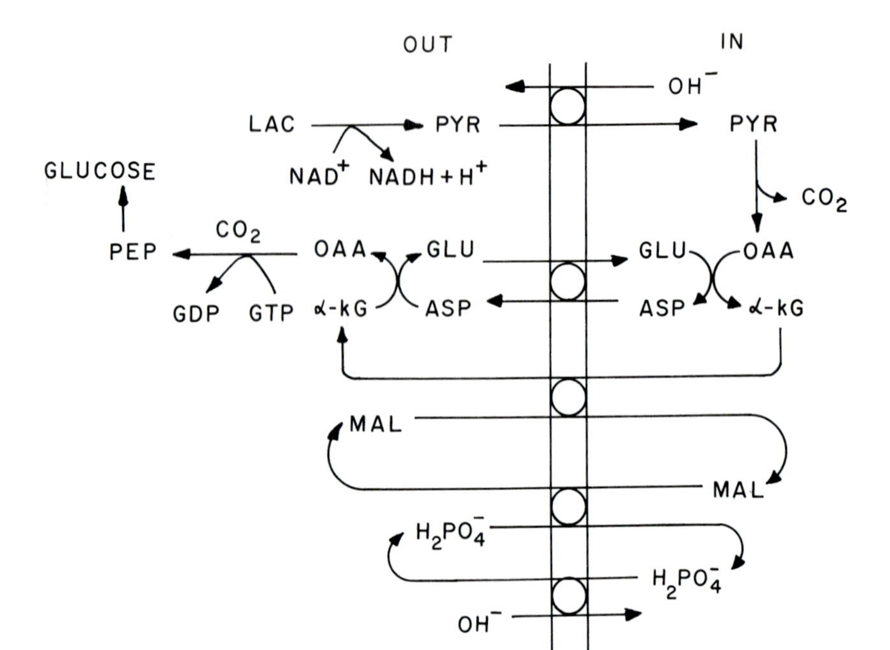

FIG. 9.15. Mitochondrial carrier systems involved in glucose synthesis from lactate. (Based on Williamson, 1976.)

for this reaction is glutamate which becomes oxidatively deaminated in mitochondria to α-ketoglutarate. The second amino group of urea is extracted from aspartate during the conversion of citrulline to argininosuccinate. This reaction takes place in the cytoplasmic compartment. Because of the compartmentation of the urea cycle enzymes, there must be a net flow of ornithine and glutamate into mitochondria and of citrulline and aspartate to the cytoplasm.

The glutamate is probably transported by both the glutamate/aspartate and glutamate/OH^- carriers. The *in vivo* operation of these transport systems, however, has not been demonstrated. There is evidence that malate transport on the dicarboxylate carrier is also important in ureogenesis. The synthesis of urea from ammonia or alanine in whole liver cells has been found to be inhibited by *n*-butylmalonate and to be stimulated by malate or conditions that raise the intracellular concentration of malate. Williamson and co-workers have proposed several anion transport schemes to explain these observations. In essence, these are aspartate shuttles in which the aspartate drained for argininosuccinate synthesis is replenished by the carboxylation of pyruvate to oxalacetate.

VII. Metal Transport

The first mitochondrial transport systems to be recognized are those responsible for the uptake of monovalent and divalent metals. Many of the techniques and approaches that have already been described in connection with substrate transport have also been used to study the mechanisms of metal transport. Ionophores, which mimic many of the properties of the natural transport systems, have also been valuable tools for mechanistic interpretations. Valinomycin, gramicidin, and nigericin in particular have been widely used in mitochondrial transport studies.

Mitochondrial transport of metals is probably important in regulating cellular metabolism. This aspect has received little attention up to now, and virtually nothing is known about the physiological significance of the metal-transporting systems.

A. Ionophores

Ionophores are chemical agents that increase the permeability of membranes to ions. The activity of an ionophore is commonly defined by its ability to enhance the rate of equilibration of an ion between two aqueous phases separated by a phospholipid film. It is still not certain whether ionophores exert their effect by acting as pores in the membrane through which selected ions can permeate or whether they are mobile carriers that transport ions by diffusion across the membrane. Although the term carrier will be used in the present discussion, it should be kept in mind that the pore hypothesis is by no means excluded at present.

1. Valinomycin

Valinomycin is a cyclic polypeptide with a repeating sequence of D-valine, D-hydroxyisovalerate, L-valine, and L-lactate (Fig. 9.16). The inner core of the peptide is a polar cage that binds alkali metals with the following order of selectivity: K^+, Rb^+, $Cs^+ \gg Na^+$, Li^+. In some systems, the ionophoric activity with K^+ has been shown to be as much as 10,000 times greater than with Na^+. Two other features of valinomycin are its hydrophobicity and absence of

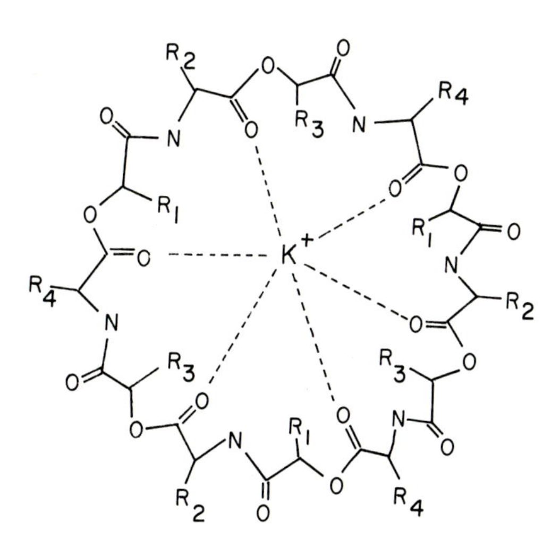

FIG. 9.16. Chemical structure of valinomycin showing the chelation of potassium through the central carbonyl oxygens of the polypeptide chain. The polypeptide chain consists of L-lactate (R_1), L-valine (R_2), D-hydroxyisovalerate (R_3), and D-valine (R_4).

charged groups. The activity of valinomycin in promoting cation exchange across synthetic phospholipid bilayers and biological membranes can be attributed to its ability to diffuse through the lipid bilayer. Since valinomycin-induced K^+ transport depends on the presence of a permeant anion, the simplest interpretation is that the positively charged complex diffuses through the bilayer in response to a membrane potential generated by the negatively charged anion. Valinomycin therefore acts as an electrophoretic carrier specific for certain alkali metals.

2. Gramicidin

Gramicidins A, B, C, and D are linear *N*-formyl pentadecapeptides. These antibiotics behave in a manner similar to valinomycin but are less specific in their binding of different cations. For example, gramicidins are equally active with Na^+ and K^+. Even though gramicidins are linear molecules, it is thought that in the complex the peptide chain wraps itself around the cation to form a cage structure. Because of the lack of discrimination between Na^+ and K^+, gramicidins can effect an exchange of both these alkali metals.

3. Nigericin

The structure of this antibiotic is not known. Nigericin forms complexes with monovalent alkali metals through the centrally oriented oxygens of the heterocyclic ring system. Since nigericin is a carboxylic acid, it can transport both cations and protons and, in some systems, behaves as a cation/proton exchange carrier. In this respect, its activity is similar to that of valinomycin when the latter is used in conjunction with an uncoupler.

4. Uncouplers

Almost all uncouplers of oxidative phosphorylation can equilibrate protons across membranes. They can therefore be considered to be ionophores or carriers of protons.

B. Mechanisms of Metal Transport

The permeability of the mitochondrial membrane to monovalent and divalent metals is too low to permit efficient diffusion of these cations into the internal compartment. Certain alkali metals, however, can be accumulated by facilitated diffusion if there is a favorable concentration gradient. Other metals can be accumulated by active transport mechanisms against their concentration gradient. A partial list of metals that have been documented to be transported by mitochondria is presented in Table 9.4.

1. Facilitated Diffusion of Na^+, Li^+, and K^+

Mitochondria undergo rapid swelling in the presence of high external concentrations of Na^+ or Li^+ and a permeant anion. If the anion is acetate and can

TABLE 9.4. Metal Transport Systems of Mitochondria

Metal	Energy requirement	Carrier	Mechanism
Na^+	−	+	Cation/proton exchange carrier
Li^+	−	+	Cation/proton exchange carrier
K^+	+	−	Electrophoretic diffusion
Mg^{2+}	+	−	Electrophoretic diffusion
Ca^{2+}	+	+	Electrogenic carrier
Sr^{2+}	+	+	Electrogenic carrier
Mn^{2+}	+	+	Electrogenic carrier

produce internal protons, no other requirements are seen for swelling. With a permeant anion such as NO_3^- or SCN^-, swelling does not occur unless an uncoupler is also added (Fig. 9.17). Both Na^+ and Li^+ are thought to penetrate mitochondria on an exchange carrier that functions as an M^+/H^+ antiport (Fig. 9.18). The function of the uncoupler is to recycle the protons which otherwise form a membrane potential positive on the outside, preventing further uptake of the cation. In the case of acetate, the need for the uncoupler is obviated, since the anion acts as an internal source of protons for the exchange carrier.

There is no swelling of mitochondria in potassium salts under similar conditions. This probably means that a K^+ carrier is absent or that its activity is too low for significant rates of transport. Mitochondria, however, can be induced to swell in potassium acetate if nigericin alone or valinomycin plus an uncoupler are added. When potassium nitrate or thiocyanate are used, the opposite requirements are found. Swelling with these anions depends on the presence of valinomycin alone or nigericin plus an uncoupler (Fig. 9.19). Based on the known ionophoric activities of valinomycin and nigericin, the swelling behavior of mitochondria in potassium acetate and nitrate can be fitted into the transport schemes shown in Fig. 9.20. In acetate, valinomycin by itself cannot effect net K^+ transport since there is no means for the extrusion of internal protons arising from the ionization of acetic acid, and therefore an uncoupler must be present. Nigericin, by virtue of being able to exchange K^+ and H^+, can support uptake of the cation in the absence of uncoupler. The same argument applies to the nitrate situation except that here an uncoupler must be combined with

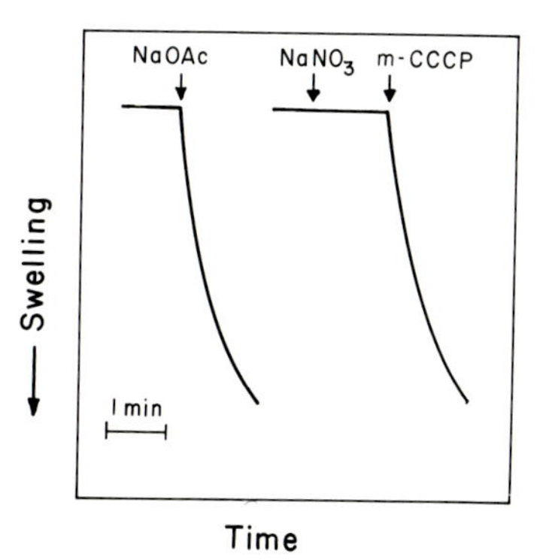

FIG. 9.17. Mitochondrial swelling induced by sodium salts of permeant anions.

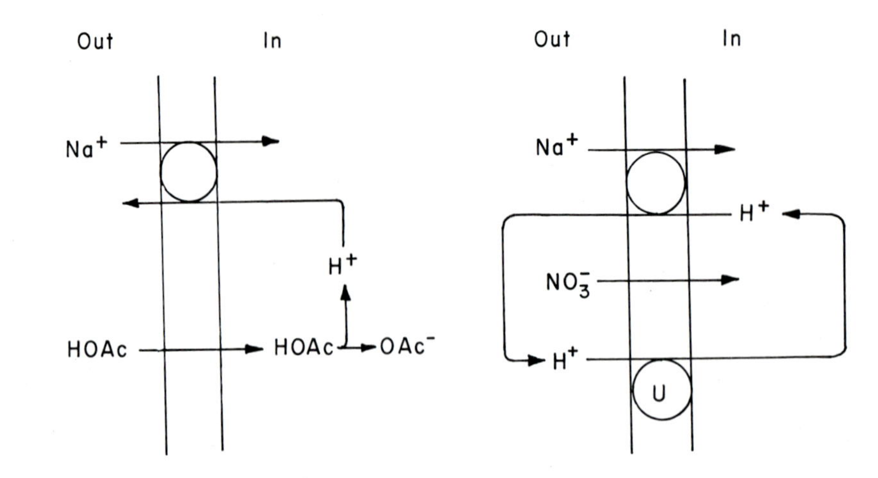

FIG. 9.18. Mechanism of sodium transport in the presence of permeant anions.

nigericin to recycle the protons extruded as a result of the obligatory exchange of K^+

2. Active Transport of K^+

Mitochondria transport K^+ in the absence of ionophores provided they are energized with a respiratory substrate or with ATP. Massive accumulation of K^+ under these conditions can be measured by the swelling technique when the permeant anions acetate or phosphate are present. The substrate-dependent swelling in isotonic potassium salts (0.12 M) is abolished by inhibitors of respiration and by uncouplers; when the energy source is ATP, swelling is prevented by uncouplers or by oligomycin. The rate of swelling is highest with substrates that are coupled to all three phosphorylation sites. These observations indicate that K^+ can be accumulated by an active transport process driven by an intermediate (or energized state) of oxidative phosphorylation. Similar

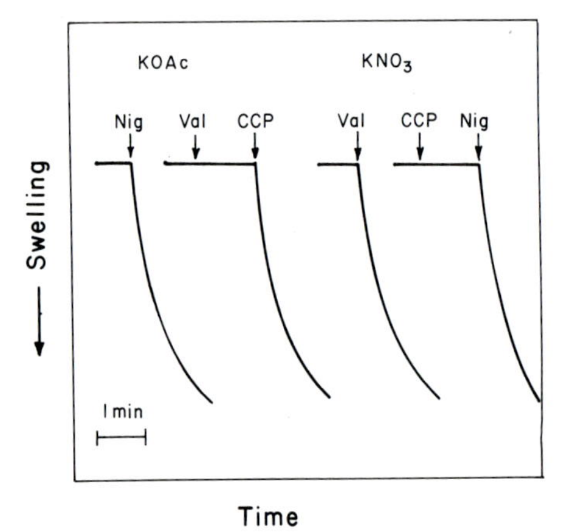

FIG. 9.19. Mitochondrial swelling induced by potassium salts of permeant anions.

requirements have been noted for the transport of a large number of other cations, e.g., Mg^{2+}, $Tris^+$, tetramethylammonium ion.

There has been considerable interest in formulating a general mechanism of mitochondrial cation transport. As with so many other questions, the interpretation of much of the known transport phenomena depends on how one chooses to think of the broader problem of energy coupling. The two mechanisms shown in Fig. 9.21 point to the basic disagreement and can be reduced to the following issue: Is the coupling event a generation of a proton gradient that can be translated into secondary ion fluxes or are there special cation pumps that are activated by a high-energy intermediate (or state) of oxidative phosphorylation? It should be obvious by now that mechanism A is a derivative of the chemiosmotic hypothesis. Here, the proton gradient formed as a result of electron transport or of ATP hydrolysis is first converted to a membrane potential negative on the inside. This occurs when phosphate is exchanged for OH^- on the phosphate carrier or when protons released from acetic acid neutralize internal hydroxide ions. The resultant membrane potential serves as the driving force that electrophoretically pulls K^+ and other cations inside. It should be pointed out that the mechanism can be adapted to include a cation carrier. For example, the transfer of cations on an electrogenic carrier might require a fairly

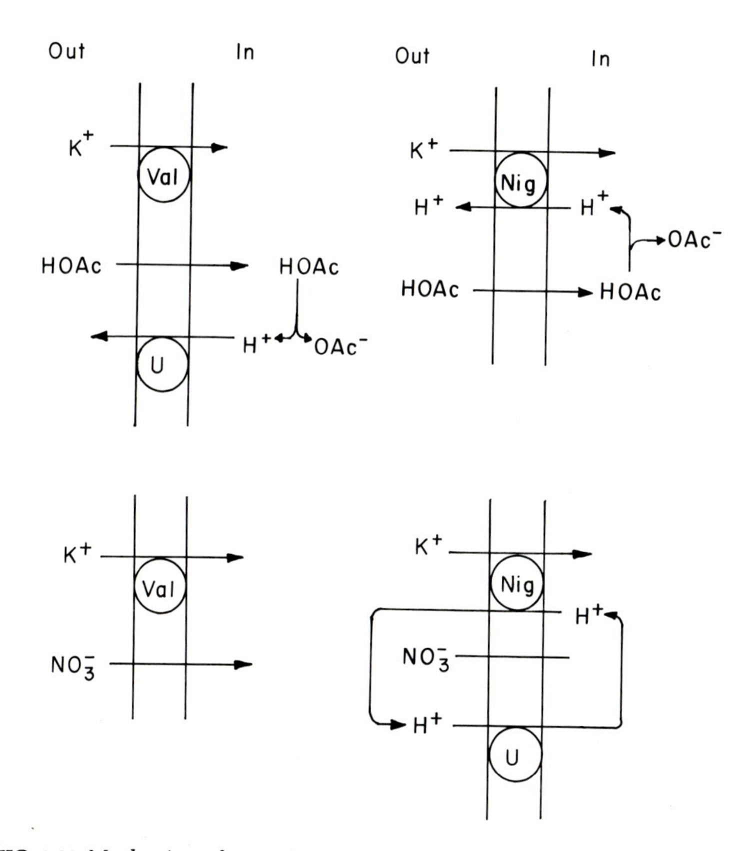

FIG. 9.20. Mechanism of potassium transport in the presence of permeant anions.

high membrane potential which can only be generated in the presence of an energy source.

In the cation pump model (mechanism B), energy is used to activate a carrier that exchanges cations for internal protons. According to this interpretation, the accumulation of anions is secondary to the movement of the cation. Energy could be used to activate the carrier by increasing its affinity for the substrate or alternatively by modifying the membrane in such a way as to allow a more rapid diffusion of the carrier.

Although both models are tenable, the existent evidence is more supportive of an electrophoretic permeation of cations in response to an anion-induced membrane potential. Brierley has pointed out that the active transport of a large number of structurally unrelated cations, many of which do not occur in nature, makes it unlikely that a carrier is involved in the transport process. Second, many reagents that modify mitochondria by making them more permeable to cations have been found to also increase substrate- and ATP-dependent cation accumulation. This observation again tends to favor a non-carrier-mediated electrophoretic mechanism of cation permeation, since an increased permeability of the membrane should cause a more rapid rate of back diffusion of cations and result in a decreased rate of net uptake according to the cation pump mechanism.

MECHANISM A

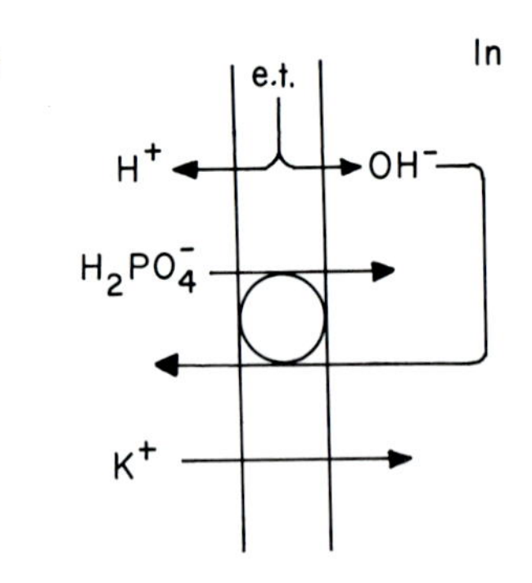

MECHANISM B

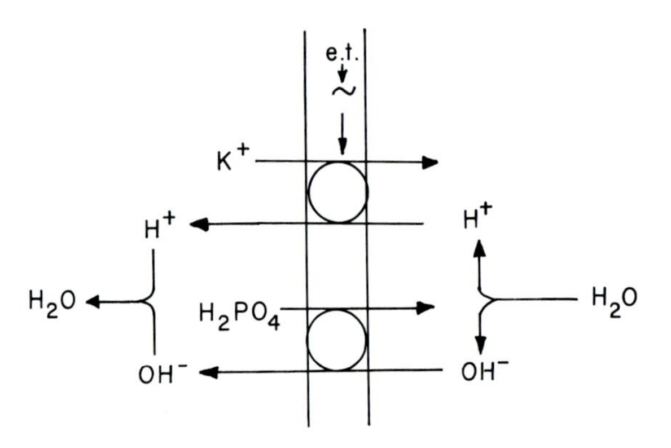

FIG. 9.21. Two mechanisms of electron-transport-driven K^+ uptake by mitochondria.

Heart, liver, and most other types of mitochondria can transport calcium at very high rates by an energy-dependent process. The active transport of Ca^{2+} has been studied in many laboratories and has been shown to have the same requirements as other energy-dependent cation transport systems. It is discussed separately because there is good evidence that the transport of this particular cation occurs on a carrier. The Ca^{2+} carrier has been shown to be specifically inhibited by low concentrations of Ruthenium Red and by La^{3+}. Competitive inhibition is also seen with Sr^{2+} and Mn^{2+} (but not Mg^{2+}), indicating that the transport of these divalent metals is probably mediated by the same carrier.

Lehninger has defined two sets of conditions for Ca^{2+} transport against an energetically unfavorable concentration gradient. "Massive loading" occurs in the presence of acetate or phosphate and relatively high concentration of Ca^{2+} (1–5 mM). When substrate or ATP is used as the energy source, as much as 2 μmoles of Ca^{2+}/mg protein penetrate into mitochondria, and in acetate there is extensive swelling. A comparable amount of Ca^{2+} is accumulated in phosphate medium except that the salt precipitates in the matrix as $Ca_3(PO_4)_2$. Mitochondria subjected to massive loading conditions become irreversibly damaged and lose their coupling functions.

The active transport of Ca^{2+} has also been studied under "limited loading" conditions where the concentration of external Ca^{2+} is low (0.1–0.2 mM). The maximal amount of Ca^{2+} accumulated is 100–150 nmoles/mg protein, and there is no detectable loss of mitochondrial functions. Chance first reported that the addition of small amounts of Ca^{2+} to state 4 mitochondria elicits the same kind of stimulation of respiration as ADP or uncouplers. From the amount of extra oxygen consumed and the amount of metal transported, the Ca^{2+}/O ratio has been determined to be 4.9 with β-hydroxybutyrate, 2.9 with succinate, and 1.8 with TMPD–ascorbate as substrates. These values yield an average stoichiometry of 1.6 Ca^{2+} transported per pair of electrons passing through each phosphorylation site. Assuming a similar efficiency for ATP-driven transport, the stoichiometry of Ca^{2+}/ATP should be the same. This has been experimentally verified.

Another important measurement is the number of protons released into the medium during Ca^{2+} transport. Initially, values of two were reported. Later, so-called "superstoichiometries" were found which could attain values as high as ten. The most recent studies indicate that there may be four protons ejected for every Ca^{2+} transported into mitochondria. The larger values have been attributed to a Ca^{2+} induced hydrolysis of endogenous ATP which could be prevented by using oligomycin. Aside from being important from a mechanistic standpoint, the number of protons ejected is also relevant to the chemiosmotic mechanism, since it is currently believed that the stoichiometry of protons transported per coupling site is two. The apparent inconsistencies and the difficulties in obtaining accurate values of the number of protons released either during substrate oxidation or cation transport seem to be caused by various fluxes of endogenous cations and anions under the conditions of the pulse experiments. Despite some of the disagreements, most of the characteristics of Ca^{2+} transport are consistent with a carrier-mediated electrophoretic process as shown in Fig. 9.21. The general features of the mechanism are similar to those already described for the energized transport of the other monovalent and divalent metals.

Selected Readings

Amoore, R. E., and Bartley, W. (1958) The permeability of isolated rat liver mitochondria to sucrose, sodium chloride and potassium chloride at 0°, *Biochem. J.* **69**:223.

Brierley, G. P. (1974) Passive permeability and energy-linked ion movements in isolated heart mitochondria, *Ann. N.Y. Acad. Sci.* **227**:398.

Brierley, G. P. (1976) Monovalent cation transport by mitochondria, in *Mitochondria: Bioenergetics, Biogenesis and Membrane Structure* (L. Packer and A. Gomez-Puyou, eds.), Academic Press, New York, pp. 3–20.

Brunnengraber, H., and Lowenstein, J. M. (1973) Effect of hydroxycitrate on ethanol metabolism, *FEBS Lett.* **36**:130.

Chance, B. (1965) The energy-linked reaction of calcium with mitochondria, *J. Biol. Chem.* **240**:2729.

Chance, B., and Montal, M. (1971) Ion translocation in energy-conserving membrane systems, in *Current Topics in Membranes and Transport* (F. Bronner and A. Kleinzeller, eds.), Vol. 2, Academic Press, New York, pp. 99–156.

Chappell, J. B. (1968) Systems for the transport of substrates into mitochondria, *Br. Med. Bull.* **24**:150.

Chappell, J. B., and Crofts, A. R. (1966) Ion transport and reversible volume changes of isolated mitochondria, in *Regulation of Metabolic Processes in Mitochondria* (J. M. Tager, S. Papa, E. Quagliarello and E. C. Slater, eds.) Elsevier, Amsterdam, pp. 293–316.

Chappell, J. B., McGivan, J. D., and Crompton, M. (1972) The anion transporting systems of mitochondria and their biological significance, in *The Molecular Basis of Biological Transport* (J. F. Woessner, Jr. and F. Huijing, eds.), Academic Press, New York, pp. 55–81.

Fonyo, A. (1968) Phosphate carrier of rat liver mitochondria. Its role in phosphate outflow, *Biochem. Biophys. Res. Commun.* **32**:624.

Gamble, J. G., and Lehninger, A. L. (1973) Transport of ornithine and citrulline across the mitochondrial membrane, *J. Biol. Chem.* **248**:610.

Grunnet, N. (1970) Oxidation of extramitochondrial NADH by rat liver mitochondria. Possible role of acetyl-SCoA elongation enzymes, *Biochem. Biophys. Res. Commun.* **41**:909.

Heldt, H. W., Klingenberg, M., and Milovancev, M. (1972) Differences between the ATP/ADP ratios in the mitochondrial matrix and extramitochondrial space, *Eur. J. Biochem.* **30**:434.

Klingenberg, M. (1970) Metabolite transport in mitochondria. An example for intracellular membrane function, in *Essays in Biochemistry* (P. N. Campbell and F. Dickens, eds.), Vol. 6, Academic Press, New York, pp. 117–159.

Klingenberg, M. (1976) The adenine nucleotide transport of mitochondria, in *Mitochondria: Bioenergetics, Biogenesis and Membrane Structure* (L. Packer and A. Gomez-Puyou, eds.), Academic Press, New York, pp. 127–150.

Lardy, H. A., Graven, S. N., and Estrada-O, S. (1967) Specific induction and inhibition of cation and anion transport in mitochondria, *Fed. Proc.* **26**:1355.

Lehninger, A. L. (1971) The transport systems of mitochondria membranes, in *Biomembranes* (L. A. Manson, ed.), Vol. 2., Academic Press, New York, pp. 147–164.

Lehninger, A. L., Carafoli, E., and Rossi, C. S. (1967) Energy-linked ion movements in mitochondria, *Adv. Enzymol.* **29**:259.

Lehninger, A. L., Brand, M. D., and Reynafarje, B. (1975) Pathways and stoichiometry of H^+ and Ca^{++} transport coupled to mitochondrial electron transport, in *Electron Transfer Chains and Oxidative Phosphorylation* (E. Quagliariello, S. Papa, F. Palmieri, E. C. Slater and N. Siliprandi, eds.) North-Holland, Amsterdam, pp. 329–334.

Meijer, A. J., and Van Dam, K. (1974) The metabolic significance of transport in mitochondria, *Biochim. Biophys. Acta* **346**:213.

Mitchell, P. (1970) Reversible coupling between transport and chemical reactions, in *Membranes and Ion Transport* (E. E. Bittar, ed.), Vol. 1, Wiley–Interscience, New York, pp. 192–256.

Palmieri, F., Quagliariello, E., and Klingenberg, M. (1970) Quantitative correlation between the distribution of anions and the pH difference across the mitochondrial membrane, *Eur. J. Biochem.* **17**:230.

Pressman, B. C. (1970) Energy-linked transport in mitochondria, in *Membranes of Mitochondria and Chloroplasts* (E. Racker, ed.), Von Nostrand Reinhold, New York, pp. 213–250.

Stein, W. D. (1967) *The Movement of Molecules Across Cell Membranes*, Academic Press, New York.

Williamson, J. R. (1976) Mitochondrial metabolism and cell regulation, in *Mitochondria: Bioenergetics, Biogenesis and Membrane Structure* (L. Packer and A. Gomez-Puyou, eds.), Academic Press, New York, pp. 79–108.

Williamson, J. R., Anderson, J., and Browning, E. T. (1970) Inhibition of gluconeogenesis by butylmalonate in perfused rat liver, *J. Biol. Chem.* **245**:1717.

Zebe, E., Delbrück, A., and Bücher, T. (1959) Uber den Glycerin-1-P Cyclus im Flugmuskel von *Locusta migratoria*, *Biochem. Z.* **331**:254.

10

Mitochondrial Biogenesis

The *in vivo* mechanism of assembly of the mitochondrion poses formidable conceptual and experimental challenges to the investigator. The morphological complexity of the organelle coupled with the fact that the synthesis of the constituent proteins and lipids occurs in two spatially separate compartments of the cell indicates a highly ordered process whose most general features remain unknown.

Not so long ago, mitochondria were considered to arise from *de novo* synthesis. Such a mechanism would require an inordinately complicated sequence in which each intermediate carries the information for and determines the next step in the assembly. Although there are examples of self-assembling biological structures (enzymes, ribosomes, viruses), these are many orders of magnitude less complex than the mitochondrion. Most of the present evidence, however, is consistent with the idea that mitochondria increase in mass and number by a process of accretion and integration of newly synthesized material into preexisting organelles.

Given the temporal continuity of mitochondria, some of the outstanding questions that need to be considered are: (1) What is the degree of the genetic and biosynthetic autonomy of the organelle? (2) How do the nucleo-cytoplasmic and mitochondrial systems mutually influence each other and coordinate their respective synthetic activities? (3) What are the mechanisms by which cytoplasmically synthesized proteins and lipids traverse the membrane barriers and recognize their proper locales within the organelle?

I. Mitochondrial Continuity

Making use of a choline auxotroph of *Neurospora crassa*, Luck did a series of ingenious experiments designed to test whether mitochondria proliferate by a process of growth and division or by *de novo* synthesis. The studies hinged on the observed differences in the buoyant density of mitochondria of cells grown in the presence or absence of choline. Since the mutant requires choline for lecithin synthesis, mitochondria of cells deprived of choline are denser because of a deficiency of lipid. Luck reasoned that if newly synthesized phospholipids are incorporated into an already mature organelle, the addition of choline to starved cells should cause the entire mitochondrial population to become less dense. Assuming, however, that phospholipids made following the shift are apportioned only to organelles undergoing *de novo* synthesis, two physically distinct populations of mitochondria would be found: high-buoyant-density mitochondria formed prior to the shift and low-density mitochondria synthesized in the presence of choline. Analysis of the mitochondria by sucrose gradient centrifugation indicated a gradual and continuous decrease in the density of the entire population during the choline growth phase of the experiment (Fig. 10.1).

Although the results with the *Neurospora* mutant convincingly showed the ability of mitochondria to accrue phospholipids, it remained uncertain whether this also holds true of protein constituents. Since yeast are capable of physiological adaptation to different growth conditions, they provide an especially favorable experimental system for probing this question. *Saccharomyces cerevisiae* is a facultative anaerobe that can grow on glycolytic substrates in an oxygen-free atmosphere. Under such conditions, cells are deficient in respiration and oxidative phosphorylation and depend on glycolysis for their source of ATP.

Despite the absence of respiratory enzymes, anaerobic yeast develop morphological entities clearly identifiable as mitochondria. The functionally immature promitochondria nonetheless contain a broad range of mitochondrial enzymes (e.g., TCA cycle) and have a protein composition remarkably similar to that of normal mitochondria. When anaerobically grown yeast are supplied with oxygen, they rapidly adapt by synthesizing the complete array of electron transfer carriers and the coupling ATPase. The aerobic adaptation of yeast has been studied in a number of laboratories, the conclusion being that it involves a direct conversion of promitochondria to respiratory functional organelles. This is based on the following evidence. Labeling studies have shown that the proteins synthesized in response to oxygen are incorporated into the promitochondrial membranes of the anaerobic cells. In addition, the relatively low turnover rates of the bulk promitochondrial proteins and electron microscopic

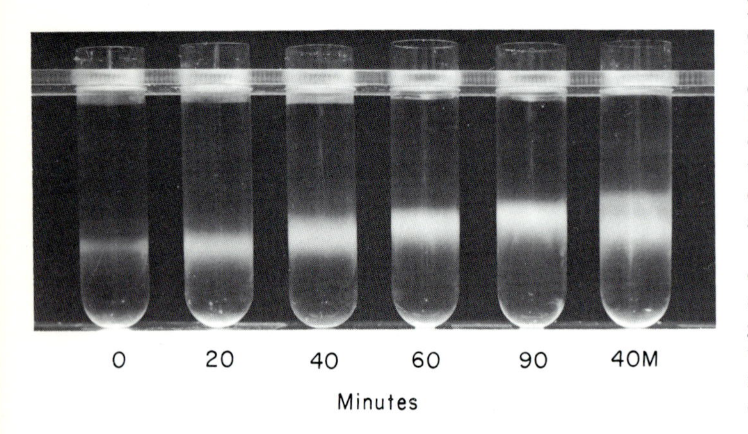

FIG. 10.1. Sucrose density banding of mitochondria of *N. crassa* following shift from low- to high-choline growth conditions. The choline auxotroph (chol-1) was pregrown on low choline (0 min) and transferred to high-choline medium. Cells were harvested after the indicated times of incubation, and mitochondria prepared. The mitochondria were subjected to isopycnic centrifugation in a 0.9–1.9 M gradient of sucrose. The 40 M sample consisted of mitochondria isolated from cells adapted for 40 min that were mixed with cells grown in high-choline medium for 15 hr. The two populations of mitochondria are clearly resolved on the gradient. (From Luck, 1965, courtesy of Dr. David Luck.)

evidence of a single population of mitochondrial particles in the fully adapted cells also argue that new catalytic proteins are added to the existent membranes.

The demonstrated physical continuity of mitochondrial lipids and proteins has provided the basis for a general concept of biogenesis the essential features of which include (1) a steady-state synthesis and turnover of all soluble and membrane-associated structural and catalytic constituents, (2) growth in mass through incorporation of new biosynthetic material into preexisting organelles, and (3) physiological control of mitochondrial number by simple division and fusion events. The ability of mitochondria to divide and fuse is well established from direct microscopic observations of live cells. The details of this process, however, are still meagerly understood. Electron microscopic studies have revealed the presence of membrane partitions in mitochondria undergoing rapid division. This situation has been described in mice fed on riboflavin-deficient diets. Liver cells of such animals acquire relatively few giant mitochondria. Return to a riboflavin-supplemented diet leads to a restoration of the normal number of mitochondria. During this transition, mitochondria presumed to be in the process of dividing have been observed to develop frequent cross membranes.

II. Components of the Mitochondrial Translation Machinery

A good deal of progress has been made in defining the extent to which mitochondria contribute towards the synthesis of their component proteins and lipids. The ability of isolated mitochondria to incorporate amino acids into a select class of proteins has been known for some time. That this activity is not the result of cytoplasmic ribosomes or bacterial contamination has been rigorously ruled out, and it is now well established that mitochondria contain ribosomes whose properties are quite different from those of their cytoplasmic counterparts. In addition to ribosomes, mitochondria contain a distinct set of transfer RNAs and the appropriate synthetases needed for the acylation reactions. Studies on the characterization of these components were crucial in demonstrating that mitochondria have a true protein synthetic machinery capable of producing finished protein products.

A. Mitochondrial Ribosomes

A characteristic of mitochondrial ribosomes is their sensitivity to antibiotics that inhibit bacterial protein synthesis. Antibacterial drugs such as chloramphenicol and erythromycin are potent inhibitors of mitochondrial ribosomes but have no effect on cytoplasmic ribosomes. The converse is true of other drugs (e.g., cycloheximide) that act specifically on cytoplasmic ribosomes. Mitochondrial ribosomes resemble bacterial ribosomes but are quite different from eucaryotic cytoplasmic ribosomes in their physical properties. The differences between the two classes of ribosomes in lower and higher eucaryotes are summarized in Table 10.1. As a rule, mitochondrial ribosomes exhibit lower sedimentation coefficients. In yeast and *N. crassa*, sedimentation values of 73–74 S have been measured; these are quite similar to the 70 S ribosomes of bacteria. Ribosomes of mammalian mitochondria have even lower sedimentation values (50–60 S). The unusual sedimentation properties reflect truly smaller ribosomes. Electron microscopic studies of isolated ribosomes from mammalian mitochondria indicate them to have smaller dimensions. This is also true of ribosomes seen in cross-sections of mitochondria (Fig. 10.2).

In all known cases, the mitochondrial ribosomes are encoded in mtDNA. The sequences of the yeast and several different mammalian ribosomal RNAs have been determined from their gene sequences. Both the fungal and mammalian RNAs are sufficiently homologous to procaryotic and eucaryotic cytoplasmic rRNAs to suggest common secondary and tertiary structural features. The 15S rRNA of yeast mitochondria has a 3′ terminal sequence compatible with a secondary structure very similar to that recently proposed by Noller and Woese for the 16S rRNA of *E. coli* (Fig. 10.3). The two rRNAs, however, have different sequences in their 3′ tails, which in the case of *E. coli* has been implicated to play an important role in the interaction of the 30S ribosomal subunit with messenger RNAs. The sequence 5′-CCUCCA-3′ at the very 3′ end of the 16S rRNA has been postulated by Shine and Dalgarno to base pair with complementary sequences occurring in the 5′ leader regions of procaryotic messengers. Although the 15S rRNA of yeast lacks the Shine and Dalgarno hexanucleotide binding sequence, it has in its stead another sequence capable of base pairing with certain sequences upstream of the initiation codons in yeast mitochondrial messengers. The putative mRNA binding site of the yeast 15S rRNA is 10 nucleotides long and has the sequence 5′-AAAUUCUAUA-3′. The extent to which this sequence can base pair with sequences present in the leaders of six different yeast mitochondrial messengers is shown in Fig. 10.4.

While the above data are consistent with a common mechanism of ribosome attachment to mRNA in *E. coli* and in *S. cerevisiae* mitochondria, this does not appear to be true of mammalian mitochondria. Evidence to be discussed in the next chapter indicate that the mRNAs of mammalian mitochondria do not have 5′ leaders. Sequence analysis of the human messengers for subunits of cytochrome oxidase has revealed that they start with the initiation codons. The absence of 5′ leaders in these mRNAs argues for some other mechanism of ribosome binding. In this context, it is significant that the 3′ tail

TABLE 10.1. Sedimentation Properties of Mitochondrial Ribosomes

		Subunits		rRNAs	
	Ribosomes	Large	Small	Large	Small
Human	60[a]	45	35	16	12
Rat	55	39	28	16	13
Plant	77	60	44	24	18
N. crassa	73	50	37	25	19
S. cerevisiae	74	50	37	21	15
Eucaryotic (cytoplasmic)	80	60	40	28	18
Bacterial (*E. coli*)	70	50	30	23	16

[a]All the values reported are in Svedberg units.

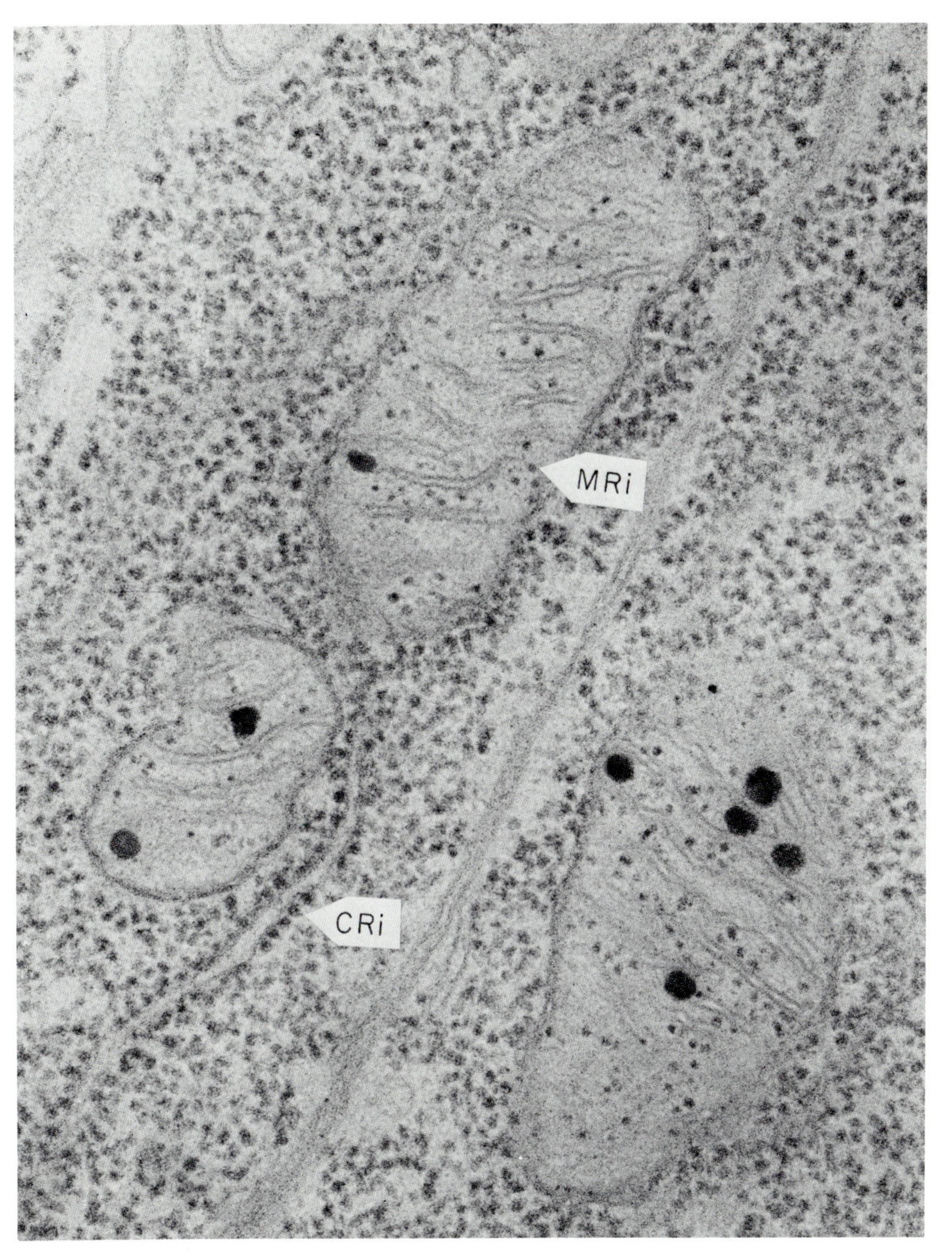

FIG. 10.2. Electron micrograph of intestinal mucosa epithelium of mouse. The cross-sections of mitochondria reveal the presence of ribosomes (MRi) substantially smaller than either free cytoplasmic ribosomes or ribosomes bound to endoplasmic reticulum (CRi). Magnification, × 80,000. (Courtesy of Dr. G. Palade.)

sequence found in *E. coli* 16S and yeast 15S rRNAs is absent in all the mammalian small subunit ribosomal RNAs.

Although the rRNAs of plant and animal mitochondria are transcribed from mtDNA, the ribosomal protein subunits are almost all products of nuclear genes. At present there are only two known instances of ribosomal proteins whose genes are located in mtDNA. The small ribosomal subunits of yeast and *N. crassa* mitochondrial ribosomes have been claimed to have one protein encoded in mtDNA. Mitochondrial ribosomes like some of the inner membrane

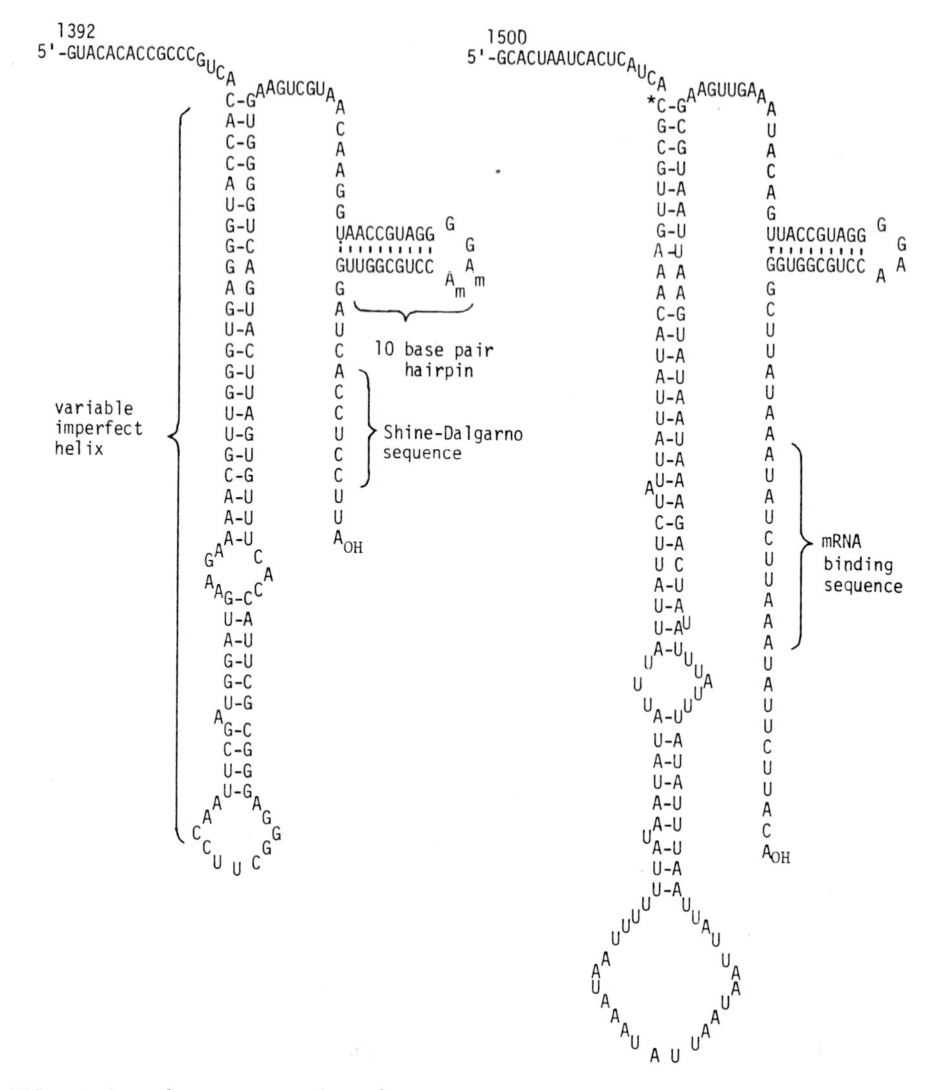

FIG. 10.3. Secondary structures of *E. coli* 16S and yeast mitochondrial 15S rRNAs in the 3′ regions. The structure of the 16S rRNA (left) is based on the model of Noller and Woese (1981). It shows two helical regions and the 3′ tail with the Shine and Dalgarno sequence. The yeast mitochondrial 15S rRNA (right) sequence has been formed to achieve maximal base pairing. Except for a longer helix and 3′ tail, it exhibits the same secondary structural features as the 16S rRNA of *E. coli*. The putative mRNA binding sequence of the yeast 15S rRNA is indicated.

respiratory complexes are hybrid structures whose biosynthesis depends on the coordinate expression of nuclear and mitochondrial genes.

B. Transfer RNAs

Mitochondrial and cytoplasmic tRNAs have different chromatographic properties and are in most instances charged only by the homologous aminoacyl synthetases. The mitochondrial tRNA species differ depending on the organism. Mitochondria of *Tetrahymena* have been claimed to contain a partial set of tRNAs, the missing species being imported from the cytoplasm. Fungal and mammalian mitochondria contain the tRNAs for all twenty amino acids and for some there are two or more isoaccepting species present. A list of the mitochondrial tRNAs of *S. cerevisiae* is presented in Table 10.2. The number of isoacceptors for some amino acids is consistent with the Wobble hypothesis. Other amino acids, however, have fewer tRNAs than are required to recognize the standard codons of the universal code. As will be discussed in Chapter 11, the genetic code of mitochondria is read by 23 tRNAs (excluding the initiator tRNA). This agrees with the recent findings of only 24 tRNA genes in yeast and human mitochondrial DNA.

The tRNAs of yeast, *N. crassa*, and human mitochondria have been either directly sequenced or their structures have been deduced from the nucleotide sequences of the genes. In almost all instances, the mitochondrial tRNAs are

TABLE 10.2. Yeast Mitochondrial tRNA Species Detected by Reverse-Phase Chromatography

Amino acid	tRNA species observed[a]	Codons	Minimum number of tRNAs required by Wobble hypothesis
Ala	2	4	2
Arg	2	6	3
Asp	1	2	1
Cys	2	2	1
Glu	1	2	1
Gln	1	2	1
Gly	1	4	2
His	3	2	1
Ile	2	3	1
Leu	1	6	3
Lys	2	2	1
Met (+fMet)	2	1	2
Phe	2	2	1
Pro	1	4	2
Ser	3	6	3
Thr	2	4	2
Trp	3	1	1
Tyr	4	2	1
Val	2	4	2
Asn	Not tested	2	1
Total	37+	61	32

[a]Unpublished data of G. Macino

characterized by an unusually low G + C content (18–35%). Although yeast and *N. crassa* mitochondrial tRNAs exhibit most of the invariant features of other procaryotic and eucaryotic tRNAs, some have rather anomalous structures. The mitochondrial initiator tRNA of *N. crassa* and the aspartic acid tRNA of yeast both lack some of the invariant bases in the TψC and D loops that are known to be important in the tertiary conformation (Fig. 10.5). The structures of mammalian tRNAs exhibit even more remarkable differences. The tRNAs of human mitochondria have either excessively large or small D and TψC loops (in one case, the D loop is absent).

In addition to the unusual structures of their tRNAs, mitochondria also show interesting deviations from the rules of the genetic code. This has necessitated some further adaptations in the anticodons of the cognate tRNAs (Chapter 11, Section VIID). A case in point is the UGA terminator which, contrary to all other known systems, in mitochondria functions as a codon for tryptophan. Accordingly, the tryptophan tRNA of mitochondria has a 3′-ACU-5′ instead of the usual 3′-ACC-5′ anticodon. The U at the Wobble position of the anticodon permits the tRNA to read both UGG and UGA as tryptophan. Normally, the Wobble position of the tryptophan anticodon is occupied by a C, thus restricting the recognition to the UGG codon.

Yeast have been shown to incorporate radioactive formate specifically into the mitochondrially translated proteins. A number of mitochondrial translation

```
3'-A C A T T C T T A T A A A T T C T A T A A A T A T T C G -

                 3'-  A A A T T C T A T A
sb2 cyt.ox.               | | | | | | |
                 5'-  G T T A A G A T T T -6

                 3'-  A A A T T C T A T A
sb3 cyt.ox.           | | | | | | |          -73
                 5'-  T T T G A G G A T A

                 3'-  A A A T T C T A T A
sb1 cyt.ox.           | | | | | | | | | |    -107
                 5'-  T T T A A G A T A T

                 3'-  A A A T T C T A T A
sb6 ATPase                | | | | | | | |    -51
                 5'-  A A T A A G A T A T

                 3'-  A A A T T C T A T A
sb9 ATPase                | | | | |          -29
                 5'-  A A T A A G A A T A

                 3'-  A A A T T C T A T A
apocyt. b                 | | | |            -100
                 5'-  A A T A A G T A T T
```

FIG. 10.4. Complementarity of the putative mRNA binding sequence in the yeast 15S rRNA and sequences upstream of the initiation codons in yeast mitochondrial messengers for subunits 1, 2, and 3 of cytochrome oxidase, subunits 6 and 9 of the ATPase, and apocytochrome *b*. The number of nucleotides intervening between the initiation codons and the proposed binding sites are indicated by the numbers.

products of yeast and *N. crassa* have, in fact, been shown to have formylmethionine at the amino terminus. Because mitochondria lack the deformylase, the formyl group remains attached on the methionine.

C. Aminoacyl Synthetases

Mitochondria contain aminoacyl synthetases capable of acylating mitochondrial but not cytoplasmic tRNAs. Similarly, cytoplasmic synthetases show a high degree of specificity towards cytoplasmic tRNAs. Although the mito-

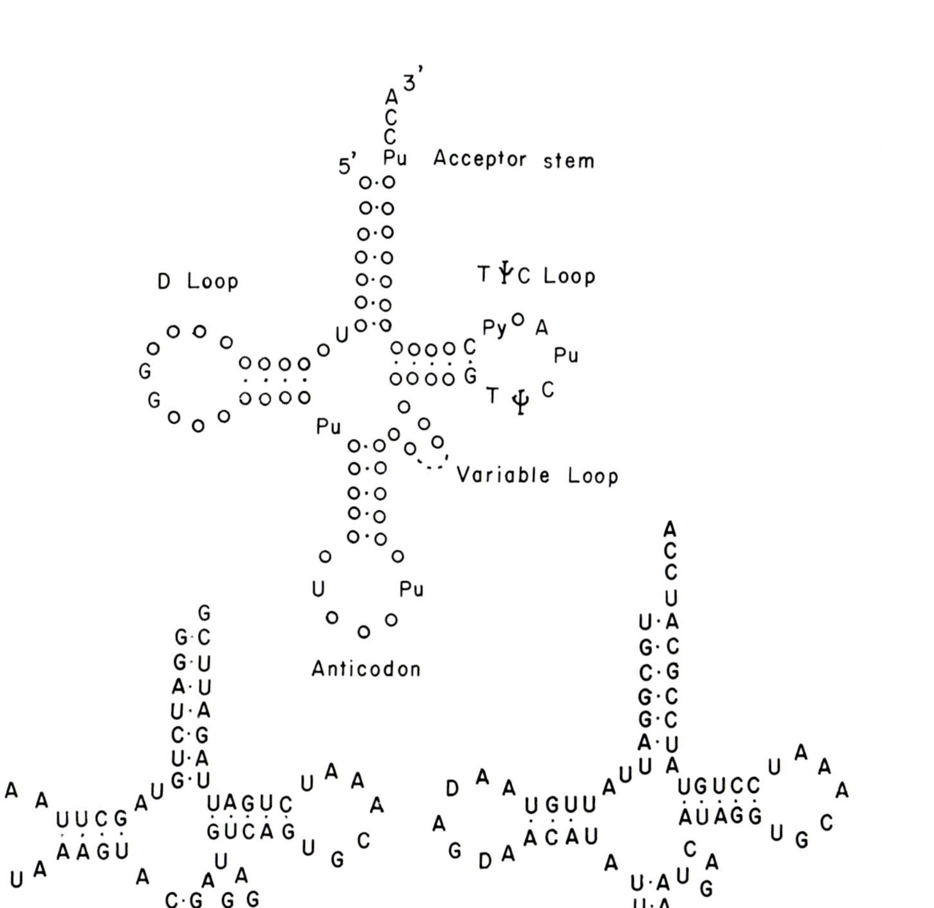

FIG. 10.5. Structures of mitochondrial tRNAs. Upper: Generalized tRNA showing the positions of the invariant bases. Lower left: Yeast mitochondrial aspartic acid tRNA. Since the secondary structure was deduced from the gene sequence, the positions of the modified bases are not known. Lower right: *N. crassa* mitochondrial initiator tRNA. (From Heckman *et al.*, 1978.)

chondrial enzymes have not been purified, they are soluble matrix proteins with cofactor requirements similar to those of the bacterial and eucaryotic cytoplasmic synthetases.

III. Mitochondrial Protein Synthesis

A. *In Vitro*

The earliest evidence that mitochondria have their own system of protein synthesis came from studies with the isolated organelle. Much of the work in this area was concerned with describing the system and demonstrating that it was a true mitochondrial activity. The properties of the protein synthetic activity of rat liver mitochondria are shown in Table 10.3. The incorporation of radioactive amino acids into acid-insoluble material is characterized by the usual requirements of other protein synthetic systems. Mitochondrial protein synthesis, however, is distinguished from cytoribosomal protein synthesis by its sensitivity to chloramphenicol and resistance to ribonuclease. Even though the activity observed with isolated mitochondria is low compared to that in whole cells, the products correspond to those made *in vivo*.

B. *In Vivo*

In eucaryotes, the vast majority of cellular proteins are made on cytoplasmic ribosomes. To study the few proteins synthesized by mitochondria, it is necessary to block the more active cytoplasmic system with a suitable inhibitor. Cycloheximide is especially useful for this purpose, since it achieves a virtually complete inhibition of cytoribosomal with little or no effect on mitochondrial translation. When intact yeast cells are incubated in the presence of cycloheximide, mitochondrial protein synthesis proceeds for 1–2 hr. The rate and net extent of incorporation of radioactive amino acids into mitochondrial

TABLE 10.3. *In Vitro* Incorporation of [^{14}C]Leucine into Rat Liver Mitochondria

Additions or omissions	Percent of complete system
Complete[a]	100
− Amino acid mix	23
− ATP	25
− KCl	7
+ Cycloheximide	100
+ Puromycin	5
+ Chloramphenicol	15

[a]Complete system consisted of 50 mM Tris, pH 6.5, 0.154 M KCl, 10 mM $MgCl_2$, ATP-generating system, 2 mM EDTA, complete mixture of amino acids, plus [^{14}C]leucine and 2–3 mg mitochondrial protein. Maximal incorporation under these conditions is approximately 4–6 pmoles of leucine per mg mitochondrial protein. (From Beattie *et al.*, 1967.)

proteins in the presence of the inhibitor can be substantially increased if the cells are first allowed to incubate in a medium containing chloramphenicol (Fig. 10.6). The stimulatory effect of a preincubation in chloramphenicol has also been observed in other organisms and suggests some regulation of the transcription or translation of mitochondrial genes by cytoplasmically synthesized proteins.

C. Number of Mitochondrial Products

The quantitative aspects of mitochondrial protein synthesis have been studied in yeast, *N. crassa*, and to some extent in animal cells. The number of proteins made by yeast mitochondria and their relative contribution to the overall mass of the organelle have been assessed in several ways. One of these has been to compare the protein compositions of mitochondria from wild type and from respiratory deficient ρ^- mutants of yeast. Since ρ^- mutants are also deficient in mitochondrial protein synthesis, proteins present in the wild type but not in the mutant cells correspond to mitochondrial translation products. In the experiment of Fig. 10.7, cells were grown in identical media except that different radioisotopes were used to label the wild type and the ρ^- mutant. All the proteins of the wild type strain were labeled with [^{14}C]leucine, and those of the mutant with [^{3}H]leucine. The two cultures were mixed, mitochondria were isolated, and the labeled proteins separated by electrophoresis on polyacrylamide gels after dissociation of the mitochondria with sodium dodecyl sulfate. An analysis of the two radioisotopes along the length of the gel shows the difference in the proteins synthesized by the two strains. When the results are plotted as the ratio of ^{14}C/^{3}H, the radioactive peaks correspond to those components present only in the wild type mitochondria and therefore derived from mitochondrial protein synthesis. An alternative means of identifying mitochondrially synthesized proteins is to label wild type cells in the presence of the cytoribosomal inhibitor cycloheximide. Under these conditions, only mitochondrially translated proteins acquire the radioactive label. Results of such exper-

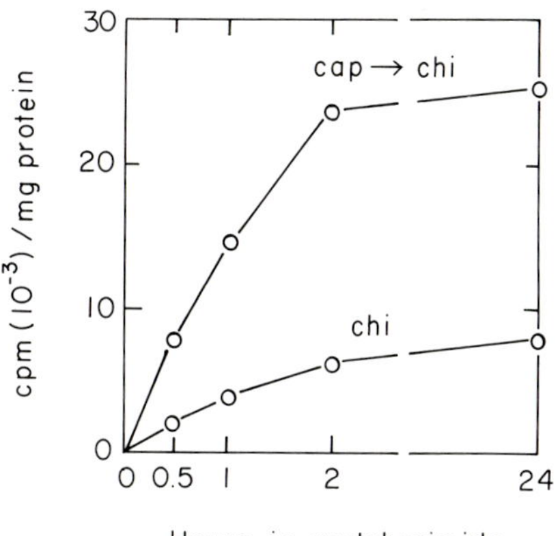

FIG. 10.6. Yeast grown under conditions of glucose repression were transferred to low-glucose medium (derepression) containing cycloheximide (chi) and [^{3}H]leucine. One-half of the glucose-repressed culture was first incubated in the presence of chloramphenicol prior to transfer to the cycloheximide medium (cap → chi). Mitochondria were isolated from each culture, and the radioactivity incorporated into the mitochondrial translation products was measured. The net incorporation is seen to be enhanced when cells are preincubated in chloramphenicol.

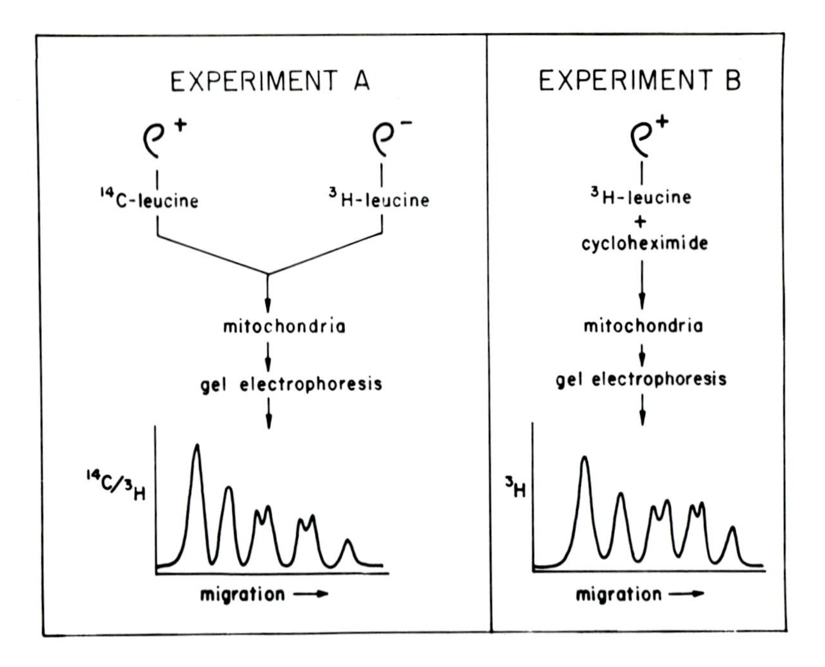

FIG. 10.7. *In vivo* labeling of mitochondrial products of yeast. In experiment A, the mitochondrially synthesized proteins are displayed as the ratio of $^{14}C/^{3}H$ counts incorporated into mitochondria of wild type (ρ^+) and mutant (ρ^-) yeast. In experiment B, specific labeling of mitochondrial products is achieved by including cycloheximide to inhibit cytoribosomal protein synthesis.

iments indicate that mitochondria synthesize eight to ten different proteins with molecular weights ranging from 45,000–8,000.*

Since mitochondria contain in excess of several hundred different proteins, it is evident that most of the organelle is derived from cytoribosomal protein synthesis. In yeast, the mitochondrially synthesized proteins have been estimated to represent only 5–10% of the total mitochondrial protein. Similar conclusions have been reached for *N. crassa* mitochondria. In higher plants and animals, the contribution by mitochondria is also nominal. As discussed below, however, not all mitochondria synthesize the same set of proteins.

D. Identification of the Mitochondrial Translation Products

The early studies with isolated mitochondria indicated that the endogenously synthesized proteins are associated exclusively with the mitochondrial inner membrane. This was also shown to be true of mitochondrial products made *in vivo* when cytoribosomal synthesis is blocked with cycloheximide. These observations were helpful in excluding the participation of mitochondria in the synthesis of a large number of enzymes known to be located in the outer membrane and the intermembrane and matrix compartments.

*These molecular weights are based on SDS gel electrophoresis and in some cases represent underestimations.

The problem of identifying the inner membrane constituents synthesized by mitochondria was first studied in yeast and *N. crassa*. It should be recalled that the four respiratory complexes and the oligomycin-sensitive ATPase constitute the five major enzyme complexes of the inner membrane. These complexes were therefore the most likely candidates to contain mitochondrially derived subunit polypeptides. Furthermore, the involvement of mitochondria in the biosynthesis of this particular set of enzymes was hinted at by other kinds of evidence. For example, it was known that cytoplasmic ρ^- mutants of yeast defective in mitochondrial protein synthesis are pleiotropically deficient in certain respiratory carriers such as cytochromes a, a_3, b, and in oligomycin-sensitive ATPase. A similar phenotype is observed in yeast grown in the presence of chloramphenicol or erythromycin, specific blocking agents of mitochondrial protein synthesis.

A more precise idea of the enzymatic lesions induced by these antibiotics came from an examination of their effects on the levels of individual enzymes. The results of such an experiment are shown in Fig. 10.8. A wild type culture of yeast pregrown on high glucose is transferred to fresh medium containing low glucose supplemented with either chloramphenicol or cycloheximide. The initial levels of respiratory and ATPase activities are low because of the high glucose in the initial growth medium which represses the synthesis of the enzymes. On transfer of the glucose-repressed cells to a low-glucose medium, the level of each enzyme increases over a period of 7 hr until the cells are fully derepressed. The increase in respiratory activity is the result of an adaptation of the yeast to aerobic metabolism of ethanol produced from the glycolytic

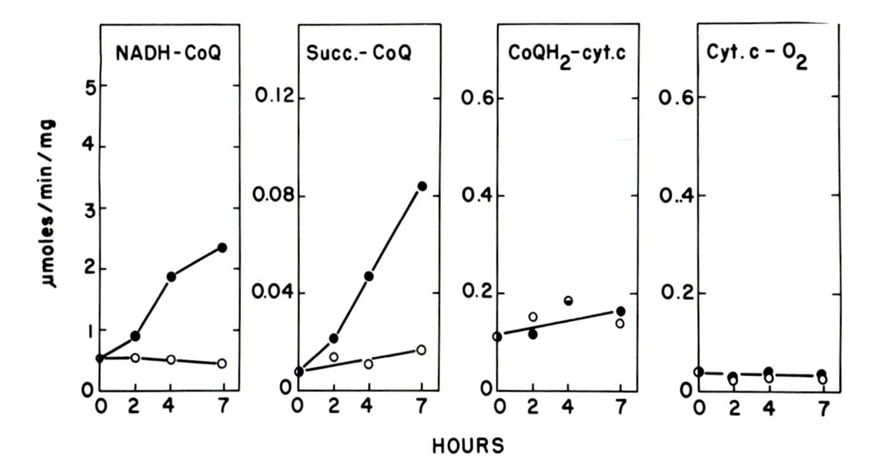

FIG. 10.8. Effect of chloramphenicol and cycloheximide on the development of respiratory activities. Yeast were grown on high glucose to induce the glucose-repressed state (low levels of respiratory complexes). The cells were transferred to low-glucose medium containing either chloramphenicol (●—●) or cycloheximide (O—O). The zero hour refers to the time of transfer from high to low glucose. Cells were collected at the indicated times, and mitochondria were prepared and assayed for each of the four respiratory complexes. The results of this experiment show that chloramphenicol only inhibits the synthesis of coenzyme QH_2-cytochrome c reductase, cytochrome oxidase, and oligomycin-sensitive ATPase (not shown) during the derepression phase of the experiment.

breakdown of glucose. When cycloheximide is added to the derepression medium, there is no increase in the activity of any of the enzymes, indicating that their biosynthesis is dependent on cytoplasmic protein synthesis. The mitochondrial inhibitor chloramphenicol, however, is seen to block only the biosynthesis of cytochrome oxidase, coenzyme QH_2-cytochrome *c* reductase, and oligomycin-sensitive ATPase. Neither NADH- or succinate-coenzyme Q reductases are affected by chloramphenicol. Such results demonstrate that the synthesis of the two dehydrogenase complexes occurs entirely on cytoplasmic ribosomes, whereas the other three complexes are derived jointly from the cytoribosomal and mitochondrial systems of protein synthesis.

The identification of the mitochondrially synthesized polypeptides of the ATPase, cytochrome oxidase, and coenzyme QH-cytochrome *c* reductase was facilitated by several technical developments. First, methods were devised for purifying the inner membrane complexes from yeast, *N. crassa*, and mammalian mitochondria. With the pure enzymes at hand, it was possible to characterize their subunit polypeptides. Second, sensitive immunochemical methods became available for the isolation of the enzymes from small samples of biological material.

The strategy used to examine the biosynthetic origin of the subunit polypeptides of the inner membrane complexes is diagrammatically illustrated in Fig. 10.9. This basic protocol was first used with yeast and *N. crassa* but was later found to be equally applicable to studies of mammalian cells in tissue cultures. In a typical experiment, cells are separately labeled under three different conditions: (1) with [^{14}C] amino acid in the absence of inhibitors, (2) with [^{3}H]amino acid plus chloramphenicol, and (3) with [^{3}H]amino acid plus cycloheximide. At the end of the labeling period, the control cells that had not seen any inhibitor are mixed with each of the other two cultures, and mitochondria are isolated. The enzyme of interest (cytochrome oxidase in the example shown in Fig. 10.9) is purified either by a standard purification method or by immunoprecipitation. The latter procedure is usually more convenient when working with small amounts of material. For the immunoprecipitation, the mitochondria are extracted with a detergent that solubilizes but does not dissociate the enzyme complex. When such extracts are treated with antisera to the holoenzyme, antibody-antigen complexes are formed that are readily sedimented from the detergent solution. The high degree of specificity of the antiserum permits a selective precipitation of the enzyme from crude extracts containing a large number of different proteins.

Immunoprecipitates obtained in this way are further depolymerized in sodium dodecyl sulfate, and the dissociated subunits are analyzed by electrophoresis on polyacrylamide gels. The subunits synthesized by mitochondria are labeled only in cells incubated in the medium containing cycloheximide. Conversely, cytoplasmically derived subunits are labeled in the presence of chloramphenicol. Since the immunoprecipitates are obtained from a mixture that includes mitochondria labeled without added inhibitors, the normal patterns of proteins present in the holoenzyme are seen on the gels and act as convenient reference markers for the identification of the mitochondrially and cytoplasmically synthesized subunits.

Labeling studies of whole yeast have made it possible to catalogue all the subunits of cytochrome oxidase, coenzyme QH_2-cytochrome *c* reductase, and the ATPase and to assign each as a product of either mitochondrial or cytoplasmic translation. Some of this information is summarized in Table 10.4. Of the seven cytochrome oxidase subunits, three have been determined to be synthesized in mitochondria (subunits 1–3) and four on cytoplasmic ribosomes (subunits 4–7). In the case of coenzyme QH_2–cytochrome *c* reductase, only one constituent polypeptide is made in mitochondria. This protein has been iden-

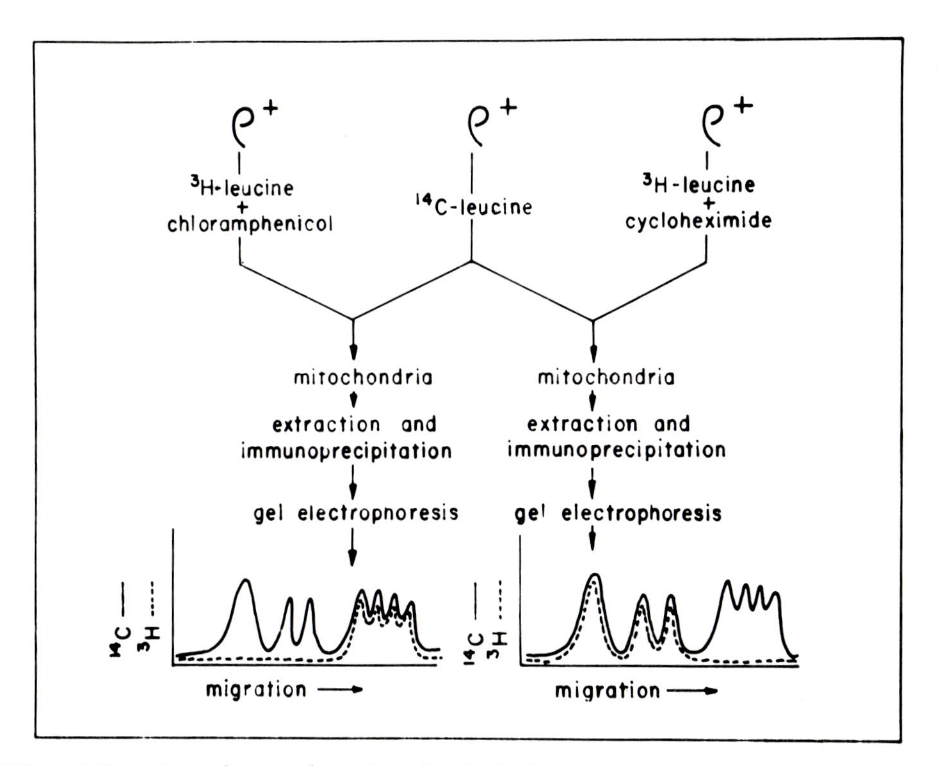

FIG. 10.9. Experimental protocols used to identify the biosynthetic origin of subunit proteins of cytochrome oxidase. The solid trace represents the total cytochrome oxidase proteins labeled with [^{14}C]leucine in wild-type (ρ^+) yeast. The dashed trace represents either mitochondrial (right) or cytoribosomal (left) products of the enzyme synthesized in the presence of the two different inhibitors.

TABLE 10.4. Products of Mitochondrial Protein Synthesis in *S. cerevisiae*

	ATPase	Cyt. oxidase	$CoQH_2$–cyt. *c* reductase
Total number of subunits	10	7	7–8
Mitochondrial products	4	3	1
Identity	Membrane factor	Cyt. *a*, a_3, Cu protein	Cytochrome *b*
Cytoplasmic products	6	4	6–7
Identity	F_1, OSCP	?	Cytochrome c_1

tified as cytochrome *b*. The remaining six or seven proteins of the complex are imported from the cytoplasm. The oligomycin-sensitive ATPase is also composed of two biosynthetically different groups of proteins. The subunits of the F_1 ATPase and OSCP are synthesized in the presence of chloramphenicol and are therefore products of cytoribosomal translation. These components of the complex, however, are not incorporated into the inner membrane when the mitochondrial system is inhibited, suggesting that the biosynthesis of the oligomycin-sensitive ATPase complex requires additional translation products of mitochondria. In fact, it has been shown that the proteins associated with the membrane factor sector of the complex are synthesized in mitochondria. Since these proteins probably function as membrane anchoring sites for F_1 and OSCP, their synthesis is necessary for proper assembly of the complex (see below).

In lower eucaryotes, some eight or nine proteins have been determined to be products of mitochondria. Although there have been claims that other minor protein species are also made in mitochondria, this has not been confirmed in other laboratories and can probably be attributed to incomplete inhibition of cytoplasmic protein synthesis by cycloheximide or other cytoribosomal inhibitors used in the experiments. With one exception, all the major products have been identified with subunits of the three complexes discussed above. The one exception is a yeast mitochondrial protein named var 1. This interesting mitochondrial product has a molecular weight of 41,000–43,000 depending on the strain of yeast. The function of var 1 is not known, but there is reasonably good evidence that it is a component of the small ribosomal subunit. Mitochondrial ribosomes of *N. crassa* have also been claimed to have one protein encoded in mitochondrial DNA and synthesized in mitochondria. Whether this protein is functionally related to var 1 needs to be verified. In these two organisms, therefore, most of the mitochondrial products are subunits of enzymes that function in respiration and oxidative phosphorylation. As will be seen in Chapter 11, the mitochondrially translated subunits of the respiratory complexes and var 1 are also gene products of mitochondrial DNA.

In *N. crassa*, the situation is quite similar but not identical with yeast. The subunit compositions and the synthetic derivations of the *N. crassa* cytochrome oxidase and coenzyme QH_2–cytochrome *c* reductase are the same as in yeast. The *N. crassa* ATPase, however, has been found to have only two proteins of mitochondrial origin. The most striking evidence for a divergence between the two organisms concerns the ATPase proteolipid (subunit 9) which has been conclusively shown in *N. crassa* to be encoded in nuclear DNA and to be synthesized on cytoplasmic ribosomes. The difference in the site of synthesis of a functionally and structurally related protein in the two organisms represents a dramatic evolutionary change. From *in vivo* labeling studies, human mitochondria appear to synthesize the same subunits of the respiratory complexes as do yeast and *N. crassa*. In addition, however, human and probably all mammalian mitochondria have five mitochondrial products not seen in lower eucaryotes. These proteins have been shown to be encoded in mitochondrial DNA, and their messages have also been detected. Although the function of the five proteins is not known, these recent findings point to a considerable diversity in the coding and biosynthetic properties of mitochondria (Table 10.5).

E. Properties of Mitochondrially Synthesized Proteins

Mitochondrial products have proven difficult to study principally because of their extreme hydrophobicity. It has been speculated that the hydrophobic character of mitochondrial products necessitated an intraorganellar site of synthesis at or near the inner membrane, thus avoiding their transport across an aqueous cytoplasmic phase. The attractiveness of this interpretation has been lessened somewhat by the already mentioned discovery that subunits 9 of yeast and *N. crassa* ATPase, despite their homologous primary structures (see Fig. 7.8), are synthesized in two different cellular compartments. The transport of the *N. crassa* proteolipid from the cytoplasm to mitochondria appears to have been successfully solved despite the hydrophobic properties of the protein.

Mitochondrial products were at one time thought to be structural proteins. This view had to be abandoned when cytochrome *b* and the heme protein of cytochrome oxidase (subunit 1) were shown to be translated on mitochondrial ribosomes. The functions of the other six to eight mitochondrial products are still obscure but will probably be better understood in the near future as more is learned about the catalytic mechanisms of the inner membrane complexes.

IV. Phospholipid Biosynthesis

Even though most of the discussion so far has centered around the protein components of mitochondria, it should be obvious that phospholipids play an equally important role in membrane assembly. Unfortunately, this aspect of biogenesis has not received the attention it deserves, and little is known of the factors controlling the coordinate synthesis of mitochondrial lipids and proteins.

Mitochondria have a limited capacity for lipid synthesis. Rat liver mitochondria are known to synthesize cardiolipin, a phospholipid unique to the organelle. The biosynthesis of cardiolipin involves the following reactions:

$$\text{phosphatidic acid} + \text{CTP} \rightarrow \text{CDP-diglyceride} \quad [10.1]$$

$$\text{CDP-diglyceride} + \text{glycerol-3-P} \rightarrow \text{phosphatidylglycerol} \quad [10.2]$$

$$\text{phosphatidylglycerol} + \text{CDP-diglyceride} \rightarrow \text{cardiolipin} \quad [10.3]$$

TABLE 10.5. Number of Mitochondrial Products in Yeast, *N. crassa*, and Mammalian Mitochondria

	Cyt. oxidase	$CoQH_2$–cyt. *c* reductase	ATPase	Ribosomes	Other
S. cerevisiae	3	1	4	1	—
N. crassa	3	1	2	1	—
Mammalian	3	1	1–2	?	4–5

The last reaction is catalyzed by a transferase found exclusively in mitochondria. The phosphatidic acid used in reaction 10.1 is probably formed internally. Most mitochondria contain an acyl transferase able to convert glycerol phosphate to lysophosphatidic acid. The lysophosphatidic acid is further acylated by a second enzyme to form phosphatidic acid. The mitochondrial enzymes are located in the outer membrane and are distinct from other acyl transferases present in the endoplasmic reticulum.

The other major mitochondrial phospholipids include phosphatidylserine, phosphatidylethanolanine, phosphatidylcholine, and phosphatidylinositol. These constituents appear to be drawn from a common pool of cellular phospholipids synthesized on endoplasmic reticulum. Their transfer to mitochondria is believed to be mediated by special carrier proteins capable of exchanging phospholipids between different membrane systems. Some four different carrier proteins have been isolated, each specific for a particular phospholipid (Table 10.6). The exchange of phosphatidylcholine between microsomes and mitochondria is shown in Fig. 10.10. In this experiment, microsomes containing radioactively labeled phospholipids were mixed with unlabeled mitochondria in the presence of a phosphatidylcholine exchange carrier. The carrier protein in the assay promotes an exchange of the radioactive phosphatidylcholine in

TABLE 10.6. Properties of Phospholipid Exchange Proteins

Source	Isoelectric point	Transfer of phospholipid[a]			
		PC	PI	PE	SPM
Bovine liver	5.8	++++	−	−	−
Bovine heart	4.7	++++	n.d.	−	+
Bovine brain	5.2	+	++++	−	−

[a]The specificity of the purified proteins was tested by measuring the exchange of the individual phospolipids between microsomes and mitochondria or liposomes and a natural membrane. PC, phosphatidylcholine; PI, phosphatidylinositol; PE, phosphatidylethanolamine; SPM, sphingomyelin. (From Wirtz and Van Deenen, 1977.)

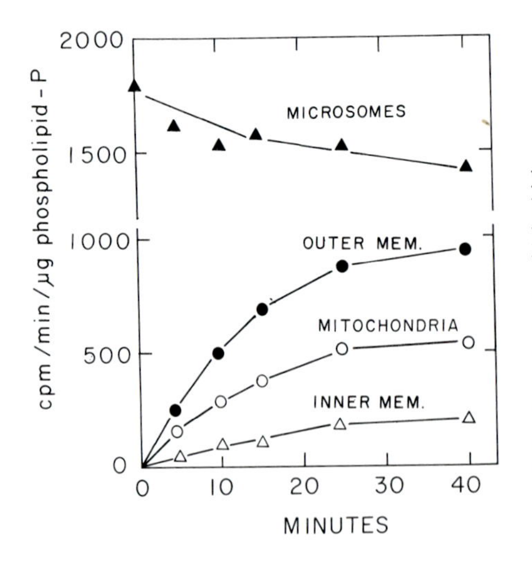

FIG. 10.10. Exchange of phosphatidylcholine between microsomes and mitochondria. [^{32}P]-labeled microsomes were incubated with unlabeled mitochondria in the presence of a rat-liver pH 5.1 supernatant fraction (crude exchange protein). Following incubation, the mitochondria and microsomes were separated, and the mitochondria fractionated into outer and inner membrane. The incorporation of microsomal phospholipids, measured as ^{32}P, was determined in the microsomes, mitochondria, and the two mitochondrial membrane fractions. (From Blok *et al.*, 1969.)

microsomes for the unlabeled phospholipid of mitochondria. It is significant that the specific activity is higher in the outer membrane. Although not shown in Fig. 10.10, the phosphatidylcholine eventually equilibrates between the two mitochondrial membranes. The transfer of lipids from the outer to the inner membrane could involve the participation of carriers endogenous to mitochondria or occur by direct transfer at points of contact between the two membranes.

The ubiquity of the lipid carriers suggests that they play an important role in intracellular partitioning of phospholipids. Whether their function is confined to preserving a steady-state balance of appropriate phospholipids among the various membrane systems or whether they are also important in active membrane growth remains to be clarified. In this context, it may be significant that in addition to promoting the exchange of phospholipids, certain carriers have been found to effect a net transfer of phospholipids. Such net transfers occur when the acceptor membrane is deficient in a given phospholipid and may be especially relevant to membrane biogenesis which requires accumulation of new lipid constituents as opposed to their mere exchange.

V. Nuclear Gene Products

The importance of nuclear genes in the biogenesis of mitochondria can be appreciated by the fact that ρ^- mutants of *S. cerevisiae* develop mitochondrial organelles lacking only those few enzymes whose synthesis is in part dependent on the endogenous system of protein translation. The biogenesis of respiratory competent mitochondria can therefore be conceived of as the product of two interdependent processes—the formation of a proorganelle by the nucleocytoplasmic system and its further modification by the mitochondrion itself. A great deal of emphasis in the past has been placed on the contribution made by mitochondria. This is understandable in view of the uniqueness of mitochondria, being one of only two organelles to participate in their own morphogenesis. The small number of mitochondrial genes and translation products has also been an important factor in attracting the attention of investigators who as a rule prefer to study problems with discernible horizons. Nonetheless, it should be obvious that not the least interesting questions deal with phenomena determined by the nuclear genome. Are there nuclear genes that influence the expression of the mitochondrial genome? What factors insure that cytoplasmically synthesized proteins find their way to the proper integration sites within the organelle? Answers to these and other related questions will ultimately depend on a complete description of the nuclear genes controlling mitochondrial biogenesis.

A. Nuclear Structural Genes

Mitochondria are composed of some 300–400 different proteins. With the few exceptions already discussed, these proteins are synthesized in the cytoplasm and are specified by nuclear genes. Nuclear mutants of yeast with defects in mitochondrial functions are referred to as *pet* mutants. Such strains cover a wide range of phenotypes. Some *pet* mutants exhibit the same pleiotropic deficiencies as do cytoplasmic ρ^- mutants. Relatively little work has been done on characterizing the primary lesions of pleiotropic *pet* mutants. Multiple deficiencies in respiratory enzymes can result from point mutations in ribosomal protein genes, in tRNA amino acyl synthetases, and in other gene functions necessary for mitochondrial protein synthesis. The absence in pleiotropic *pet* strains of cytochromes *a*, a_3, *b*, and oligomcyin-sensitive ATPase is most likely a secondary effect of a primary lesion in a component essential for protein synthesis, DNA replication or processing of mitochondrial RNA.

The *pet* mutants also include strains deficient in a single enzyme such as ATPase, cytochrome oxidase, cytochrome *c*, etc. Cytochrome *c* mutants in particular have been extensively studied. There are two chromatographically different cytochrome *c*s in yeast. Iso-1-cytochrome *c* is the major electron transport carrier constituting some 95% of the total cytochrome *c*. The less abundant iso-2-cytochrome *c* makes up 5%. The two proteins have identical spectral properties but differ in their primary structures (80% homology) and are encoded by two separate genes.

Sherman and co-workers have carried out a detailed genetic and biochemical analysis of a large number of cytochrome *c*-deficient strains. Based on complementation analysis, the mutants have been assigned to nine genetic loci (*cyc1–cyc9*). Two loci, *cyc1* and *cyc7* have been shown to be the structural genes of iso-1- and iso-2-cytochrome *c*, respectively. *Cyc1* mutants have been used to derive a fine genetic map of the iso-1-cytochrome *c* gene. It has also been possible to deduce the nucleotide substitutions in *cyc1* strains by correlative biochemical studies of the amino acid sequences of the mutant proteins. The other seven loci are in genes required for heme biosynthesis and regulation of cytochrome *c* synthesis.

Pet mutants with lesions in respiratory carriers other than cytochrome *c* have also been reported. Such strains are selected either for a particular enzyme or cytochrome deficiency. Mutations in the respiratory complexes are often expressed by the absence of cytochromes. For example, many nuclear mutants deficient in cytochrome oxidase also lack cytochromes *a* and a_3. Similarly, mutants deficient in coenzyme QH_2–cytochrome *c* reductase sometimes lack cytochrome *b* (Fig. 10.11). In a single study of 19 cytochrome oxidase mutants, nine were found to be in different complementation groups. In the same study, mutations affecting the coenzyme QH_2–cytochrome *c* reductase complex were associated with ten complementation groups. The large number of complementation groups in an analysis of a relatively small group of mutants indicates a complex array of nuclear genes of which only a few are likely to code for structural components of the enzymes.

Another interesting class of *pet* strains is characterized by an impaired mitochondrial ATPase. F_1 mutants have been isolated by selecting for strains resistant to ATPase inhibitors. Certain aurovertin-resistant mutants have been shown to produce an F_1 ATPase whose activity is not affected by concentrations of aurovertin that are inhibitory to wild type F_1. Other mutations result in improper assembly of the ATPase or in an enzymatically inactive F_1.

Studies on the biogenesis of complex viruses have been greatly aided by the availability of mutants defective in different steps of the assembly process.

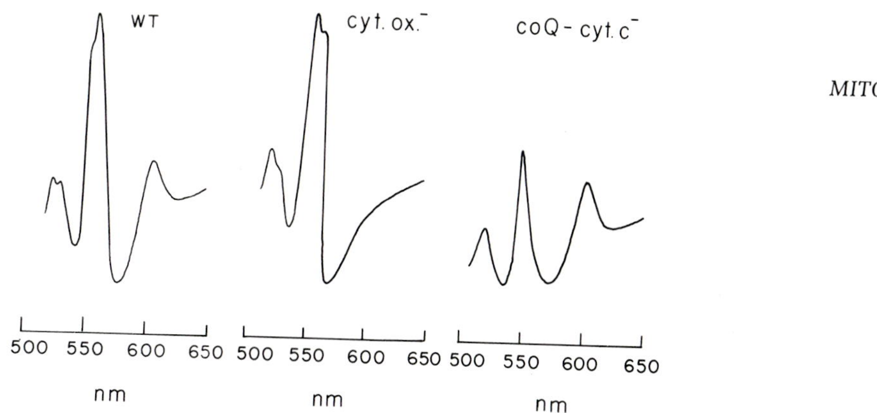

FIG. 10.11. Visible absorption spectra of yeast mitochondria from wild-type and nuclear mutants deficient in cytochrome oxidase or coenzyme QH_2–cytochrome *c* reductase. The cytochrome oxidase mutant lacks the α bands of cytochromes $a + a_3$; the coenzyme QH_2–cytochrome *c* reductase mutant lacks the α band of cytochrome *b*.

The substantial number of *pet* mutants with lesions in mitochondrial inner membrane complexes offers a similar opportunity to reconstruct the *in vivo* sequence of assembly of these enzymes. In addition, *pet* mutants can be used to clarify the functions of the subunit polypeptides encoded in the nuclear genome. Undoubtedly, this will become one of the major areas of mitochondrial research in the near future.

B. Nuclear Regulation of the Mitochondrial Genome

The mitochondrial and cytoplasmic translation products of cytochrome oxidase and other dually derived enzymes of the inner membrane are synthesized in a stoichiometric fashion. The expression of the nuclear and mitochondrial genomes must therefore be coordinated through cellular signals insuring a proportional and orderly synthesis of the two sets of proteins. Mitochondria appear to play a passive role in this process. This is clearly evident from the fact that ρ^0 mutants of yeast lacking mitochondrial DNA nonetheless exhibit normal rates of sythesis of mitochondrial constituents including the proteins normally associated with products of the organellar genome.

In searching for regulatory phenomena, we must therefore look at the nucleo-cytoplasmic system. There is a fair amount of evidence pointing to profound effects of nuclear genes on the expression of mitochondrial DNA. Nuclear mutations have been found to turn off the synthesis of single mitochondrial gene products. For example, certain *pet* mutants are deficient in subunit 2* of cytochrome oxidase, even though the other mitochondrial products are expressed normally. Other *pet* mutations selectively affect the synthesis of cytochrome *b*,* and others still of the ATPase proteolipid.*

*In yeast, these proteins are specified by mitochondrial DNA (see Chapter 11).

The mechanisms controlling the transcription of mitochondrial DNA are only now beginning to be investigated. In view of what has been learned from bacterial systems, it is reasonable to suppose that certain cytoplasmic factors may activate or repress the transcription and perhaps translation of mitochondrial genes. Mitochondrial genes with intervening sequences such as the genes for cytochrome *b* and subunit 1 of cytochrome oxidase (see Chapter 11, Section VIIC) offer alternative means for nuclear control, since the processing enzymes involved in the splicing of the messenger RNA are likely to be encoded in nuclear DNA.

VI. Transport of Proteins into Mitochondria

The synthesis of the vast majority of mitochondrial proteins on cytoplasmic ribosomes necessitates their transport from the cytoplasm to the internal compartments of the organelle. This process probably involves two events: recognition of the target membrane and translocation of the protein across the membrane(s).

The orderly segregation of proteins among the various cellular organelles is a general problem of membrane biogenesis that has been the subject of much speculation. The mere fact that each membrane is composed of a unique set of proteins implies the existence of membrane receptors capable of binding only a specific subset of proteins from the total cytoplasmic pool. In the case of mitochondria, some of the receptors, whether protein, lipid, or a combination of both, must reside in the outer membrane, since it is the immediate barrier separating the cytoplasm from the rest of the organelle. It should be remembered that passive diffusion of solutes across the outer membrane is restricted to compounds with molecular weights of 2000 or less. Consequently, most proteins must enter mitochondria by a facilitated mechanism, i.e., direct interaction of the protein with some component of the membrane.

The number of different receptors present in the outer membrane is not known. Because of the many proteins transported, it is unlikely that a separate receptor exists for each protein. A reasonable assumption is that there are a limited number of receptors, each recognizing some structural feature common to a group of proteins.

The inner membrane acts as a second physical barrier, and it also must exercise a discriminatory function. Thus, it accommodates the transfer of certain proteins into the matrix while confining other proteins either to the intermembrane space or to the inner membrane itself. The apportionment of proteins to these compartments could be signaled by another set of receptors located in the inner membrane.

A. Transport of Cytochrome *c*

The existence of outer membrane receptors is supported by studies on the synthesis and transport of cytochrome *c*. This respiratory carrier is synthesized on cytoplasmic ribosomes and eventually becomes integrated into the inner membrane. Apocytochrome *c* has been synthesized in cell-free translation systems programmed either with *N. crassa* or with rat liver messenger RNA. The primary translation product has a size identical to the mature cytochrome. The absence of an amino terminal extension is also supported by recent data on the nucleotide sequence of the gene.

When the *in vitro* translation system is supplemented with mitochondria, most of the newly synthesized apocytochrome *c* is transferred to the intermembrane space where it is enzymatically converted to functional cytochrome *c*. Several observations made with this reconstituted system suggest that the apoprotein binds to a receptor on the outer membrane. (1) The kinetics of apocytochrome *c* incorporation into mitochondria exclude simple passive diffusion of the protein. (2) Transport is restricted to apocytochrome *c* and does not occur with the mature cytochrome. (3) The transport is blocked by certain peptide fragments of cytochrome *c*. Apparently, the receptor is capable of discriminating between the tertiary structures of the apo- and mature forms of the cytochrome.

B. Internalization of Mitochondrial Proteins

Two mechanisms have been proposed to account for the transport of proteins across biological membranes. Co-translational transport, more generally referred to as the signal hypothesis, involves a transfer of the polypeptide chain through the membrane during the elongation phase of translation. This mechanism has been shown to operate in the transport of secretory proteins from the cytoplasmic to the lumen side of the endoplasmic reticulum. In co-translational transport, the amino terminal sequence of the nascent polypeptide chain (the signal portion of the protein) interacts with the membrane and acts as an anchoring or attachment site for the ribosome. This is followed by penetration of the growing polypeptide chain through the membrane (Fig. 10.12). The amino

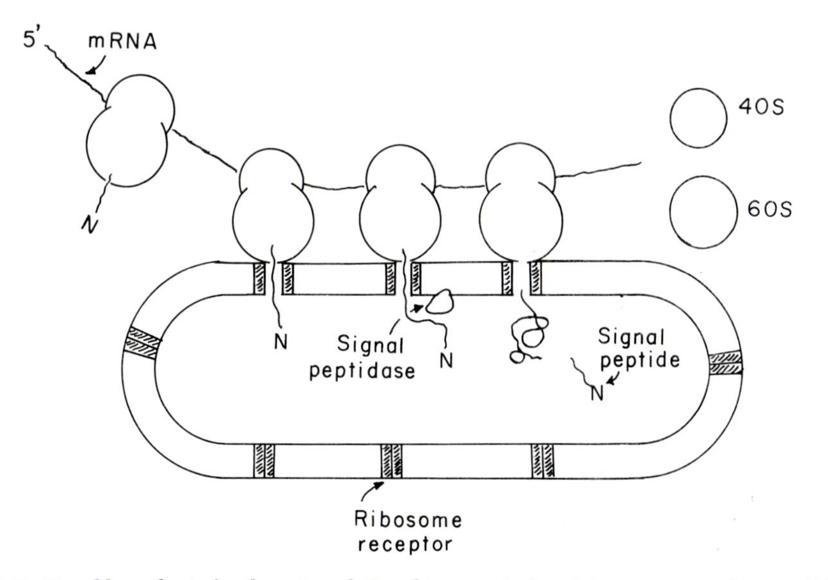

FIG. 10.12. Signal hypothesis for the cotranslational transport of proteins across membranes. (After Blobel and Dobberstein, 1975.)

terminal extension of secretory proteins is cleaved on the lumen side of the membrane, thus providing the driving force for the translocation.

Co-translational transport has been implicated in the transfer of certain bacterial proteins from the cytoplasmic side of the inner membrane to the periplasmic space. The integration of some hydrophobic proteins into membranes has also been found to occur by a mechanism incorporating most of the features of the signal hypothesis. The G component of vesicular stomatitis virus is a glycoprotein whose insertion into the plasma membrane is coupled to translation. The asymmetric integration and proteolytic cleavage of the first 16 amino acid residues of the protein have been shown to be dependent on the presence of the receptor membrane during translation. As with secretory proteins, the integration and post-translational maturation of the G protein fail to occur when the completed polypeptide chain is added to the membrane. Unlike secreted proteins which completely traverse the membrane during translation, integral membrane proteins, because of their hydrophobic properties, partition in the lipid phase of the membrane—in effect, their translocation is arrested.

Although the transfer of proteins from the cytoplasm to mitochondria was thought to occur by a co-translational mechanism, this has not been borne out by recent studies on the synthesis and transport of some well-defined soluble and inner membrane proteins. The situation in mitochondria appears to be complex. Although certain cytoplasmic translation products are synthesized as higher-molecular-weight precursors with amino terminal extensions, others have sizes identical to the mature proteins. Table 10.7 summarizes some of the information currently available on the precursors of cytoplasmic and mitochondrial translation products. Studies on the subunits of F_1 ATPase and cytochrome oxidase translated in a cell-free system have shown that the primary

TABLE 10.7. Precursor Forms of Mitochondrial Proteins

	Size (daltons)	
	Mature	Precursor
Cytoplasmically synthesized		
Yeast F_1-ATPase subunit		
α	58,000	64,000
β	54,000	56,000
γ	34,000	40,000
N. crassa ATPase subunit 9	7,500	13,500
Yeast cytochrome oxidase subunit		
4	14,000	16,000
5	13,000	15,000
6	10,000	12,000
$CoQH_2$–cyt. *c* reductase subunit 5	25,000	27,000
Carbamyl phosphate synthetase	160,000	165,000
Ornithine transcarbamylase	39,000	43,000
Mitochondrially synthesized		
N. crassa cytochrome oxidase subunit		
1	56,000	58,000
2	26,500	28,500
Yeast cytochrome oxidase subunit 2	26,500	28,500

products are incorporated into mitochondria post-translationally. In the case of F_1, the signal peptides of the α, β, and γ subunits are removed during or following their transfer to the matrix where they are assembled into the polymeric protein. Similar results have been obtained with some of the cytoplasmically translated subunits of cytochrome oxidase (Fig. 10.13). The mechanism of transport of these proteins also differs from that of secretory proteins in its requirement for ATP.

A post-translational mode of transport of mitochondrial proteins is also indicated by *in vivo* studies. In yeast, it is possible to completely deplete mitochondria of endogenous ATP. This is achieved by growing a ρ^- mutant in the presence of atractyloside which inhibits the adenine nucleotide carrier and prevents the transport of glycolytically made ATP into mitochondria. Since ρ^- mutants are deficient in respiration, there is also no internal generation of ATP from oxidative phosphorylation. Under these conditions, the precursor forms of cytochrome oxidase subunits 4, 5, and 6 have been found to accumulate in the cytoplasmic fraction. Presumably, the absence of mitochondrial ATP prevents the translocation and maturation of these proteins. It is of interest that the transport of chloroplast proteins has also been found to be an energy-coupled process.

The mechanism of post-translational transport of mitochondrial proteins is not understood at present. Most likely, the signal peptide or, more generally,

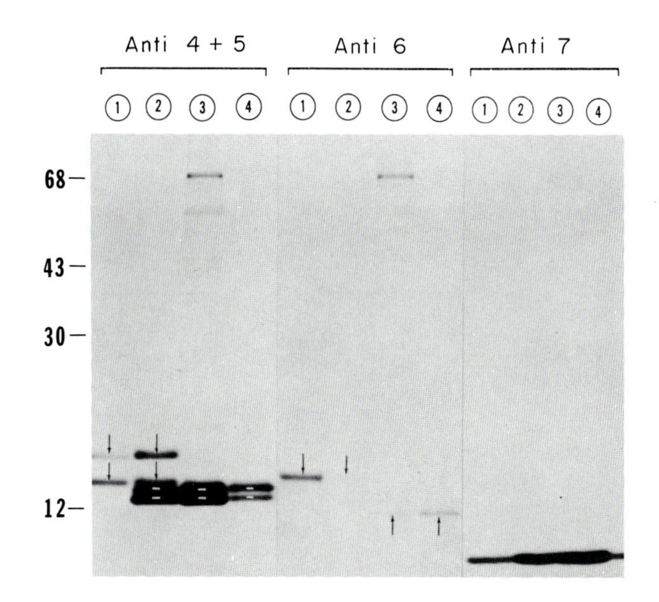

FIG. 10.13. Cytochrome oxidase subunits 4, 5, 6, and 7 synthesized *in vitro* and *in vivo*. The subunits were precipitated with antibodies to a mixture of subunits 4 and 5, to subunit 6, and to subunit 7. The cytochrome oxidase proteins were synthesized under to following conditions: (1) *in vitro* translation with poly-A-containing yeast mRNA; (2) *in vivo* with pulse-labeled yeast spheroplasts; (3) *in vivo* by pulse-chase labeling of spheroplasts; (4) continuous *in vivo* labeling of cells. Arrows pointing up or white bars show the mature subunits; arrows pointing down show the precursors. The molecular weights are indicated in the left-hand margin. (From Mihara and Blobel, 1980, courtesy of Dr. Günter Blobel.)

some aspect of the tertiary structure of the protein is required for binding to a membrane receptor. Once the receptor–protein complex is formed, there is an energy-dependent change in the conformation of the complex leading to internalization and, in the case of the precursors, cleavage of the amino terminal extension (Fig. 10.14).

The fact that certain mitochondrial products (Table 10.7) are synthesized as precursors suggests that signal peptides also function in the integration of this class of proteins into the inner membrane. Again, however, the situation is far from simple. Not all mitochondrial products are synthesized as precursors. Even more puzzling are recent findings that, depending on the mitochondria, the same protein may be synthesized with or without an extension. This is true of subunits 1 and 2 of cytochrome oxidase. The primary translation products of yeast and *N. crassa* subunit 2 are 14 amino acid residues longer than the mature protein. The mammalian subunit 2 lacks the signal sequence. The significance of this evolutionary divergence is not clear but probably indicates that proper integregation of subunit 2 in higher eucaryotes is determined by some new structural feature of the protein.

VII. Biosynthesis of Inner Membrane Enzymes

Most of the structural and functional properties of the inner membrane can be accounted for by the orientation and topology of its constituent enzymes in the lipid bilayer. The biogenesis or, more appropriately, growth of the mem-

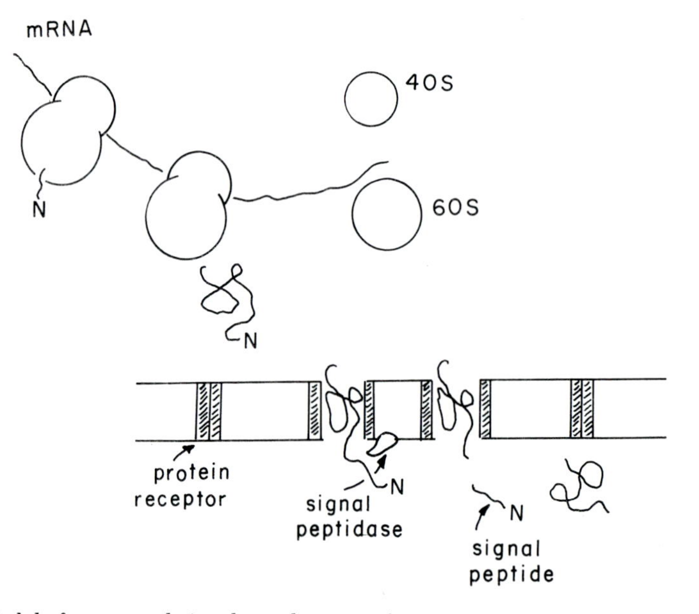

FIG. 10.14. Model of post-translational translocation of a protein across the membrane. The completed polypeptide chain with the signal sequence forms a complex with a membrane receptor. Following cleavage of the signal peptide, the protein undergoes a conformational change resulting in its release from the receptor.

brane can therefore be viewed as a function of the mechanisms of biosynthesis and integration of the enzymes themselves. These enzymes are generally lipoprotein complexes consisting of multiple subunit polypeptides. Does the assembly of the complexes from the subunits take place prior to membrane integration or does it occur concomitant with the deposition of the newly synthesized polypeptides into the lipid bilayer? Of equal importance is the question of what determines the proper orientation of asymmetric enzymes with respect to the two sides of the membrane.

Earlier in this chapter, we saw that whereas the majority of mitochondrial proteins are synthesized in the cytoplasm, a limited number of inner membrane enzymes such as cytochrome oxidase and the oligomycin-sensitive ATPase have subunits derived from both protein-synthesizing systems. A cogent argument can be made that at least part of the assembly of these complexes proceeds within the boundaries of the inner membrane.

A. Oligomycin-Sensitive ATPase

The biosynthesis of the ATPase complex has been examined when the production of the cytoplasmically synthesized polypeptides is arrested with cycloheximide. Under these conditions, the mitochondrial products are found exclusively in the inner membrane even though there is no final assembly of the complex. This can be demonstrated by assaying the membrane either for the physical or functional presence of the membrane factor proteins (Fig. 10.15). The converse, however, is not true. Although normal levels of F_1 and OSCP are synthesized by cells subjected to a chloramphenicol block, neither of these ATPase components enters into a stable association with the inner membrane. Since F_1 and OSCP accumulated in cells grown in the presence of chloramphenicol can be used to form new oligomycin-sensitive ATPase subsequent to removal of the inhibitor, assembly of the complex does not depend on the simultaneous synthesis of both sets of proteins.

The following conclusions can be drawn from these observations.

1. Synthesis and membrane integration of the mitochondrial products proceed in the absence of cytoplasmic protein synthesis. The assembly

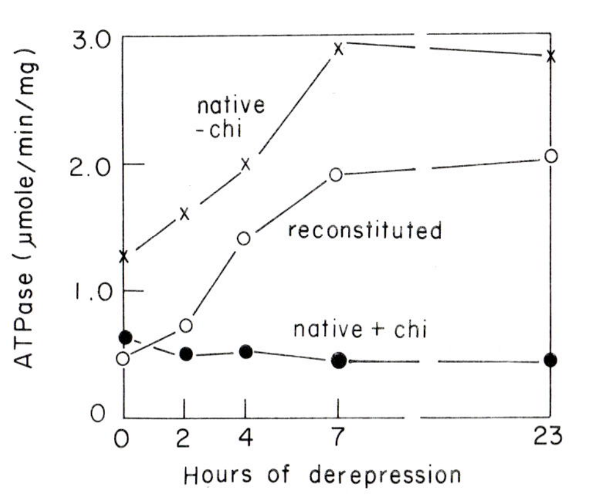

FIG. 10.15. Synthesis of membrane factor. Glucose-repressed yeast were transferred to a derepression medium without (X—X) or with cycloheximide (●—●). The oligomycin-sensitive ATPase activity of mitochondria was measured after the cells had been allowed to derepress for the indicated periods of time. The ATPase activity was also measured in mitochondria of cells incubated in the presence of cycloheximide after they had been reconstituted with F_1 and OSCP (O—O). This experiment shows that the synthesis of the membrane factor, as reflected in the ability of the particles to bind F_1 and OSCP, is not inhibited by cycloheximide.

of the membrane factor proteins into a functional unit occurs in the membrane and is not obligatorily coupled to the synthesis of OSCP or F_1. Whether any of the mitochondrial subunits are synthesized with signal peptides that guide their insertion into the membrane is still an open question.

2. Although not required for the synthesis of the cytoplasmic products, the membrane factor serves a critical function in the assembly of the complex. The membrane factor subunits must provide the appropriate attachment sites for both OSCP and F_1. This set of proteins probably determines the eventual orientation of the enzyme in the membrane.
3. Three of the F_1 subunits (α, β, and γ) have been shown to be made as precursors with amino terminal extensions. Most likely, the signal sequences are important for the transport of the individual subunits into the matrix rather than for their assembly into a polymeric F_1 or their association with the membrane factor. This conclusion stems from the fact that mitochondria of chloramphenicol-grown yeast or of ρ^- mutants contain a catalytically active F_1 polymer whose subunits have sizes indistinguishable from those of the normal wild type enzyme. The above evidence also suggests that the assembly of F_1 occurs in the soluble rather than membrane phase.

According to the model of ATPase biosynthesis proposed here (see Fig. 10.16), there are two separate assembly sequences, one giving rise to the membrane factor and the other to F_1. The hydrophobic subunits of the membrane factor are synthesized on mitochondrial ribosomes, probably at the membrane surface. The insertion of these proteins from the internal side of the membrane may be a sufficient condition to establish the eventual orientation of the complex. The interaction of the membrane factor proteins to form a unit capable of binding OSCP and F_1 occurs in the lipid phase of the membrane as a

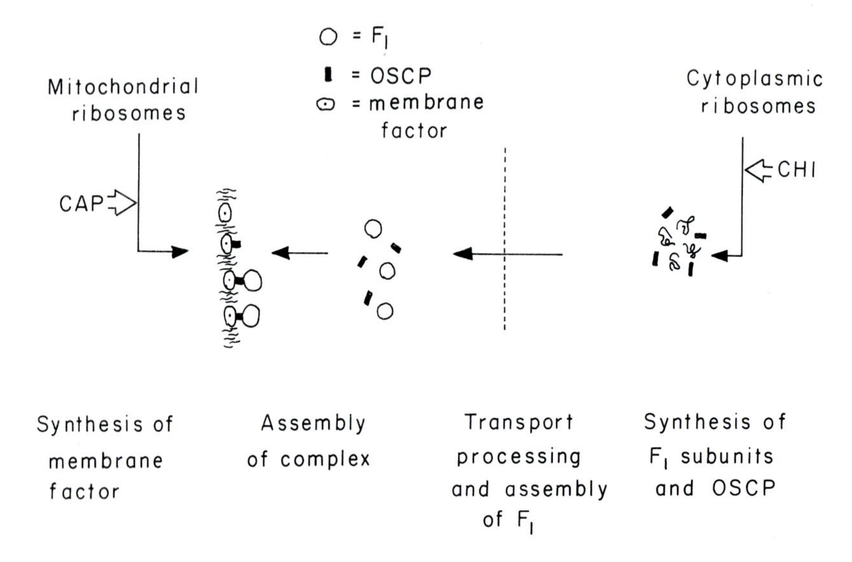

FIG. 10.16. Synthesis and assembly of the mitochondrial ATPase complex.

consequence of their intrinsic affinities for one another. Assembly of F_1 is an independent event that takes place in the soluble matrix phase. The individual subunits are transported into the matrix after termination of the polypeptide chains. For some subunits, transport is accompanied by proteolytic cleavage of the signal peptides. Once processed, the mature subunits polymerize to form F_1. The final assembly may involve a spontaneous association of OSCP and F_1 with the membrane factor.

B. Cytochrome Oxidase

The biosynthetic route leading to the formation of functional cytochrome oxidase is probably similar to that described for the ATPase. In this case also, synthesis and insertion of the mitochondrially derived subunits into the inner membrane must be early events preceeding the integration of the four cytoplasmically translated polypeptides. The latter subunits are transported into mitochondria post-translationally. Subunits 4, 5, and 6 are synthesized as individual precursors and are processed down to their mature size during translocation. The integration of the processed subunits with the membrane depends on the presence of the mitochondrial translation products. Although the broad outlines of the assembly of the complex are beginning to be understood, there are still many details that need to be clarified. For example, it is not known if the cytoplasmic subunits form intermediate assemblies prior to association with the membrane. Nor is it known whether they are all transported to the matrix. It is conceivable that the assembly is bidirectional; i.e., some subunits may interact with the membrane from the intermembrane space and others from the matrix side.

C. Other Membrane Constituents

The inner membrane serves a multitude of different functions. In addition to housing the electron transport chain, it contains most if not all of the transport systems of the organelle as well as the various dehydrogenase complexes linking the oxidation of citric acid cycle and fatty acid intermediates to the main respiratory pathway. These enzymes often have complex compositions with subunit polypeptides exhibiting a wide range of different properties. For instance, NADH–coenzyme Q reductase consists of the primary dehydrogenase plus at least seven to ten other proteins. When separated from the rest of the complex, the flavoprotein dehydrogenase is perfectly water soluble. Other subunits of the complex, however, including those containing the iron–sulfur centers are extremely hydrophobic and belong to the integral class of membrane proteins.

Are there some rules that apply for the assembly of inner membrane complexes whose synthesis is governed entirely by the nucleo-cytoplasmic system? As a working hypothesis, it may be profitable to generalize from the models proposed for the assembly of the ATPase and cytochrome oxidase. Such a general model postulates that the assembly process is initiated by the insertion of the hydrophobic proteins into the membrane. The translational origin of the integral proteins may only be important in determining the direction of insertion. If translation occurs on mitochondrial ribosomes, the polypeptide chain must be inserted from the matrix side. Synthesis on cytoplasmic ribosomes implies insertion from the side of the intermembrane space. Once the hydrophobic core of the complex is formed, assembly is completed either through a sequential integration of individual polar subunits or of a preassembled unit of such subunits. The direction from which the polar subunits are inserted may depend on the compartment (intermembrane or matrix) to which they are transported. According to this model, assembly is guided primarily by specific protein–protein interactions. The membrane localization of the complex is a consequence of the initial deposition of the hydrophobic polypeptides in the membrane during their translation (mitochondrial) or transport (cytoplasmic). Finally, the orientation of the complex is determined by the direction from which different polypeptides are inserted.

Selected Readings

Agsteribbe, E., Datema, R., and Kroon, A. M. (1974) Mitochondrial polysomes from *Neurospora crassa*, in *The Biogenesis of Mitochondria* (A. M. Kroon and C. Saccone, eds.), Academic Press, New York, pp. 305–314.

Attardi, G., and Ojala, D. (1971) Mitochondrial ribosomes in HeLa cells, *Nature* [*New Biol.*] **229**:133.

Barrell, B. G., Bankier, A. T., and Drouin, J. (1979) A different genetic code in human mitochondria, *Nature* **282**:189.

Barrell, B. G., Anderson, S., Bankier, A. T., de Bruijn, M. H. L., Chen, E., Coulson, A. R., Drouin, J., Eperon, I. C., Nierlich, D., Staden, R., and Young, I. G. (1980) Different pattern of codon recognition by mammalian mitochondrial tRNAs, *Proc. Natl. Acad. Sci. U.S.A.* **77**:3164.

Beattie, D. S., Basford, R. E., and Koritz, S. B. (1967) The inner membrane as the site of the *in vitro* incorporation of L(^{14}C)leucine into mitochondrial proteins, *Biochemistry* **6**:3099.

Blobel, G., and Dobberstein, B. (1975) Transfer of proteins across membranes I. Presence of proteolytically processed and unprocessed nascent immunoglobulin light chains on membrane-bound ribosomes of murine myeloma. II. Reconstitiution of functional rough microsomes from heterologous components, *J. Cell Biol.* **67**:835, 852.

Blok, M. C., Wirtz, K. W. A., and Scherphof, G. L. (1969) Exchange of phospholipids between microsomes and inner and outer membranes of rat liver, *Biochim. Biophys. Acta* **233**:61.

Bonitz, S. G., Berlani, R., Coruzzi, G., Li, M., Macino, G., Nobrega, F. G., Nobrega, M. P., Thalenfeld, B., and Tzagoloff, A. (1980) Codon recognition rules in yeast mitochondria, *Proc. Natl. Acad. Sci. U.S.A.* **77**:3167.

Borst, P. (1972) Mitochondrial nucleic acids, *Annu. Rev. Biochem.* **41**:333.

Chua, N.-H., and Schmidt, G. W. (1979) Transport of proteins into mitochondria and chloroplasts, *J. Cell Biol.* **81**:461.

Conboy, J. G., Kalousek, F., and Rosenberg, L. E. (1979) *In vitro* synthesis of a putative precursor of mitochondrial ornithine transcarbamylase, *Proc. Natl. Acad. Sci. U.S.A.* **76**:5724.

Cote, C., Solioz, M., and Schatz, G. (1979) Biogenesis of the cytochrome bc_1 complex of yeast mitochondria. A precursor form of the cytoplasmically made subunit V, *J. Biol. Chem.* **254**:1437.

Davidson, J. B., and Stanacev, N. Z. (1971) Biosynthesis of cardiolipin in mitochondria isolated from guinea pig liver, *Biochem. Biophys. Res. Commun.* **6**:1191.

Grivell, L. A., Reijnders, L., and Borst, P. (1971) The effect of temperature and ionic strength on the electrophoretic mobility of yeast mitochondrial RNA, *Eur. J. Biochem.* **19**:64.

Hallermayer, G., Zimmermann, R., and Neupert, W. (1977) Kinetic studies of the transport of cytoplasmically synthesized proteins into mitochondria of intact cells of *N. crassa*, *Eur. J. Biochem.* **81**:523.

Hare, J. F., Ching, E., and Attardi, G. (1980) Isolation, subunit composition and site of synthesis of human cytochrome *c* oxidase, *Biochemistry* **19**:2023.

Heckman, J. E., Hecker, L. I., Schwartzbach, S. D., Barnett, W. E., Baumstark, B., and RajBhandary, U. L. (1978) Structure and function of initiator tRNA from the mitochondria of *Neurospora crassa, Cell* **13**:83.
Heckman, J. E., Sarnoff, J., Alzner-Deweerd, B., Yin, S., and RajBhandary, U. L. (1980) Novel features in the genetic code and codon reading patterns in *Neurospora crassa* mitochondria based on sequences of six mitochondrial tRNAs, *Proc. Natl. Acad. Sci. U.S.A.* **77**:3159.
Koch, G. (1976) Synthesis of the mitochondrial inner membrane in cultured *Xenopus laevis* oocytes, *J. Biol. Chem.* **251**:6097.
Kovac, L. (1974) Biochemical mutants: An approach to mitochondrial energy coupling, *Biochim. Biophys. Acta* **346**:101.
Lamb, A. J., Clark-Walker, G. D., and Linnane, A. W. (1968) The biogenesis of mitochondria. The differentiation of mitochondrial and cytoplasmic protein synthesizing systems *in vitro* by antibiotics, *Biochem. Biophys. Acta* **161**:415.
Lambowitz, A. M., Chua, N.-H, and Luck, D. J. L. (1976) Mitochondrial ribosome assembly in *Neurospora*. Preparation of mitochondrial ribosomal precursor particles, sites of synthesis of mitochondrial ribosomal proteins and studies on the poky mutant, *J. Mol. Biol.* **107**:223.
Leaver, D. J., and Harmey, M. A. (1972) Isolation and characterization of mitochondrial ribosomes from higher plants, *Biochem. J.* **129**:37p.
Lewin, A. S., Gregor, I., Mason, T. L., Nelson, N., and Schatz, G. (1980) Cytoplasmically made subunits of yeast mitochondrial F_1-ATPase and cytochrome *c* oxidase are synthesized as individual precursors, not as polyproteins, *Proc. Natl. Acad. Sci. U.S.A.* **77**:3998.
Lizardi, P. M., and Luck, D. J. L. (1972) The intracellular site of synthesis of mitochondrial ribosomal proteins in *N. crassa, J. Cell. Biol.* **54**:56.
Luck, D. J. L. (1963) Formation of mitochondria in *Neurospora crassa*. A quantitative radioautographic study, *J. Cell Biol.* **16**:483.
Luck, D. J. L. (1965) Formation of mitochondria in *Neurospora crassa*. A study based on mitochondrial density changes, *J. Cell Biol.* **24**:461.
Maccecchini, M.-L., Rudin, Y., Blobel, B., and Schatz, G. (1979) Import of proteins into mitochondria. Precursor forms of the extramitochondrially made F_1-subunits in yeast, *Proc. Natl. Acad. Sci. U.S.A.* **76**:343.
Martin, N. C., Rabinowitz, M., and Fukuhara, H. (1977) Yeast mitochondrial DNA specific tRNAs for 19 amino acids. Deletion mapping of the tRNA genes, *Biochemistry* **21**:4672.
Mason, T. L., and Schatz, G. (1973) Cytochrome oxidase of Baker's yeast II. Site of translation of the protein components, *J. Biol. Chem.* **248**:1355.
McLean, J. R., Cohn, G. L., Brandt, I. K., and Simpson, M. V. (1958) Incorporation of labeled amino acids into the protein of muscle and liver mitochondria, *J. Biol. Chem.* **223**:657.
Mihara, K., and Blobel, G. (1980) The four cytoplasmically made subunits of yeast mitochondrial cytochrome *c* oxidase are synthesized individually and not as a polyprotein, *Proc. Natl. Acad. Sci. U.S.A.* **77**:4160.
Neher, E.-M., Harmey, M. A., Hennig, B., Zimmermann, R., and Neupert, W. (1980) Post-translational transport of proteins in the assembly of mitochondrial membranes, in *The Organization and Expression of the Mitochondrial Genome* (A. M. Kroon and C. Saccone, eds.), North-Holland, Amsterdam, pp. 413–422.
Noller, H. F., and Woese, C. R. (1981) Secondary structure of 16 S ribosomal RNA, *Science* **212**:403.
O'Brien, T. W. (1971) The general occurrence of 55 S ribosomes in mammalian liver mitochondria, *J. Biol. Chem.* **246**:3409.
Plattner, H., Salpeter, M. M., Saltzgaber, J., and Schatz, G. (1970) Promitochondria of anaerobically grown yeast IV. Conversion into respiring mitochondria, *Proc. Natl. Acad. Sci. U.S.A.* **66**:1252.
Pring, D. R. (1974) Maize mitochondria. Purification and characterization of ribosomes and ribosomal ribonucleic acid, *Plant Physiol.* **53**:677.
Raymond, Y., and Shore, G. C. (1979) The precursor for carbamyl phosphate synthetase is transported to mitochondria via a cytosolic route, *J. Biol. Chem.* **254**:9335.
Robberson, D., Aloni, Y., and Attardi, G. (1971) Expression of the mitochondrial genome in HeLa cells. VI. Size determination of mitochondrial ribosomal RNA by electron microscopy, *J. Mol. Biol.* **60**:473.
Roodyn, D. B., Reix, P. J., and Work, T. S. (1961) Protein synthesis in mitochondria. Requirements for the incorporation of radioactive amino acids into mitochondrial protein, *Biochem. J.* **80**:9.

Sager, R. (1972) *Cytoplasmic Genes and Organelles*, Academic Press, New York.

Schatz, G. (1968) Impaired binding of mitochondrial adenosine triphosphatase in the cytoplasmic "petite" mutant of *Saccharomyces cerevisiae*, *J. Biol. Chem.* **243**:2192.

Schatz, G., and Mason, T. L. (1974) The biosynthesis of mitochondrial proteins, *Annu. Rev. Biochem.* **43**:51.

Schatz, G., Groot, G. S. P., Mason, T. L., Rouslin, W., Wharton, D. C., and Saltzgaber, J. (1972) Biogenesis of mitochondrial inner membrane in baker's yeast, *Fed. Proc.* **31**:21.

Schwab, A. J., Sebald, W., and Weiss, H. (1972) Different pool sizes of the precursor polypeptides of cytochrome oxidase from *Neurospora crassa*, *Eur. J. Biochem.* **30**:511.

Sebald, W. (1977) Biogenesis of mitochondrial ATPase, *Biochim. Biophys. Acta* **463**:1.

Sebald, W., Hoppe, J., and Wachter, E. (1979) Amino acid sequence of the ATPase proteolipid from mitochondria, chloroplasts and bacteria, in *Function and Molecular Aspects of Biomembrane Transport* (E. Quagliariello, F. Palmieri, S. Papa and M. Klingenberg, eds.), North-Holland, Amsterdam, pp. 63–74.

Sevarino, K. A., and Poyton, R. O. (1980) Mitochondrial membrane biogenesis. Identification of a precursor to yeast cytochrome *c* oxidase subunit II, an integral polypeptide, *Proc. Natl. Acad. Sci. U.S.A.* **77**:142.

Sherman, F., and Slonimski, P. P. (1964) Respiratory deficient mutants in yeast II. Biochemistry, *Biochim. Biophys. Acta* **90**:1.

Sherman, F., and Stewart, J. W. (1971) Genetics and biosynthesis of cytochrome *c*, *Annu. Rev. Genet.* **5**:257.

Shine, J., and Dalgarno, L. (1974) The 3′-terminal sequence of *Escherichia coli* 16 S ribosomal RNA: Complementarity to nonsense triplets and ribosome binding sites, *Proc. Natl. Acad. Sci. U.S.A.* **71**:1342.

Shore, G. C., Carignan, P., and Raymond, Y. (1979) *In vitro* synthesis of a putative precursor to the mitochondrial enzyme, carbamyl phosphate synthetase, *J. Biol. Chem.* **254**:3141.

Slonimski, P. P. (1953) *La Formation des Enzymes Respiratoire chez la Levure*, Masson, Paris.

Tzagoloff, A. (1969) Assembly of the mitochondrial membrane system II. Synthesis of the mitochondrial adenosine triphosphatase F_1, *J. Biol. Chem.* **244**:5027.

Tzagoloff, A., Akai, A., and Needleman, R. (1975) Assembly of the mitochondrial membrane system. Isolation of nuclear mutants of *S. cerevisiae* with specific defects in mitochondrial functions, *J. Biol. Chem.* **250**:8228.

Tzagoloff, A., Macino, G., and Sebald, W. (1979) Mitochondrial genes and translation products, *Annu. Rev. Biochem.* **48**:419.

Utter, M. F., Duell, E. A., and Bernofsky, C. (1968) Alterations in the respiratory enzymes of the mitochondria of growing and resting yeast, in *Aspects of Yeast Metabolism* (A. K. Mills, ed.) Blackwell, Oxford, pp. 197–212.

Weiss, H., and Ziganke, B. (1974) Cytochrome *b* in *Neurospora crassa* mitochondria. Site of translation of the heme protein, *Eur. J. Biochem.* **41**:63.

Wickner, W. (1979) The assembly of proteins into biological membranes. The membrane trigger hypothesis, *Annu. Rev. Biochem.* **48**:23.

Wirtz, K. W. A., and van Deenen, L. L. M. (1977) Phospholipid-exchange proteins: A new class of intracellular lipoproteins, *Trends Biochem. Sci.* **2**:49.

Wirtz, K. W. A., Kamp, H. H., and van Deenen, L. L. M. (1972) Isolation of a protein from beef liver which specifically stimulates the exchange of phosphatidylcholine, *Biochim. Biophys. Acta* **274**:607.

Zimmermann, R., Paluch, U., and Neupert, W. (1979) Cell-free synthesis of cytochrome *c*, *FEBS Lett.* **108**:141.

11

Mitochondrial Genetics

A remarkable feature of mitochondria is their possession of an independent genetic system, now known to be necessary for the morphogenesis of a respiratory-competent organelle. This aspect of mitochondria has attracted the attention of geneticists and molecular biologists, and, as a result, much progress has been made in our understanding of this rather unique genetic system. The foundations of mitochondrial genetics were laid in the late 1940s by Ephrussi and his collaborators who discovered a mutation that abolished the capacity of yeast to grow on nonfermentable substrates such as ethanol or glycerol whose utilization depends on a functional respiratory chain. Even though such respiratory-deficient strains of yeast are capable of growing on glucose and other sugars, they form small colonies and for this reason are called "petite" mutants (Fig. 11.1). Ephrussi's studies showed that mutations resulting in the petite phenotype are inherited in a non-Mendelian fashion, and he therefore postulated the lesions to be in an extrachromosomal or cytoplasmic element. This element was designated as the rho (ρ) factor, and hence, cytoplasmic petite mutants are also referred to as ρ^- mutants. Although this was not known at the time of the discovery, the ρ factor was subsequently shown to be identical to mitochondrial DNA.

Virtually all of our current knowledge of mitochondrial genetics is based on studies of the unicellular yeast *Saccharomyces cerevisiae;* most of the discussion in this chapter will therefore be devoted to a description of the yeast mitochondrial genome.

I. Mitochondrial DNA

Mitochondrial DNA (mtDNA) is a closed circular duplex molecule with superhelical twists (Fig. 11.2). The molecular weight of the *S. cerevisiae* genome is estimated to be 50×10^6, corresponding to a 25-μm circle. The size of mtDNA in other yeasts and in multicellular fungi is somewhat smaller. In general, the largest genomes are found in plant and the smallest in animal mitochondria (Table 11.1). In addition to the disparities in molecular weight, there is recent evidence of diversity in genetic content of various mtDNAs. The number of mitochondrial genes, however, is not necessarily related to the size of the DNA.

The exact number of copies of mtDNA is not known, but in most cells it probably ranges from 50 to 100. An important question is whether all the copies are identical. At present, all the evidence is consistent with the presence of a genetically homogeneous population of circles. As will be seen, two circumstances—a pool consisting of multiple and identical copies of mtDNA—are of paramount importance in understanding the population aspect of mitochondrial genetics.

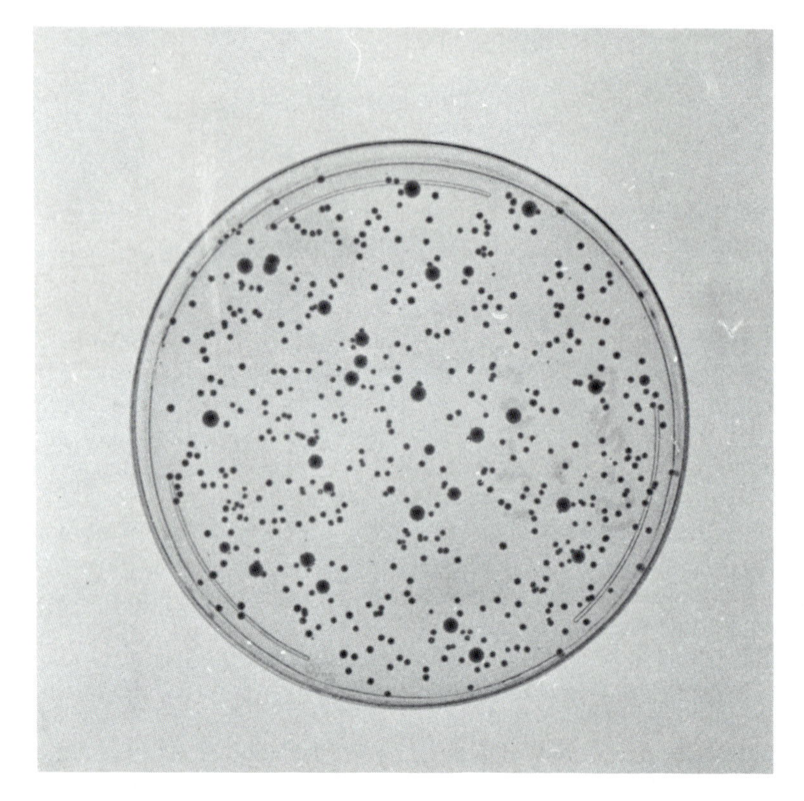

FIG. 11.1. Colony morphology of wild-type and respiratory-deficient mutants of *Saccharomyces cerevisiae*.

TABLE 11.1. Properties of Mitochondrial DNA

Organism	Density (g/cm^3)	Contour length (μm)	Molecular weight ($\times 10^6$)
Saccharomyces cerevisiae	1.682	25	50
Neurospora crassa	1.701	20	40
Drosophila melanogaster	1.680	6	12
Pisum sativum	1.708	30	74
Rat	1.701	5	10.3
Man	1.708	5	10

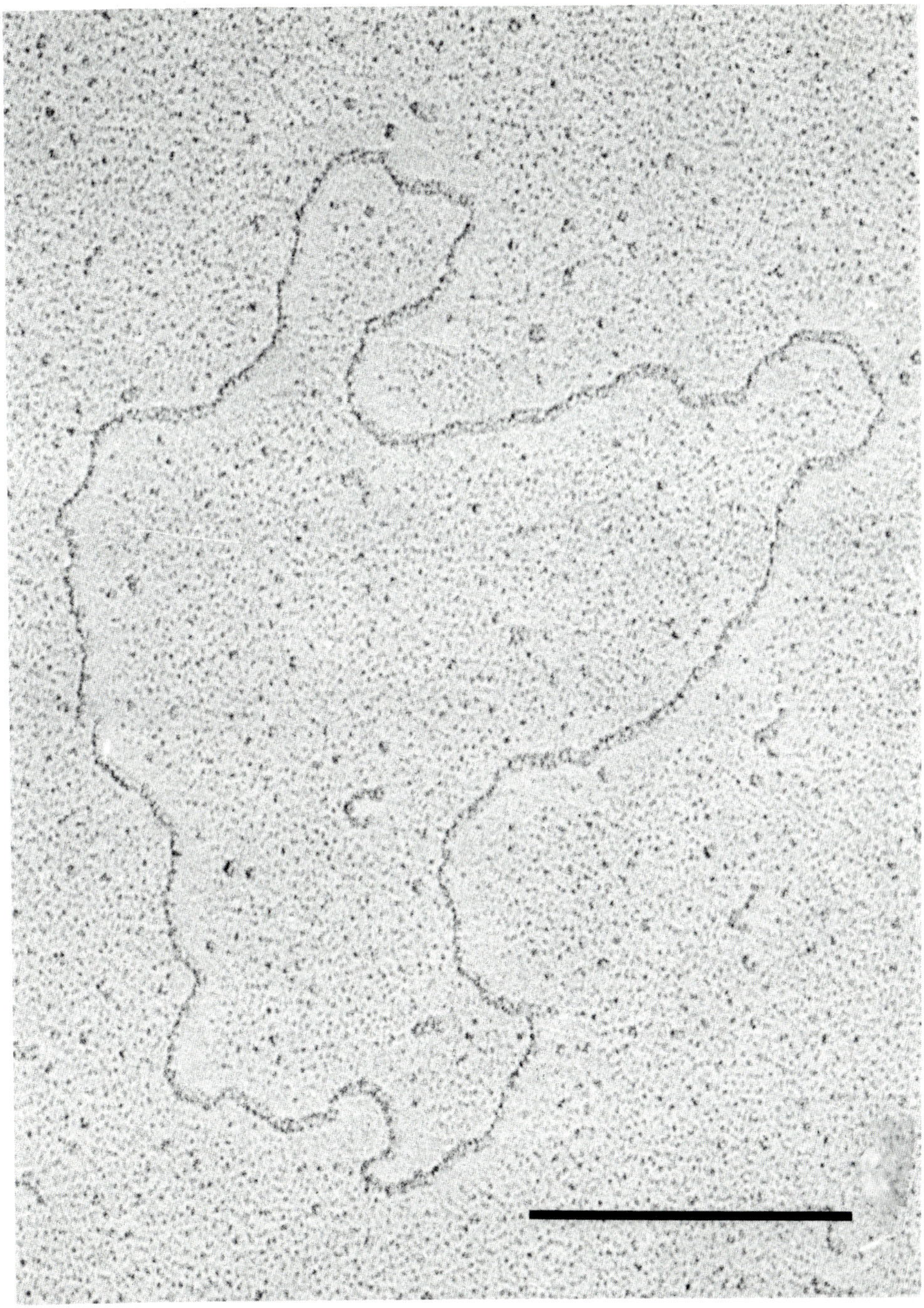

FIG. 11.2. Electron micrograph of human mitochondrial DNA. The contour length of the molecule is 5.6 μm. The scale bar represents 0.5 μm. (Courtesy of Dr. David Clayton.)

Because of its low G + C content, yeast mtDNA has a lighter buoyant density than nuclear DNA. This is of some practical advantage, since it allows the two species to be separated on cesium chloride gradients. Although mtDNA represents only a small percentage of the total DNA content of most cells, it can be purified by procedures entailing an initial extraction of mitochondria with detergents followed by centrifugation of the extracts on cesium chloride density gradients or chromatography on hydroxylapatite. Preparations of yeast mtDNA obtained in this way are free of nuclear DNA but contain only a small number of intact circles. Most of the DNA consists of a heterogeneous collection of linear molecules whose average molecular weight is approximately one-half that of the native DNA. The presence of single-stranded nicks in yeast mtDNA is another disadvantage that has hampered certain types of studies.

The degradation of mtDNA during purification seems to be most serious with organisms having a large genome size (yeast, fungi, plants) but has not been a problem with animal mtDNA. Preparations of mtDNA from rat liver mitochondria, for example, consist of intact circles with relatively few single-stranded nicks.

Substantial progress has been made in the past few years in defining the nature of the genetic information encoded in the mitochondrial genome. Two functionally different classes of genes have been identified in yeast. The first class includes genes necessary for the biogenesis of the mitochondrial protein synthetic apparatus. The second class consists of genes coding for protein products required in a structural capacity for the biosynthesis of a limited number of mitochondrial enzymes. These two classes of genes will be discussed in more detail in a later section. The point to be emphasized here is that the mitochondrial genome is a relatively simple molecule compared to other eucaryotic DNAs. The simplicity stems from its small size and limited informational content. This, in turn, has facilitated its description in great detail both from the standpoint of the identity of the genes and their molecular organization.

II. Life Cycle of *Saccharomyces cerevisiae*

In order to discuss mitochondrial genetics of *S. cerevisiae,* it is necessary first to review the life cycle of this organism. *Saccharomyces cerevisiae* is a unicellular yeast that multiplies vegetatively either in the haploid or diploid states. During vegetative growth, the yeast produces buds that eventually separate from the mother cell. Each bud receives a complete complement of nuclear genes as well as part of the cytoplasm of the mother cell including mitochondria and other subcellular organelles.

Haploid strains of *S. cerevisiae* occur in one of two different sexual mating types designated as "α" and "a." When two haploid strains of opposite mating type are mixed in liquid or solid media, fusion occurs, giving rise to a zygote with a diploid number of chromosomes in the nucleus and a hybrid cytoplasm containing cytoplasmic components from each of the two haploid parents. In a rich medium, the zygotes form buds by the same process as do haploid yeast, the only difference being that the buds have a diploid number of chromosomes in their nuclei. The new diploid cells continue to divide vegetatively by budding as long as they are not deprived of substrate. Under conditions of limiting substrate, a diploid cell undergoes sporulation, a process involving a meiotic event which results in the formation of four spores, each having a haploid number of chromosomes. The four spores are enclosed by a spherically shaped ascus coat and, in this dormant state, can remain viable for long periods of time. If, however, they are placed in a rich medium, germination ensues, and a new cycle of growth and mating is initiated. The different stages of the life cycle of *S. cerevisiae* are shown in Fig. 11.3.

Saccharomyces cerevisiae is an ideal eucaryotic cell for genetic studies, since most of the standard techniques used in bacterial genetics are equally applicable to the handling of this yeast. The fact that diploid cells can be induced to undergo meiosis and give rise to haploid clones is very useful in some types of work. For example, it provides an easy way to construct genetically defined strains. Haploid yeast with different genetic markers can be crossed, and meiotic segregants with the desired combinations of the input markers isolated after sporulation.

III. Genetic Criteria for Distinguishing Nuclear and Mitochondrial Mutations

The development of functional mitochondria depends on a normal complement of nuclear and mitochondrial genes. Mutants with lesions in mitochondrial functions can, therefore, arise from mutational events in either of the two genomes. There are three genetic tests commonly used to distinguish nuclear and mitochondrial mutations. These will be described here since they introduce some useful concepts in understanding later sections dealing with the mapping of mitochondrial mutations.

A. Restoration of Wild-Type Growth by a ρ^0 Tester

Ethidium bromide is a DNA-intercalating dye capable of inducing the loss of long segments of mitochondrial DNA in *S. cerevisiae*. Extensive treatment

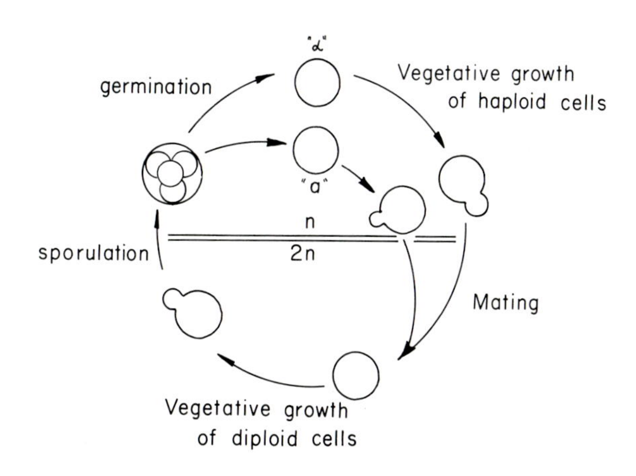

FIG. 11.3. Life cycle of *Saccharomyces cerevisiae*.

of *S. cerevisiae* with this chemical results in strains with no detectable mitochondrial DNA. Such strains are referred to as ρ^0 mutants to indicate the total absence of mtDNA. Since ethidium bromide has no effect on nuclear DNA, ρ^0 mutants derived from a respiratory competent yeast will have a normal complement of nuclear genes including those necessary for the development of functional mitochondria. The ρ^0 strains are useful for testing whether a mutant defective in some mitochondrial function has a genetic lesion in nuclear or mitochondrial DNA. If the mutation is in nuclear DNA, diploid cells obtained from a cross of the mutant to the ρ^0 tester should have a wild-type phenotype because of the presence of a normal complement of nuclear genes (ρ^0 input) and of mitochondrial genes (from the mutant). On the other hand, if the mutation is in mitochondrial DNA, the diploid progeny resulting from the cross will exhibit the mutant phenotype since the ρ^0 parent, having no mitochondrial DNA, cannot supply the wild-type copy of the mutated gene (Fig. 11.4). This genetic test is very easy to perform, particularly when the mutation is in a gene necessary for a respiratory function. In this case, the diploid cells are simply scored for growth on a medium containing a nonfermentable substrate (glycerol or ethanol). If growth is observed, the mutation is most likely to be in a nuclear gene. Absence of growth usually indicates a mitochondrial mutation.

B. Meiotic Segregation

The segregation ratios observed in tetrad analysis form a second useful criterion with which to discriminate between nuclear and mitochondrial mutations. In this test, the mutant is crossed to a wild-type haploid strain of the opposite mating type. The diploid cells are forced to sporulate, and the four spores from individual tetrads are dissected, grown up as separate clones, and checked for the presence or absence of the mutant phenotype. Since diploid yeast have two sets of nuclear genes, one from each parent, any mutation in nuclear DNA will segregate in a Mendelian or 1:1 ratio in the meiotic spore progeny; i.e., the mutant gene will be expressed in only two of the four spores

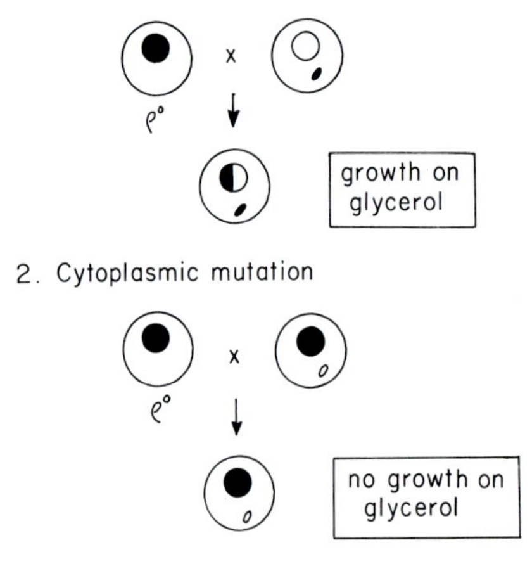

FIG. 11.4. Complementation test with ρ^0 tester. Mutations in nuclear DNA are represented by the large open circle (nucleus) and in mitochondrial DNA by the open oval-shaped mitochondrion. Wild-type nuclear and mitochondrial genomes are shown as solid black organelles.

of each tetrad. If, however, the mutation is in mtDNA, non-Mendelian segregation ratios of the mutant phenotype will be found among the spore progeny.

Non-Mendelian ratios arise from the segregation of mitochondrial DNA during vegetative growth of the diploid cells. The zygotes formed in the cross of the wild type and mutant haploid strains will presumably have an equal number of copies of the two parental mtDNAs. During vegetative growth there is a random assortment and segregation of DNA, and after a sufficient number of generations (usually 20–30), most of the diploid cells will be pure or homoplastic with respect to their content of mtDNA. Two types of clones will be present in a fully segregated population, cells having pure wild-type DNA or pure mutant DNA. When such a mixed culture is induced to sporulate, most of the tetrads will be of the 0:4 and 4:0 types (Fig. 11.5). If the segregation at the time of sporulation is not complete, other ratios (3:1, 1:3, 2:2) will be observed.

C. Mitotic Segregation

In this test, the mutant is mated to a wild-type haploid strain, and the diploid progeny grown for a sufficient number of generations to yield pure clones. Since most nuclear mutations are recessive, the presence of clones with the mutant phenotype in the diploid population is indicative of a mutation in cytoplasmic or mitochondrial DNA. Nuclear recessive mutations will yield diploid cells with a wild-type phenotype. In a sense, this test, although operationally different, is based on the same principle as tetrad analysis. Both depend on the segregation of mitochondrial DNA during vegetative growth of diploid cells.

IV. Mitochondrial Genes and Mutants

A. Genes

Mitochondria contain mit and syn genes. Syn genes code for components of the mitochondrial protein synthetic machinery. This class includes the genes

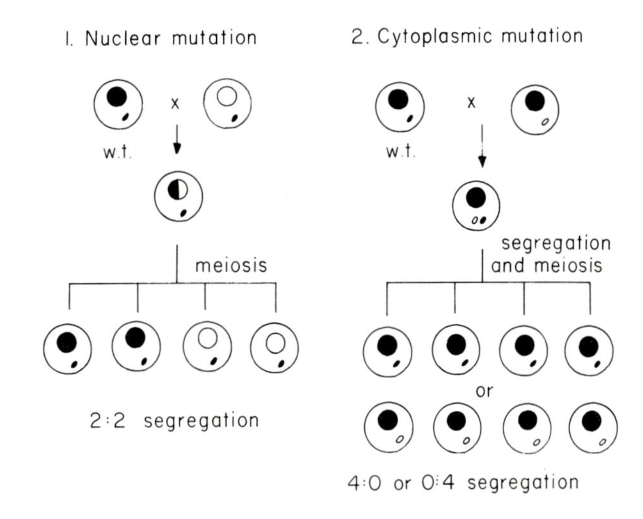

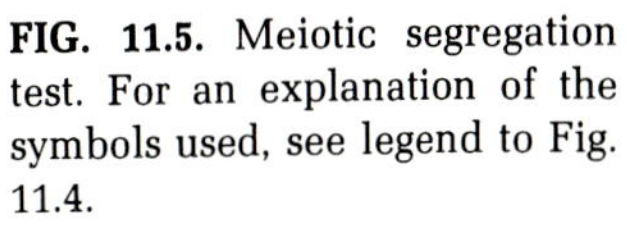
FIG. 11.5. Meiotic segregation test. For an explanation of the symbols used, see legend to Fig. 11.4.

of the mitochondrial ribosomal RNAs and a complete set of transfer RNA genes. Other protein factors essential for mitochondrial translation may also be encoded in the genome, but as yet none except one ribosomal protein subunit has been identified.

The mit genes code for proteins required for the development of the terminal respiratory chain. At present, it is known that this class consists of structural genes involved in the biosynthesis of cytochrome oxidase, coenzyme QH_2–cytochrome *c* reductase, and oligomycin-sensitive ATPase. No other functions have been found to be controlled by mit genes.

B. Mutants

One of the goals of mitochondrial genetics has been to identify all of the genes present on the genome and to establish their locations on the circular map. This information is vital for an understanding of the mechanisms regulating the expression of mitochondrial genes. Are there operator and repressor genes? Do nuclear genes control the transcription and/or translation of mitochondrially encoded genes? These questions bear directly on the extent of the biogenetic autonomy of the organelle and, in principle, should be answerable by purely genetic approaches. We now consider the properties and methods of isolation of the different types of mitochondrial mutants of yeast listed in Table 11.2.

1. ρ^- Mutants

This class includes an almost infinite number of mutations and therefore strains. All ρ^- mutants studied so far share certain properties that permit them to be defined operationally. The ρ^- mutants arise from long deletions in mitochondrial DNA. The nondeleted segments of DNA retained by various ρ^- strains may represent different parts and lengths of the genome, and most independently isolated ρ^- mutants differ in their retained segments. The loss of a large part of the genome, whether by a single continuous deletion or by multiple deletions, has several consequences which have been experimentally ver-

TABLE 11.2. Phenotypes of Yeast Mitochondrial Mutants

Mutant	Nature of mutations	Phenotype
ρ^-	Large deletions	Deficiencies in mitochondrial protein synthesis, respiration, and oxidative phosphorylation
Ant^R	Point mutations	Resistance to antibiotics that inhibit ribosomes, ATPase, or respiratory complexes
Mit^-	Point mutations or small deletions	Lesions in cytochrome oxidase, coenzyme QH_2–cytochrome *c* reductase, or ATPase
Syn^-	Point mutations or small deletions	Deficiencies in mitochondrial protein synthesis, respiration, and oxidative phosphorylation

ified for a large number of different mutants. First is the absence of mitochondrial protein synthesis. This is because extended deletions have a high probability of causing the loss of some syn genes which are scattered on the genome. As a consequence of the loss of protein synthesis, ρ^- mutants fail to synthesize enzymes with subunit proteins made on mitochondrial ribosomes, viz, cytochrome oxidase, coenzyme QH_2-cytochrome *c* reductase, and ATPase (see Chapter 10, Section IIID). The ρ^- mutants, therefore, exhibit pleiotropic deficiencies in respiration and oxidative phosphorylation. A second important property is that no two ρ^- mutants have yet been found to yield wild-type recombinants when crossed to each other. This property will be discussed in more detail when we consider criteria for distinguishing ρ^- from syn^- mutants.

Another interesting finding that has emerged from studies of well-characterized ρ^- mutants is that the retained segment of mitochondrial DNA is amplified by tandem head-to-tail or palindromic repeats of the DNA segment. The mitochondrial genome of ρ^- mutants is a circular DNA whose dimensions are similar to the wild-type genome but consists of numerous repeats of the retained segment (Fig. 11.6). This, of course, means that the smaller the retained sequence, the more times it will be repeated in the genome of the mutant. For example, a segment 1 μm long may be repeated up to 25 times. Thus, ρ^- mutants can be used to enrich for mitochondrial genes and, as will be seen, have become important tools for sequencing mitochondrial DNA.

The mitochondrial DNA of *S. cerevisiae* is unusually labile, and in most haploid strains, 0.1–1% of the population consists of spontaneously induced ρ^- mutants. The frequency of ρ^- production can be increased substantially by various chemical agents or physiological conditions. The most commonly used reagents are ethidium bromide and acridine dyes which intercalate with mitochondrial DNA and, at relatively low concentrations, induce ρ^- mutants with as high as 90–100% efficiency.

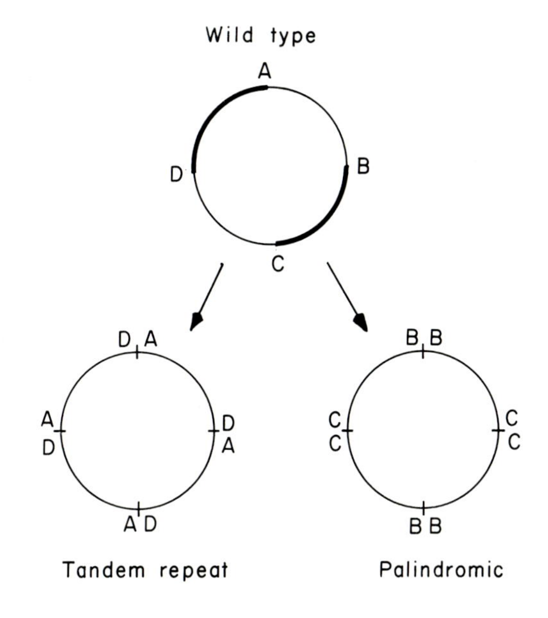

FIG. 11.6. Organization of the mitochondrial genomes in ρ^- clones with tandem (head-to-tail) and palindromic (head-to-head) arrangements of the repeat unit. The segments of mtDNA retained in the two mutants are shown as the heavier arcs on the wild-type genome.

2. Antibiotic-Resistant (ant^R) Mutants

A large number of antibiotics and drugs inhibit the growth of *S. cerevisiae* on nonfermentable substrates but do not affect growth on sugars. These agents act either as selective inhibitors of mitochondrial protein synthesis (mitochondrial ribosomes), of the terminal respiratory chain, or of the oligomycin-sensitive ATPase. Some of the better-known inhibitors and their targets are listed in Table 11.3. Respiratory and translational inhibitors act by either directly or indirectly interfering with the synthesis of ATP by oxidative phosphorylation and thereby prevent growth of yeast on substrates whose metabolism depends on functional mitochondria.

In the 1960s, several laboratories began isolating mitochondrial mutants that showed an increased resistance to antibiotics. The first such strains to be reported were resistant to erythromycin and chloramphenicol. In subsequent studies, other mutants resistant to oligomycin and antimycin were also isolated. Spontaneous or mutagen-induced antibiotic-resistant mutants are easily obtained, since the method depends on a positive selection. A commonly used procedure involves plating a heavy lawn of the sensitive haploid strain on a medium containing the antibiotic and glycerol or some other oxidizable substrate. After a period of several days to a week, resistant clones appear as single colonies on the plate.

Antibiotic-resistant strains probably have point mutations or small deletions in mit or syn genes which lower the binding affinity of the affected component for the antibiotic. For example, the binding site of oligomycin is most probably on subunit 9 of the ATPase complex. In oligomycin-resistant mutants, the binding site is modified through a change in the amino acid sequence of the protein. Although antibiotic-resistant mutants are recognized by their response to the inhibitors both *in vivo* and *in vitro*, the functions of the mutated components are not noticeably modified. Thus, chloramphenicol- or paromomycin-resistant strains carry out mitochondrial protein synthesis at optimal rates, and oligomycin- or antimycin-resistant mutants synthesize ATP at nearly the same efficiency as wild-type cells.

TABLE 11.3. Antibiotic Inhibitors of Mitochondrial Protein Synthesis and Respiratory Functions

Antibiotic	Locus	Site of inhibition
Chloramphenicol	*cap*	Large ribosomal subunit
Erythromycin	*ery*	Large ribosomal subunit
Paromomycin	*par*	Small ribosomal subunit
Oligomycin	*oli* 1,2,3,4	ATPase
Venturicidin	*ven*	ATPase
Ossomycin	*oss*	ATPase
Antimycin	*ana* 1,2	Coenzyme QH_2–cytochrome *c* reductase
Funiculosin	*fun*	Coenzyme QH_2–cytochrome *c* reductase
Mucidin	*muc*	Coenzyme QH_2–cytochrome *c* reductase
Diuron	*diu*	Coenzyme QH_2–cytochrome *c* reductase

3. Mit$^-$ Mutants

This class of mutants is characterized by their inability to grow on nonfermentable substrates because of point mutations or intragenic deletions in mit genes. Although mit$^-$ mutants have the same gross phenotype as ρ^- mutants and appear as small colonies on glucose medium, they differ from the latter in several important respects. Unlike ρ^- mutants which are defective in mitochondrial protein synthesis and are pleiotropically deficient in respiratory functions, mit$^-$ mutants retain the capacity for mitochondrial protein synthesis, and their deficiency is restricted to a single enzyme activity.

To date, only three types of mit$^-$ mutants have been isolated; their enzymatic phenotypes and absorption spectra are summarized in Fig. 11.7. The first type consists of strains that have a normal spectrum but are defective in oligomycin-sensitive ATPase activity. Mutants of the second type are deficient in cytochrome oxidase activity and show a spectral absence of cytochromes *a* and a_3. The third type is deficient in coenzyme QH_2-cytochrome *c* reductase and generally also lacks spectral cytochrome *b*. It is noteworthy that each of the three enzymes affected by mit$^-$ mutations has one or more subunit proteins translated on mitochondrial ribosomes (Chapter 10, Section IIID).

Three methods have been used to isolate mit$^-$ mutants. The most general of these is outlined in Fig. 11.8. A haploid strain of *S. cerevisiae* is mutagenized with Mn^{2+} to induce mutations predominantly in mitochondrial DNA. The culture is plated on a medium containing a low concentration of glucose and a high concentration of glycerol. On this medium, wild-type or respiratory-competent cells form large colonies, whereas respiratory-deficient mutants (ρ^-, syn$^-$, and mit$^-$) stop growing once the glucose is exhausted and appear as small colonies. Since most of the small colonies that are collected consist of ρ^- strains, the second step in the selection procedure is to assay the individual glycerol-negative clones for their ability to carry out mitochondrial protein synthesis. Strains that have a functional system of mitochondrial protein synthesis consist

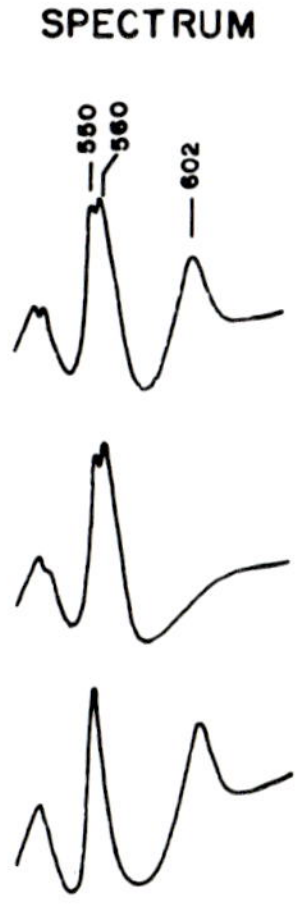

SPECTRUM (550, 560, 602)

Class	ATPase	Cyt. ox.	CoQ-c
I	−	+	+
II	+	−	+
III	+	+	−

FIG. 11.7. Spectral properties and enzymatic deficiencies of mit$^-$ mutants. The cytochromes are identified by the absorption maxima of the α bands. Cytochromes $a + a_3$ (603 nm); cytochrome *b* (560 nm); and cytochromes $c + c_1$ (550 nm).

largely of mit$^-$ mutants with enzymatic lesions in one of the three aforementioned enzymes. The protein-synthesis-negative strains are predominantly ρ^- mutants but also include syn$^-$ mutants.

Another means of selecting for mit$^-$ mutations is to use a nuclear mutant (*opl*) in which the ρ^- condition is lethal. Following mutagenesis with Mn^{2+}, ρ^- mutants are killed, and the surviving glycerol-negative strains are enriched in mit$^-$ mutants. A third method consists of crossing glycerol-negative strains with a set of ρ^- testers, each containing a different segment of mitochondrial DNA. The segments of DNA in the testers should cover the entire genome. Strains whose growth on glycerol is not restored by any testers consist of ρ^- mutants, whereas those that form respiratory-competent diploids when crossed to any of the testers are usually either mit$^-$ or syn$^-$ mutants.

4. Syn$^-$ Mutants

Syn$^-$ mutations result in an impaired system of mitochondrial protein synthesis. Mutants of this class, therefore, do not grow on nonfermentable substrates and are similar to ρ^- strains in being pleiotropically deficient in the respiratory and ATPase complexes. They can, however, be distinguished from ρ^- mutants on the basis of their genetic properties. Since the loss of the protein synthetic activity in syn$^-$ mutants is caused by point mutations rather than large deletions of DNA, their growth on glycerol can be restored when crossed with ρ^- strains whose retained segments of mtDNA contain the wild-type allele of the mutated gene.

In the screening procedure outlined in Fig. 11.8, the strains defective in mitochondrial protein synthesis consist of ρ^- and syn$^-$ mutants. There are sev-

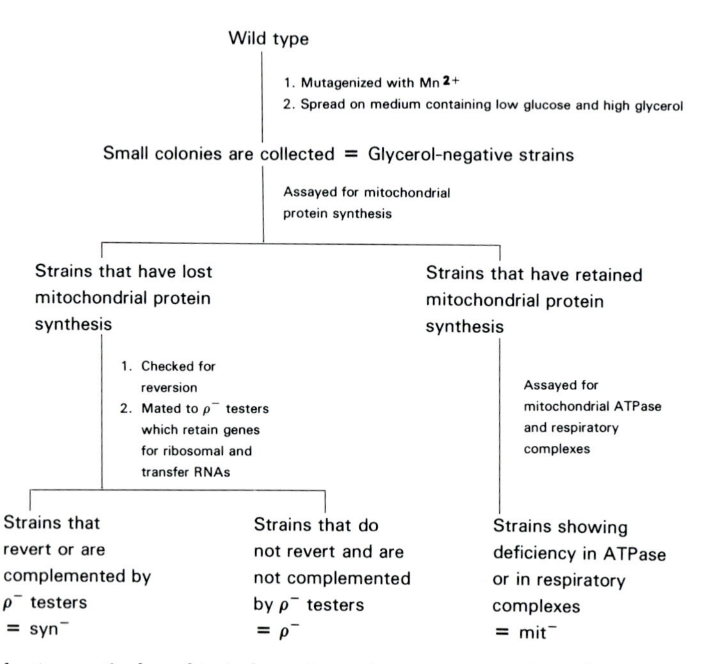

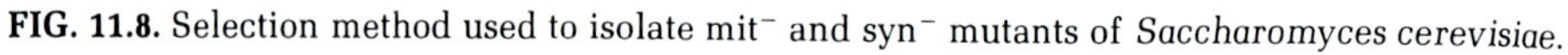

FIG. 11.8. Selection method used to isolate mit$^-$ and syn$^-$ mutants of *Saccharomyces cerevisiae*.

eral ways in which it is possible to distinguish between the two. The first is a reversion test. The rationale of this test is that since ρ^- mutants do not revert to wild type, strains with a measurable reversion frequency are likely to have point mutations in genes specifying components of the mitochondrial protein synthetic machinery. The second test is based on the previously mentioned fact that wild-type recombinants are not observed in crosses of two ρ^- mutants. Strains that yield respiratory-competent diploids when crossed to ρ^- testers are likely to be syn$^-$ mutants, particularly if the testers used have regions of mitochondrial DNA containing syn genes.

V. Genetic Methods for Mapping Miıochondrial Mutations

When two haploid strains are mated, the resultant zygotes acquire a common pool of the two parental mtDNAs. It has been experimentally shown that recombination of genetic markers following zygote formation is a continuous process that proceeds until the population is fully segregated and consists of pure diploid clones. Recombination of mtDNA is not restricted to crosses involving strains with a full complement of mitochondrial genes but occurs with equal efficiency when one or both partners are ρ^- mutants, even though the retained DNA segments may be relatively short.

Because of the recombination and segregation of pure recombinant clones, there is no simple test for genetic complementation* in mitochondrial genetics. In the absence of a convenient complementation test, it has been difficult to assign mitochondrial mutations to genes, and the looser concept of a *locus* has been introduced. The term locus can be operationally defined in the following way: two mutations are considered to be in the same locus if they have the same phenotype (e.g., resistance to an antibiotic or a defective enzyme) and if by genetic or other criteria they can be shown to be closely linked. The value of 1% recombination can be taken as an arbitrary but acceptable indication of close linkage. From this definition, it follows that a locus must contain at least one gene but does not exclude the possibility of two or more genes. By the same token, it is conceivable that two or more loci may be in the same gene. In fact, it is now known that the same gene may have several different antibiotic-resistance and mit$^-$ markers.

With these introductory remarks, we can now proceed to consider some of the methods currently being used to assign mutations to genetic loci.

A. Loci on Mitochondrial DNA

1. Antibiotic-Resistance Loci

Mutations in alleles that confer resistance to antibiotics are quite easily assigned to loci based on their phenotypes (each unrelated antibiotic defines a new locus) and their recombination frequencies. The known antibiotic-resis-

*It is possible to test for complementation when the mutations affect respiratory activity. The emergence of respiration in zygotes formed in a cross of two respiratory-deficient mutants has been shown to be a valid test for intergenic complementation.

tance loci on the *S. cerevisiae* genome are listed in Table 11.3. Of these *par*, *cap*, *oli* 1, *oli* 2, and *ana* 1 are the most useful as genetic markers, since they are unlinked and occur on widely separated parts of the genome. Other loci exhibit variable degrees of linkage to the above five. Mutations in *oli* 1 and *oli* 3 recombine at low frequency; nonetheless, they are viewed as separate loci because of the cross resistance of *oli* 3 but not *oli* 1 mutants to venturicidin.

It should be noted that mutations within a single locus are not necessarily allelic. For example, independently isolated oligomycin-resistant strains in *oli* 1 give recombination values ranging from 0–0.1% when crossed with each other. This indicates that there are a number of closely linked mutational sites or alleles in the *oli* 1 locus. A similar argument applies to the other resistance loci.

2. Mit Loci

There are currently two methods for assigning mit$^-$ mutations to loci. The first depends on the frequency of wild-type recombinants produced in pairwise crosses of different mutant strains (mit$_1^-$ × mit$_2^-$). Such crosses can be done either qualitatively by a patch test or by quantitating the percentage recombination. The method is illustrated in Table 11.4 which shows the results of pairwise crosses of a series of independent cytochrome *b*-deficient strains. In the qualitative patch test, the diploids replicated on glycerol show either confluent growth, papillae type of growth, or no growth at all. Confluent growth usually corresponds to recombination values of 1% or more, whereas crosses giving rise to papillae indicate 0.01–0.1% wild-type recombinants. Based on the results of Table 11.4, the cytochrome *b* mutants can be divided into two groups designated as *cob* 1 and *cob* 2. Each group satisfies the definition of a locus since crosses involving pairs of mutants within a group exhibit low recombination frequencies. Conversely, in crosses involving combinations of mutants from different groups, the frequency of wild-type recombinants is high.

An analysis of all mit$^-$ mutants isolated to date indicates that they fall into one of seven different loci. Three of these are associated with a deficiency in cytochrome oxidase, two in coenzyme QH_2–cytochrome *c* reductase (or cytochrome *b*), and two in oligomycin-sensitive ATPase (Table 11.5).

Another way of classifying mit$^-$ mutations is by means of ρ^- testers whose mtDNAs are capable of discriminating different mit loci. The principle of the method is depicted in Fig. 11.9. In this hypothetical example, a group of mit$^-$ mutants (A–H) is mated to four different testers, each with a different retained segment of mitochondrial DNA, and the diploids issued from the crosses are scored for glycerol growth. Since neither the mit$^-$ nor the ρ^- testers grow on glycerol, the appearance of respiratory-competent diploids in any given cross indicates that the ρ^- tester contains the wild-type allele of the mit$^-$ mutant. The results of the crosses suggest that the mutations fall into four separate groups or loci. Mutants A–D form a cluster that is restored by tester 1 only; a second cluster is represented by mutants G–H; the other two mutants, E and F, are seen to be different from each other and from the first two groups. Although the initial classification of mit$^-$ mutants is done with many different ρ^- clones, once the loci are defined, a few testers can be used to assign new mutations to the pre-

TABLE 11.4. Results of Pairwise Crosses of Cytochrome *b*-Deficient Yeast Mitochondrial Mutants[a]

	cob 1								*cob 2*				
	M6-200	M7-40	M8-53	M8-181	M13-101	M15-207	M17-231	M24-241	M10-152	M17-162	M18-68	M33-119	M21-71
M6-200	—	P	P	P	P	P	—	P	+	+	+	+	+
M7-40		—	—	+	—	P	P	P	+	+	+	+	+
M8-53			—	+	—	P	P	—	+	+	+	+	+
M8-181				—	+	P	+	P	+	+	+	+	+
M13-101					—	—	P	—	+	+	+	+	+
M15-207						—	P	P	+	+	+	+	+
M17-231							—	P	+	+	+	P	+
M24-241								—	+	P	+	+	+
M10-152									—	—	+	+	+
M17-162										—	P	—	—
M18-68											—	—	P
M33-119												—	P
M21-71													—

[a]Confluent growth of diploid patches on glycerol is indicated by +, papillae growth by P, and absence of growth by –.

viously established loci. These testers are chosen for their genetic stability and for the lengths of their retained segment of DNA which preferably should include only one locus.

In general, there is good agreement in the assignment of mutations by pairwise crosses of mit^- mutants and by crosses of mit^- to ρ^- testers.

3. Syn Loci

Mutations in the transfer and ribosomal RNA genes can be analyzed in the same way as mit^- mutations. Unfortunately, only relatively few syn^- strains have been isolated. Those currently available have mutations in tRNAs. Pre-

TABLE 11.5. Mit^- Loci in Yeast Mitochondrial DNA

Locus	Enzyme deficiency	Gene
oxi 1	Cytochrome oxidase	Subunit 2
oxi 2	Cytochrome oxidase	Subunit 3
oxi 3	Cytochrome oxidase	Subunit 1
cob 1	Coenzyme QH_2–cyt. *c* reductase	Apocytochrome *b*
cob 2	Coenzyme QH_2–cyt. *c* reductase	Apocytochrome *b*
box 4/5	Coenzyme QH_2–cyt. *c* reductase	Apocytochrome *b*
box 3	Coenzyme QH_2–cyt. *c* reductase	Apocytochrome *b*
box 8	Coenzyme QH_2–cyt. *c* reductase	Apocytochrome *b*
box 10	Coenzyme QH_2–cyt. *c* reductase	Apocytochrome *b*
box 1/9	Coenzyme QH_2–cyt. *c* reductase	Apocytochrome *b*
box 7	Coenzyme QH_2–cyt. *c* reductase	Apocytochrome *b*
box 2	Coenzyme QH_2–cyt. *c* reductase	Apocytochrome *b*
box 6	Coenzyme QH_2–cyt. *c* reductase	Apocytochrome *b*
pho 1	ATPase	Subunit 6
pho 2	ATPase	Subunit 9

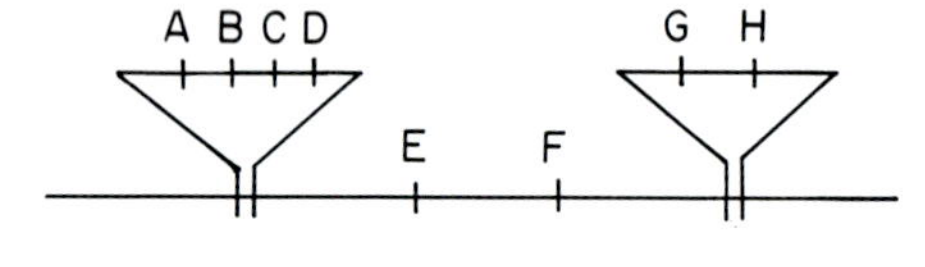

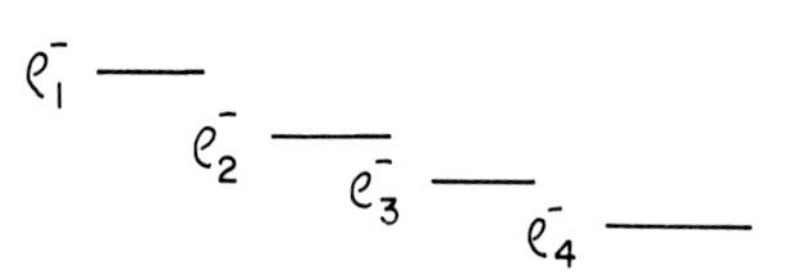

RESTORATION OF RESPIRATION

	A	B	C	D	E	F	G	H
ρ_1^-	+	+	+	+	–	–	–	–
ρ_2^-	–	–	–	–	+	–	–	–
ρ_3^-	–	–	–	–	–	+	–	–
ρ_4^-	–	–	–	–	–	–	+	+

Fig. 11.9. Test for assigning mit^- mutations to genetic loci on mtDNA. Mutational sites or alleles are designated by the letters A–H. The segments of mtDNA present in the ρ^- testers are indicated by the length of the bars. The genetic test involves crosses of the mit^- mutants with lesions in the different alleles to the ρ^- testers and scoring of the diploids for growth on glycerol (= respiration). The results of such tests for the hypothetical series of mutations are shown in the lower part of the figure.

sumptive tRNA mutants are characterized by examining the ability of the different tRNA species to be charged by their respective amino acids. The few mutants studied have been found to have a species of tRNA that fails to be acylated by its appropriate amino acid. This is shown in Fig. 11.10 for a mutant in threonine tRNA. *Saccharomyces cerevisiae* has two isoaccepting species of threonine tRNA. In the mutant, one of the two species fails to be charged with radioactive threonine. The mutation, therefore, defines a single gene, $tRNA_1^{Thr}$. It should be noted that in the case of tRNA mutations, we can speak of genes rather than loci, since each tRNA can be assumed to be specified by a single gene.

B. Mapping of Mitochondrial Markers by Recombination

Mitochondrial ant^R and mit loci have been located on the circular map by recombinational analysis and deletion mapping with ρ^- clones. The latter method is unique to yeast mitochondrial genetics and has become a powerful tool suitable both for localization of genetic loci relative to other markers and for fine mapping of mutations within loci.

1. Recombination of Mit and Ant^R Loci

The antibiotic-resistance loci are useful reference markers for the localization of mit^- and syn^- mutations on mtDNA. In some instances, genetic linkage established by recombination has permitted mit loci to be mapped in the vicinity of antibiotic-resistance markers. Recombination has also been used to map mit^- and syn^- mutations in spans of mitochondrial DNA that are bordered by antibiotic-resistance markers.

As in classical recombinational mapping, the degree of linkage of mitochondrial markers or their proximity is related to the frequency with which they recombine. This rule, however, does not apply for mutations that exceed a certain distance on the DNA. For reasons not entirely understood, distant markers on mtDNA exhibit an upper recombination limit of 20–25%. This

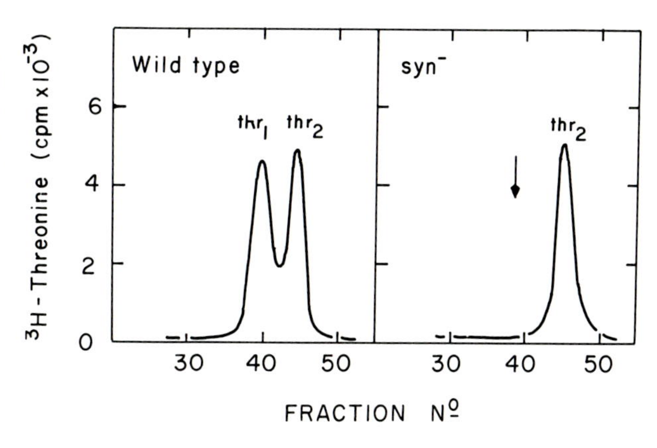

FIG. 11.10. Analysis of a syn^- mutant by reverse-phase chromatography. Total mitochondrial tRNAs were prepared from wild-type and a presumptive syn^- mutant of *S. cerevisiae*. The tRNAs were charged with individual radioactive amino acids, and the amino-acylated products separated on a reverse-phase RPC5 column. The results of this assay are shown when the tRNA mixture was charged with [^{3}H]threonine. It is seen that the mutant lacks the radioactive peak corresponding to $tRNA_1^{Thr}$. Since all other tRNAs in the mutant charged normally, the mutation was identified to be in the threonine-tRNA (Macino and Tzagoloff, 1979).

value is observed for all markers separated by a distance of 10% or more of the total genome length—in physical terms, this corresponds to 2.5 μm of DNA. In the case of more closely spaced mutations, the percent recombination is a good index of the degree of linkage.

The simplest test for genetic linkage is to cross an antibiotic-resistant strain (mit^+ant^R) to a mit^- mutant (mit^-ant^S). The diploids issued from the cross are grown for 20–30 generations to obtain pure clones which are scored for their phenotypes. In this type of cross, four mitotic segregants are produced: the two parental classes (mit^+ant^R, mit^-ant^S) and the two recombinants (mit^+ant^S, mit^-ant^R). Since the antibiotic-sensitive or -resistant alleles can only be scored on the respiratory-competent (mit^+) half of the population, the percent recombination is equal to the ratio of mit^+ant^S to mit^+ant^R + mit^+ant^S. Based on the observed recombination frequencies of various antibiotic-resistance and mit markers, close linkage has been established for the following pairs: *pho* 1/*oli* 2, *pho* 2/*oli* 1, *cob* 2/*ana* 1, and *cob* 2/*ana* 2. Since the recombination values for these markers are 1% or less, there is a good probability that the mutated alleles are within the same gene. In addition to the above, a lesser degree of linkage has been observed for *cob* 1 and *oli* 1 mutants. All other mit loci appear to be unlinked to the currently known antibiotic-resistance markers.

Another type of recombinational analysis is done by crossing a mit^- mutant to a respiratory-competent strain with three different antibiotic-resistance markers. By analyzing the different recombinant classes produced in the cross, it is possible to locate the mit mutation in one of the three spans of mtDNA defined by the ant^R markers. This test works only if all the input markers (mit^- and ant^R) are unlinked to each other. The rationale of the method is illustrated in Fig. 11.11. Let us assume that the antibiotic-resistant tester has the three

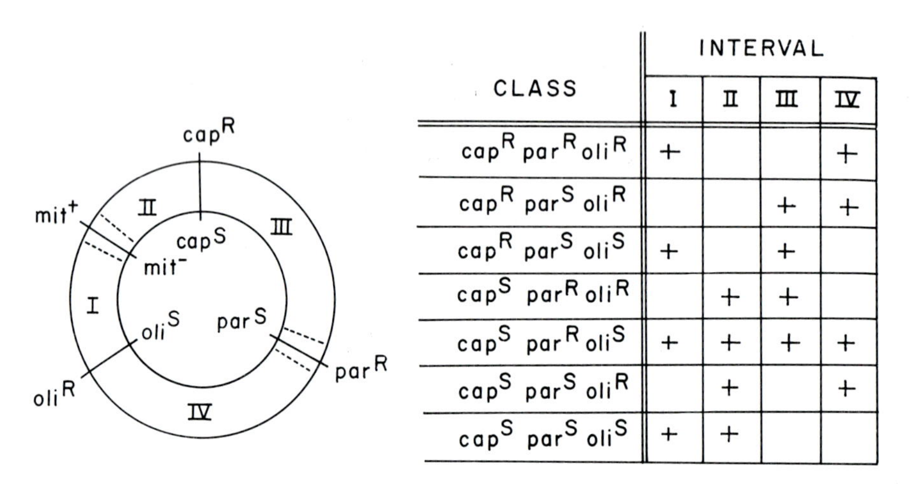

CLASS	INTERVAL I	II	III	IV
$cap^R\ par^R\ oli^R$	+			+
$cap^R\ par^S\ oli^R$			+	+
$cap^R\ par^S\ oli^S$	+		+	
$cap^S\ par^R\ oli^R$		+	+	
$cap^S\ par^R\ oli^S$	+	+	+	+
$cap^S\ par^S\ oli^R$		+		+
$cap^S\ par^S\ oli^S$	+	+		

FIG. 11.11. Model showing exchanges necessary for the various mit^+ recombinant classes produced from a mit^+*cap*^{R}*par*^{R}*oli* 1^R × mit^-*cap*^{S}*par*^{S}*oli* 1^S cross. The outer circle represents the mitDNA of the mit^+*cap*^{R}*par*^{R}*oli* 1^R, and the inner circle that of the mit^-*cap*^{S}*par*^{S}*oli* 1^S haploids, respectively. The dotted lines show the four exchanges that produce the *cap*^{S}*par*^{R}*oli*^{S} discriminating class. The intervals I–IV correspond to the spans in the diagram, and the plus signs indicate the intervals in which the exchanges must occur to produce the indicated recombinants. The results of an actual cross are shown in Table 11.9 (taken from Slonimski and Tzagoloff, 1976).

markers, *oli* 1^R, par^R, cap^R, and that we are interested in determining whether a certain mit^- mutation is localized in the *cap–par*, *par–oli* 1 or *oli* 1–*cap* span. In a cross of the mit^- mutant (mit^-, *oli* $1^S par^S cap^S$) to the antibiotic-resistant tester (mit^+ *oli* $1^R par^R cap^R$), eight different recombinant classes will be produced which can be distinguished by their resistances to the various antibiotics. Depending on which span the mit^- mutation is located in, only one recombinant class (discriminating class) will arise from four exchanges in mtDNA—the remaining seven classes each require only two exchanges. For example, if the mit^- mutation is in the *oli* 1–*par* span, the discriminating class is mit^+ *oli* $1^S par^S cap^R$. By analogy, the discriminating classes for the *oli* 1–*cap* and *cap–par* spans are mit^+ *oli* $1^S par^R cap^S$ and mit^+ *oli* $1^R par^S cap^S$, respectively. In any given cross, the least frequent type of recombinant should correspond to one of the three discriminating classes and, depending on which of the three it is, determines the localization of the mutation. An actual example of this method is shown in Table 11.6—the data were obtained from a cross of an *oxi* 1 mutant to a respiratory-competent haploid with the resistance markers *oli* 1^R, cap^R, and par^R. The least frequent recombinants scored in the cross were of the *oxi* 1^+, *oli* $1^R par^S cap^S$ type, indicating that the *oxi* 1 locus is in the *cap–par* span of mtDNA. This approach has been used to localize *oxi* 1, *oxi* 2, and some transfer RNA mutations in the *cap–par* and the *oxi* 3 locus in the *par–oli* 2 segments.

2. Recombination of Mit and Syn Markers

Since all the mit loci are genetically unlinked to each other, interlocus crosses of mit^- mutants yield recombination values that range from 20–25%, the usual recombination values of nonlinked mitochondrial markers. The relative distances of mit^- mutations within single loci, however, can be estimated by the frequency of wild-type recombinants issued from $mit_1^- \times mit_2^-$ crosses. In such intralocus crosses, mutational sites separated by 30–40 base pairs recombine with a recombination frequency of 0.1%. Although there are experimental problems in obtaining accurate recombination values, this method of mapping has been useful in ordering a series of mutations in the cytochrome *b* region of the genome. Quantitative recombinational analysis can also be performed in crosses involving $mit^- \times syn^-$ or $syn_1^- \times syn_2^-$ mutations. At present, the mapping of syn genes by this method has been limited by the small number of syn^- mutants available.

TABLE 11.6. Recombinant Classes Measured in a Cross of an *oxi* 1 to a Triple Antibiotic-Resistant Respiratory-Competent Strain

Number of $\rho^+ mit^+$ colonies scored[a]								
$C^R O_I^R P^R$	$C^R O_I^R P^S$	$C^R O_I^S P^R$	$C^R O_I^S P^S$	$C^S O_I^R P^R$	$C^S O_I^R P^S$	$C^S O_I^S P^R$	$C^S O_I^S P^S$	Total recombinants
1446	247	106	115	61	22	52	68	671

[a] The $C^R O_I^R P^R$ genotype is not considered a recombinant class since it corresponds to one the parental types used in the cross. The three discriminatory classes in this cross are $C^S O_I^R P^S$, $C^R O_I^S P^S$, and $C^S O_I^S P^R$. The smallest number of colonies were scored in the $C^S O_I^R P^S$ class, thus placing the *oxi* 1 mutation in the *cap–par* span of the genome.

C. Mapping of Mitochondrial Mutations with ρ^- Clones

The ρ^- mutants have become very useful for locating and mapping mutations on mtDNA. All of the methods currently in use are a form of deletion mapping based on the loss or retention of markers in independently isolated ρ^- clones. The presence or absence of mit, syn, or ant^R markers in the mtDNA of ρ^- clones can be tested in a relatively simple way. The mit$^-$ or syn$^-$ mutants are crossed with the collection of ρ^- clones. Crosses giving rise to respiratory-competent diploids indicate the retention of the wild-type mit or syn allele in the mtDNA segment of the ρ^- mutant; conversely, crosses that fail to produce respiratory-competent diploids indicate the loss of the allele in the ρ^- mutant. The presence of ant^R alleles in the mtDNA of ρ^- mutants is also easily determined. If the ρ^- mutants are derived from an antibiotic-resistant parent, the retention of the ant^R marker in the ρ^- mtDNA is ascertained by crossing the ρ^- mutants to a wild-type sensitive tester. Both sensitive and resistant diploids will be issued from crosses in which the ρ^- mutants have retained the resistance allele; ρ^- mutants that have lost the resistance marker will give rise to antibiotic-sensitive diploid cells only. The same analysis can be done if the ρ^- clones are derived from an antibiotic-sensitive parent except that in this case, the wild-type tester used should have the antibiotic-resistance alleles.

Let us now consider a few of the methods that have been devised to order mutations on the circular map.

1. Coretention and Codeletion Analysis

Schweyen and colleagues have developed a simple method for establishing the order of any four unlinked markers by determining the frequency of coretention or loss of two markers in a random collection of ρ^- clones. As shown in Fig. 11.12, there are three possible orders in which four markers can

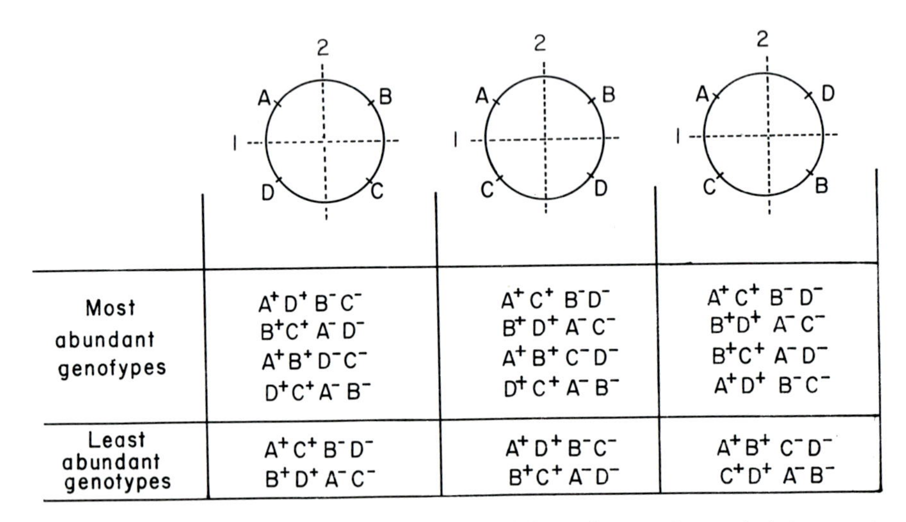

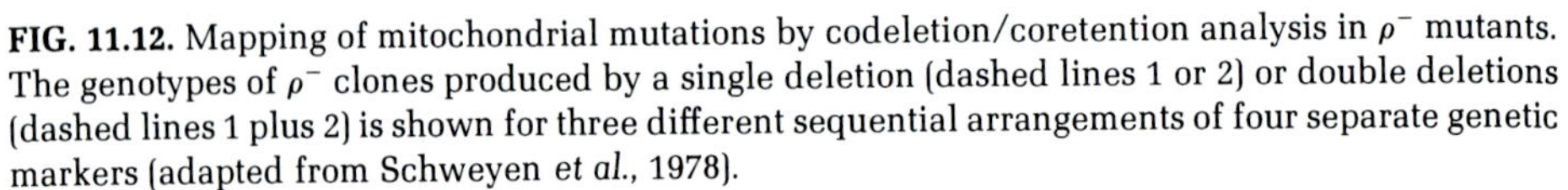

FIG. 11.12. Mapping of mitochondrial mutations by codeletion/coretention analysis in ρ^- mutants. The genotypes of ρ^- clones produced by a single deletion (dashed lines 1 or 2) or double deletions (dashed lines 1 plus 2) is shown for three different sequential arrangements of four separate genetic markers (adapted from Schweyen *et al.*, 1978).

be arranged on a circular DNA molecule. A choice of the most likely order is possible because the number of deletions necessary for the simultaneous loss or retention of different pairs of markers depends on the order in which they occur on the circle. Given the order ABCD, the retention or loss of the pairs AB, BC, CD, and DA requires only one continuous deletion, whereas the pairs AC and DB require two independent deletions. The pairs requiring two and four deletions will be different depending on the relative arrangement of markers. Since ρ^- clones having two deletions are less frequent than those with a single continuous deletion, it is possible by analyzing a collection of ρ^- clones for the coretention or codeletion of markers to assign a unique order for any four markers.

The frequency of coretention or deletion of markers can also serve as an index of their physical separation on the DNA. This is reasonable, since the probability with which a deletion will be initiated or will terminate in a segment of DNA bordered by two different markers will be higher as the length of the segment or the distance between the two markers increases. The probability of coretention (or codeletion) is therefore related to map distances and can be expressed as a coefficient that takes into account the total frequency with which two markers are retained in a random collection of ρ^- clones.

Although the coretention/codeletion method originally made use of stable secondary clones, it works equally well with primary ρ^- clones (heteroplasmic mtDNA) obtained by ethidium bromide mutagenesis. This is of some advantage when the analysis is done on markers that show linkage, and a large number of ρ^- clones are needed in order to obtain statistically significant numbers.

2. Mapping with Physically Characterized ρ^- Clones

Map distances can be assigned to mitochondrial markers by using ρ^- mutants whose retained segments of mtDNA are characterized in terms of length and genotype. This method was introduced in Linnane's laboratory to map both mit and antibiotic-resistance loci. The procedure involves the following steps.

1. Stable ρ^- clones are isolated and genetically verified to have retained the various markers of interest. From the previously established order of markers on the genome, clones are chosen if they satisfy the criterion of having arisen from a single continuous deletion.
2. The mtDNA of the ρ^- clones is isolated and hybridized to wild-type DNA in order to determine the fraction of the wild-type genome retained in each clone. The extent of sequence overlaps in the mtDNAs of the different ρ^- clones can be deduced by hybridization to the mtDNAs of a series of reference ρ^- clones.
3. With the information obtained in steps 1 and 2, it is possible to assign map distances to markers. This is shown in Fig. 11.13. In this example, the reference ρ^- mutant (U4) was found to retain 36% of the wild-type sequence. The percentage of DNA retained by the other ρ^- mutants as well as their markers is also indicated. A previous knowledge of the order in which the five markers are arranged in the genome allows maximal limits to be set for the distances intervening between each set of

markers. For example, the DNA sequence of U3 (15% of the wild-type DNA) is virtually completely included in the sequence of U4. The difference in length between U3 and U4 (21% of the genome) is the maximal separation of *oli* 1 and *cap* since they are the two outside markers lost by U3.

It is apparent that the precision with which markers can be positioned depends on having a number of closely neighboring markers and the appropriate ρ^- testers capable of discriminating among them. The limitations of this method lie in the accuracy of the hybridization method as a means of establishing the size of the retained segment.

3. Deletion Mapping in a Confined Region of mtDNA

The ρ^- mutants are potentially capable of discriminating among closely linked mutations. It is therefore possible to use ρ^- mutants to order a series of mutations in a relatively limited span of mitochondrial DNA. This approach has been used to map mutations in the *cob* 1 and *cob* 2 loci. In this instance, the ρ^- clones were selected for their ability to restore glycerol growth to two different cytochrome b mutants, one from the *cob* 1 and the other from the *cob*

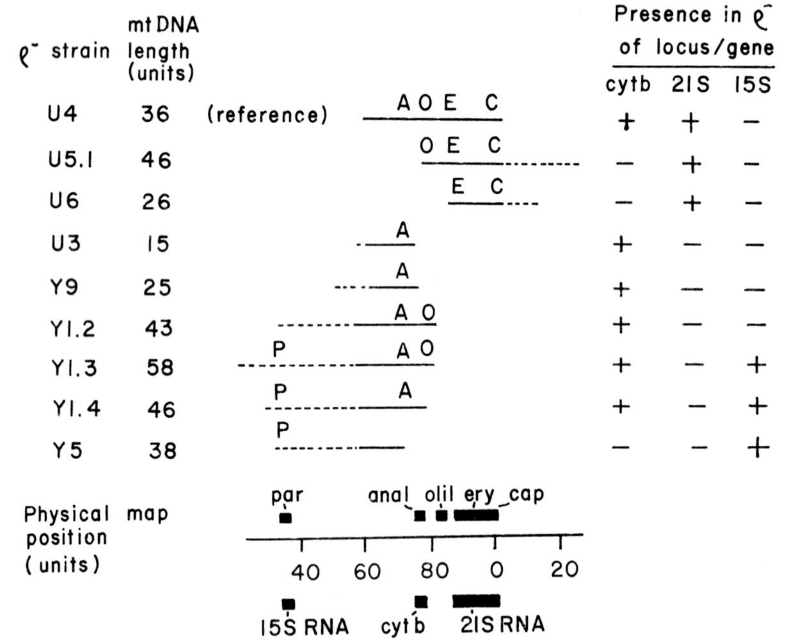

ρ^- strain	mt DNA length (units)		Markers	Presence in ρ^- of locus/gene: cytb	21S	15S
U4	36	(reference)	A O E C	+	+	−
U5.1	46		O E C	−	+	−
U6	26		E C	−	+	−
U3	15		A	+	−	−
Y9	25		A	+	−	−
Y1.2	43		A O	+	−	−
Y1.3	58		P A O	+	−	+
Y1.4	46		P A	+	−	+
Y5	38		P	−	−	+

Physical map position (units)

par ana1 oli1 ery cap

40 60 80 0 20

15S RNA cyt b 21S RNA

FIG. 11.13. Mapping of mitochondrial mutations with ρ^- mutants of known complexity. The sequence length of mitDNA is shown by the length of the solid and dashed lines and is based on the hybridization to wild type and the reference U4. The retention of the cytochrome *b* (cytb), 21 S (large ribosomal RNA), and 15S (small ribosomal RNA) is indicated by the plus signs. Other genetic markers are: A, *ana* 1; O, *oli* 1; E, *ery*; C, *cap*; P, *par*. The approximate positions of the antibiotic-resistance and mit loci are shown on the physical map in the lower part of the figure. The entire wild-type genome is set at 100 units.

2 locus. Out of approximately 119 stable secondary clones collected, 15 were able to discriminate among the different cytochrome *b* mutants studied. The other 104 clones restored all the cytochrome *b*-deficient strains and were therefore not useful for mapping purposes. The genotypes of the 15 discriminatory clones in terms of their retention of the various *cob* 1, *cob* 2, and the two resistance loci, *oli* 1 and *oli* 2, are shown in Table 11.7. By analyzing the different markers retained in the ρ^- clones, the relative order of the antibiotic-resistance and cytochrome *b* alleles indicated in Fig. 11.14 was deduced.

This method was also used by Fukuhara to map tRNA genes in the *cap–par* span of the genome except that the retention or loss of the genes was established not genetically but by hybridization of the tRNAs to isolated mtDNA from the ρ^- clones.

VI. Physical Mapping of Mitochondrial Genes

Restriction endonucleases have been used to construct physical maps of various viral, plasmid, and, more recently, mitochondrial DNAs. These genomes are relatively small (10^6–10^8 daltons), and maps showing the distribution of restriction sites can be derived without too much difficulty, particularly for enzymes that introduce a limited number of cuts in the DNA.

TABLE 11.7. Genotypes of ρ^- Mutants with Segments of mtDNA in the *cob* Region[a]

	Markers										
			cob 1				*cob* 2				
ρ^-	*ery*	*oli* 1	M7-40	M8-53	M6-200	M8-181	M33-119	M17-162	M21-71	M10-152	*oli* 2
RP-5B	−	+	+	+	+	+	+	−	−	−	−
-7B	−	−	+	+	+	+	−	−	−	−	−
-40B	−	+	+	+	+	+	−	−	−	−	−
-59B	−	+	+	+	+	+	−	−	−	−	−
-78B	−	+	+	+	+	−	−	−	−	−	−
-86B	+	+	+	+	+	+	−	−	−	−	−
-95B	+	+	+	+	+	+	−	−	−	−	−
-104B	+	+	+	+	+	+	−	−	−	−	−
-125B	−	−	+	+	+	+	−	−	−	−	−
-14B	−	−	−	−	+	+	+	+	+	+	−
-27B	−	−	−	−	−	−	+	+	+	+	+
-45B	−	−	−	−	−	−	+	+	+	+	+
-57B	−	−	−	−	−	+	+	+	+	+	−
-76B	−	−	−	−	−	+	+	+	+	+	+
-84B	−	−	−	−	−	−	+	+	+	+	+
-96B	−	−	−	−	−	−	+	+	+	+	+
-100B	−	−	−	−	+	+	+	+	+	+	+
-119B	−	−	−	−	−	+	+	+	+	+	+
-123B	−	−	−	−	+	+	+	+	+	+	−

[a]To test for the presence of the mit⁻ markers in the ρ^- clones, the cytochrome *b*-deficient strains (M7-40, etc.) were mated to the ρ^- clones, and the diploids issued from the crosses were scored for growth on glycerol. Plus indicates growth and minus absence of growth on glycerol.

As might be anticipated, the restriction maps of mtDNA from organisms belonging to different taxonomic groups bear little similarity to each other. This is true both for the number and relative positions of the restriction sites. Among more related organisms (e.g., mammals), the restriction maps are also quite different even though the genes are probably the same. Restriction analysis is sufficiently sensitive to detect sequence divergences even in genomes of individuals of the same species. Substantial polymorphism has been reported to exist among human individuals and in closely related strains of rats. Such polymorphism is also evident in the genomes of commonly used laboratory strains of *S. cerevisiae*. Although the physical maps of the strains of yeast studied until now are generally quite similar, certain regions of the DNA exhibit marked differences. The most noteworthy of these occur in the regions coding for apocytochrome *b* and subunit 1 of cytochrome oxidase. As will be described below, both genes contain intervening sequences some of which are lacking in certain strains of yeast.

Mitochondrial genes have been localized by direct hybridization of RNA transcripts to restriction fragments of wild-type DNA. This method is especially useful for mapping ribosomal and transfer RNA genes. The class of messenger RNAs can also be mapped by hybridization to restriction fragments of known origin on the genome. This approach has been used to obtain transcriptional maps of yeast and animal mtDNAs. Since the protein products encoded by the

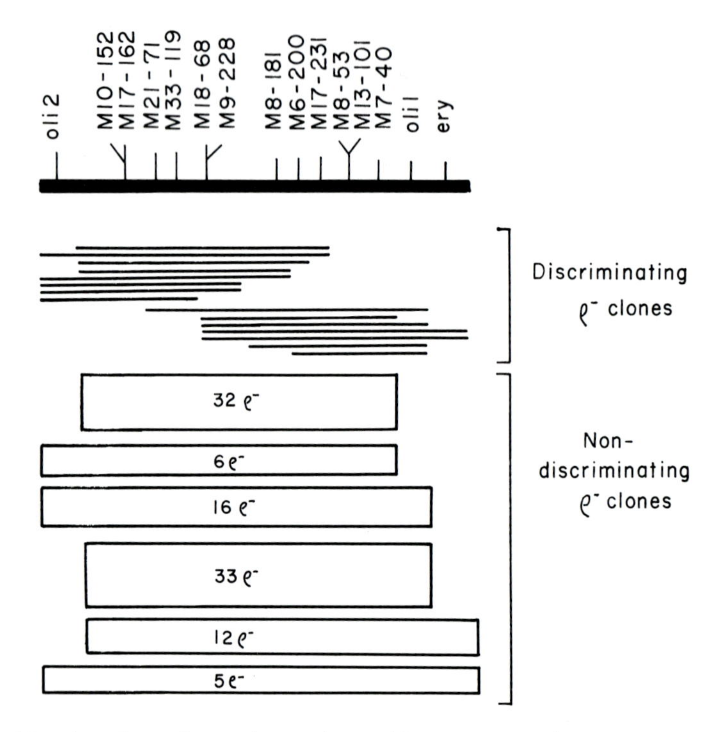

FIG. 11.14. Mapping of cytochrome *b* mutations with ρ^- mutants. The retention of the different cytochrome *b* alleles by the ρ^- clones is indicated by overlap of the lines and areas denoting the segments of mitochondrial DNA in the clones and the dark line showing the positions of the cytochrome *b* and antibiotic-resistance alleles.

messengers are not always known, the transcriptional maps do not by themselves provide information about the identity of the genes.

A. Yeast Mitochondrial Genes

Mit$^-$ mutations in *S. cerevisiae* constitute important markers in genes that code for messenger RNAs. This is also true of certain antibiotic-resistance loci (*oli* 1, *ana* 1, etc.). Most of the known mit$^-$ and ant^R loci of yeast have now been mapped by genetic as well as physical methods. The physical mapping of such markers has been achieved by comparative restriction analysis of the mitochondrial genomes present in genetically defined ρ^- clones. This will be illustrated for the *oli* 1 and *pho* 2 loci, both of which were known to be markers in the structural gene of the ATPase proteolipid.

A library of independent ρ^- clones was analyzed for their mitochondrial genotypes by crosses to appropriate ant^R and mit$^-$ testers. Out of 200 clones screened, several were ascertained to have retained a segment of mtDNA containing only the *cob*, *oli* 1, *oli* 3, and *pho* 2 loci. One of these clones (DS400) was mutagenized with ethidium bromide, and new clones were isolated with less complex genotypes. The marker compositions of some representative secondary clones are presented in Table 11.8. The three clones DS401, DS400/A3, and DS400/A4 had lost the *cob* loci as a result of the mutagenesis but still retained some of the ATPase markers. The repeat lengths of the mtDNAs of these clones ranged from 1 to 5.4 kbp. An analysis of the restriction sites present in the mtDNA segments of the three clones permitted these genomes to be aligned relative to one another and to the wild-type restriction map. As shown in Fig. 11.15, the most complex clone, DS401, has a segment of mtDNA extending from approximately 79–87.7 map units. The genomes of the lower-complexity clones, DS400/A3 and DS400/A4, represent smaller spans within this region. Based on the genotypes and the restriction maps of ρ^- mtDNAs, the *oli* 1 and *oli* 3 markers were localized between 80.9 and 82.6 map units (the deletion endpoints of DS400/A4). Since the *pho* 2 marker was present in DS400/A3

TABLE 11.8. Genetic Markers and Physical Sizes of mtDNA in ρ^- Clones Containing the ATPase Proteolipid Gene[a]

Clone	*oli* 1	*oli* 3	*pho* 2	*cob* 1	*cob* 2	Size of mtDNA repeat in base pairs
DS400	+	+	+	+	+	10,000
DS401	+	+	+	−	−	5,400
DS400/A3	+	+	+	−	−	1,800
DS400/A4	+	+	−	−	−	1,080

[a]The ρ^- clones were derived from *S. cerevisiae* D273-10B/A21. This respiratory-competent haploid strain carries a mutation in the *oli* 1 locus conferring resistance to oligomycin. To test for the retention of the *oli* locus, the clones were crossed to a wild-type sensitive haploid strain. The diploid progeny were replicated on glycerol medium containing rutamycin (oligomycin). Growth on this medium indicates the presence of the *oli* 1 resistance allele in the mtDNA of the clone. The presence of mit loci was checked by crossing the clones to mit$^-$ testers with mutations in *cob* 1, *cob* 2, *pho* 2 as well as other mit loci. The diploid cells were replicated on glycerol medium. Growth on glycerol in these tests indicates the presence of the wild-type mit allele in the ρ^- clone. Only the results with one antibiotic and three mit$^-$ testers are shown in the table. All other mit and antibiotic-resistance loci were found to be absent in the clones.

but not DS400/A4, it was most likely to lie between 80.4 and 80.9 map units. The ATPase proteolipid is only 76 residues long and, assuming a co-linear gene (this was subsequently verified by DNA sequencing), should be encoded by 228 nucleotides. The clustering of the markers in the vicinity of 81 map units therefore suggested this to be the most likely position for the gene. It is obvious that the accuracy with which genetic markers can be mapped by this procedure depends on the repeat lengths of the ρ^- genomes used in the analysis. Although very simple clones with repeat lengths of several hundred base pairs have been obtained from some regions of wild-type mtDNA, other regions tend to produce clones with rather long segments. In the latter case, the localization of markers tends to be imprecise.

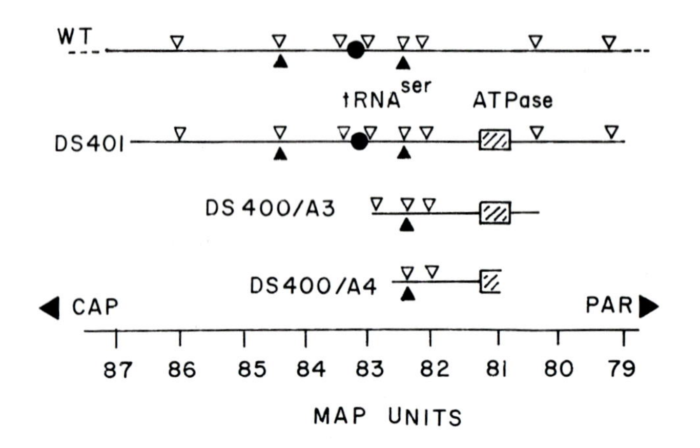

FIG. 11.15. Restriction maps of DS401, DS400/A3, and DS400/A4. The restriction maps of the ρ^- and wild-type DNAs have been aligned to indicate correspondence of the restriction sites for HpaII (△), HincII (●), and HhaI (▲). The position of the HincII site at 83 units is based on the physical map of D273-10B (see Fig. 11.16). Each map unit is equivalent to 700 nucleotides. The HincII site marks the position of the serine tRNA. The structural gene of the ATPase proteolipid is depicted by the hatched box. The direction of transcription is *par* → *cap* for both the serine tRNA and the ATPase gene.

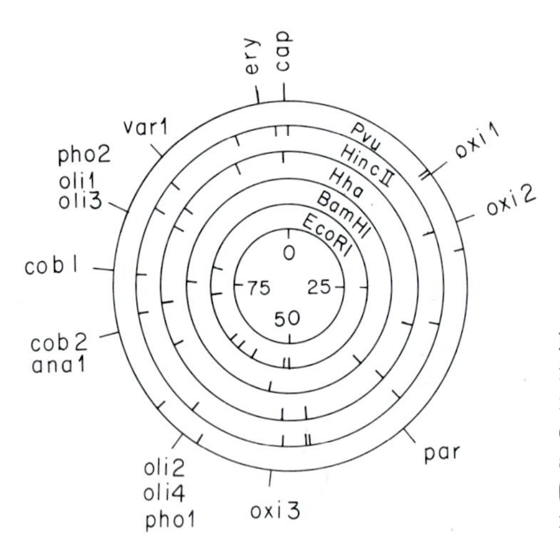

FIG. 11.16. Physical and genetic map of mtDNA in *S. cerevisiae* D273-10B. The restriction sites for five endonucleases are shown on the inner circles. The location of the antibiotic-resistance and mit loci are indicated on the outer circle. (Based on the studies of Morimoto and Rabinowitz, 1979.)

Almost all known genetic loci of the *S. cerevisiae* have been mapped within fairly accurate physical limits defined by the wild-type restriction map. In general, the locations of markers deduced from physical and genetic mapping are in very good agreement. The genetic and physical maps of the mitochondrial genome in *S. cerevisiae* D273-10B are presented in Figs. 11.16 and 11.17. Similar maps have been obtained for other strains that differ only in the presence of several long insertions in the *cob* and *oxi* 3 regions but otherwise have the same order and approximate location of the markers. The combined genetic and physical maps reveal a number of interesting features about yeast mtDNA. Most noticeable is the scattering of functionally related genes. For example, the three cytochrome oxidase loci, *oxi* 1, *oxi* 2, and *oxi* 3, are separated by either tRNA genes or genes coding for protein constituents of other mitochondrial enzymes. A similar disposition is seen in the case of the ATPase loci, *pho* 1 and *pho* 2, and the two ribosomal RNA genes. It is also obvious from an examination of the map that some 40% of the genome lacks genetic markers. Since most of the yeast mitochondrial genes have been saturated by mutations, the absence of genetic markers in long stretches of DNA suggested that a sub-

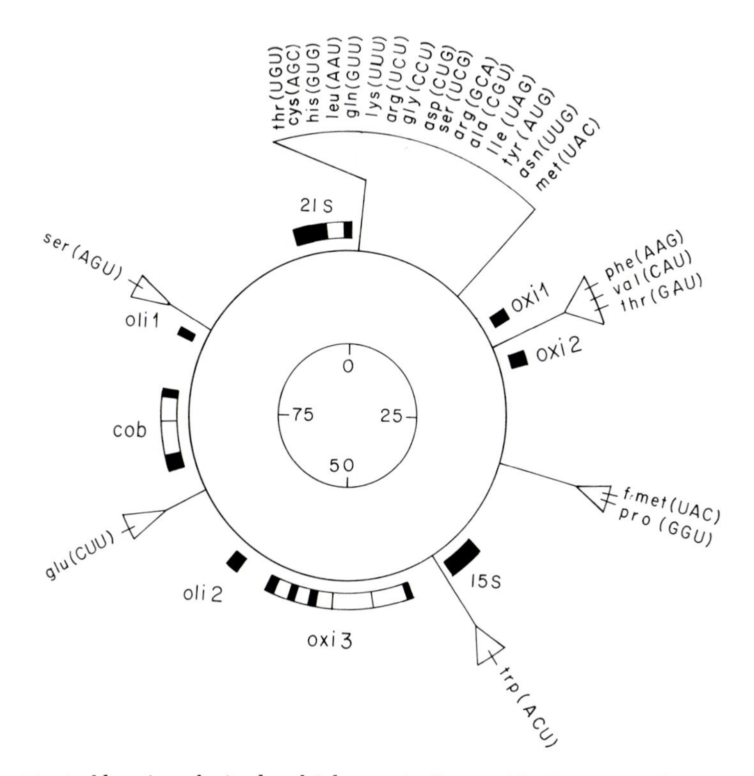

FIG. 11.17. Physical location of mitochondrial genes in *S. cerevisiae* D273-10B. The map shows the genes that have been sequenced. The dark portions of the 21S rRNA, *cob*, and *oxi* 3 loci denote the exon coding regions. The genes encoded in the various loci are as follows: 21S (large rRNA), 15S (small rRNA), *oxi* 1 (subunit 2 of cytochrome oxidase), *oxi* 2 (subunit 3 of cytochrome oxidase), *oxi* 3 (subunit 1 of cytochrome oxidase), *cob* (apocytochrome b), *oli* 1 (subunit 9 of ATPase), *oli* 2 (subunit 6 of ATPase). The locations of the tRNA genes and their anticodons (3′ → 5′) are also indicated.

stantial percentage of the genome consists of noninformational DNA. This has been substantiated by the transcriptional maps and even more conclusively by DNA sequencing. Thus, regions with no genetic markers are transcriptionally silent and are composed of very (A + T)-rich sequences. The yeast mitochondrial genome is, therefore, extremely specialized, its main function being to provide the organelle with the RNA components of the protein translational machinery and a small number of messenger RNAs vital to the development of respiration and oxidative phosphorylation.

B. Identity of Yeast Mitochondrial Genes

All of the ant^R, syn$^-$, and mit$^-$ loci of yeast mtDNA have been assigned to specific genes (Table 11.9). The genes have been identified (1) by analysis of defective proteins or tRNAs in mutants, (2) by physical mapping of genetic loci in physical spans of DNA with known structural gene sequences, e.g., *cap* and *ery* in the 21S rRNA, and (3) by DNA sequence analysis of the genes.

1. Syn Genes

The two ribosomal RNA genes appear to be part of the genetic makeup of all mtDNAs. It was already mentioned that yeast mitochondria also code for the tRNAs of the 20 amino acids and the initiator tRNA. There is some evidence that at least one ribosomal protein subunit is encoded in the yeast genome. This protein carries the genetic marker *var* 1.

2. Mit Genes

These markers have been shown to occur in genes coding for proteins that are translated on mitochondrial ribosomes. The identification of the specific gene products was initially based on the production of novel polypeptides in certain mit$^-$ mutants. For example, mutations in the *cob* or *box* loci frequently result in the synthesis of truncated forms of apocytochrome *b*. Similarly, certain *oxi* 1 mutants have been shown to make new low-molecular-weight products

TABLE 11.9. Genetic Loci and Gene Products of Yeast Mitochondrial DNA

Gene	Loci
21S rRNA	*ery*, *cap*
16S rRNA	*par*
ATPase subunit 6	*oli* 2, *oli* 4, *ven*, *pho* 1
ATPase subunit 9	*oli* 1, *oli* 3, *pho* 2
Apocytochrome b	*cob* 1, *cob* 2, *box* 1–10, *ana* 1, *ana* 2, *muc*, *diu*
Cytochrome oxidase	
Subunit 1	*oxi* 3
Subunit 2	*oxi* 1
Subunit 3	*oxi* 2
Ribosomal subunit var 1	*var* 1

antigenically related to subunit 2 of cytochrome oxidase. Although the earlier conclusions about the nature of the gene products encoded in yeast mtDNA were based on somewhat indirect evidence, they have been fully confirmed by more recent DNA sequence data.

C. Other Mitochondrial Genomes

At present, the best-studied genomes are those of yeast, *N. crassa*, human, and bovine mitochondria. Even though information about the genetic content and organization of other mtDNAs (plants) is still fragmentary, there is reason to believe that this situation will not persist for long. Recent studies indicate that yeast mitochondrial DNA probes can cross-hybridize with high specificity to homologous sequences in other types of mtDNA. Defined DNA probes of *S. cerevisiae* have already been used to identify and map mit genes in *N. crassa* and other yeast species. Since mitochondrial gene products appear to have highly conserved amino acid sequences, the use of heterologous DNA probes may prove to be a general and rapid means for delineating the gene compositions and genetic maps of mtDNAs from widely different sources.

The organization of human mtDNA will be described in more detail later in this chapter. It is worth emphasizing that despite being only one-fifth the size of the yeast genome, it has a greater coding capacity. This is because of its extremely compact arrangement of genes.

The genome of *N. crassa* is not unlike that of yeast. In addition to the ribosomal and tRNA genes, *N. crassa* mitochondria code for the three large subunits of cytochrome oxidase, apocytochrome *b*, and at least one subunit of the ATPase (subunit 6). In contrast to yeast, however, the proteolipid subunit of *N. crassa* ATPase is encoded in nuclear DNA*. There exists the intriguing possibility that even though the expressed proteolipid gene of *N. crassa* is in chromosomal DNA, a silent copy of it is also present in mtDNA. Thus, yeast probes with sequences of the gene have been found to hybridize to a specific region of *N. crassa* mtDNA. The region of hybridization has not been sequenced, and it is therefore still uncertain whether the cross-hybridization occurs because of the presence of a homologous sequence.

From the relatively few genomes studied up to now, it is already becoming clear that mtDNA has undergone profound changes during evolution. A comparison of the yeast and human DNAs points to differences not only in codon utilization but also in the structures of the tRNAs and even the number of genes. One of the more exciting prospects is that future studies may reveal concrete examples of how genes evolve and are transferred from one genome to another.

VII. Sequencing of Mitochondrial DNA

The development of rapid methods for sequencing of DNA has opened the way to answer many longstanding questions about the organization and regu-

*This is also true of the human and bovine proteolipids.

lation of transcription of genes in procaryotic and eucaryotic genomes. Because of its relatively small size, mtDNA is one of the few eucaryotic genomes (chloroplast DNA is another example) that can be sequenced in its entirety. This goal has already been achieved for the human genome and is not too far from completion in yeast.

A. Chemical Method of DNA Sequencing

One of the more popular methods of DNA sequencing will be briefly described to help further discussion of this topic. This method, devised by Maxam and Gilbert, is based on the chemical cleavage of DNA at positions corresponding to each of the four bases. A variation of this method is schematically depicted in Fig. 11.18.

Double-stranded DNA is labeled at the 5′ end with ^{32}P. The labeling is done enzymatically with [$\gamma^{32}P$]ATP in the presence of polynucleotide kinase. The labeled DNA is denatured and separated into single strands which are then subjected to the chemical modification reactions. The single strands are

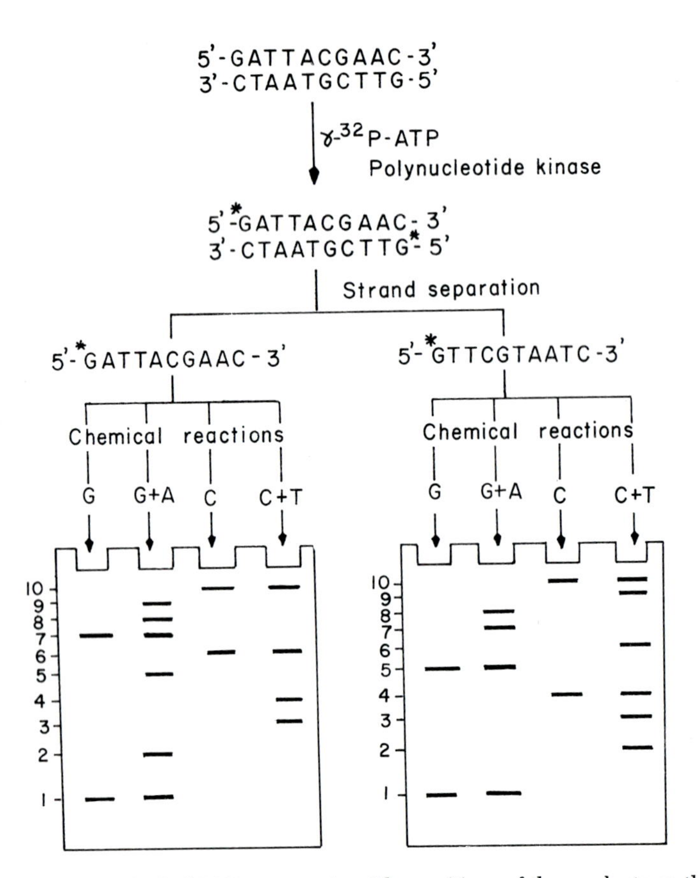

FIG. 11.18. Chemical method of DNA sequencing. The positions of the products on the polyacrylamide gel are shown by the solid lines. The numbers on the left-hand side indicate the lengths (in nucleotides) of the various products. (Adapted from Maxam and Gilbert, 1977.)

cleaved at positions corresponding to one or two of the four bases. A number of different reactions have been used to effect such specific cleavages. The following four are sufficient to illustrate the general principle.

1. G + A: Depurination at acid pH followed by alkaline hydrolysis causes cleavage at adenine and guanine.
2. G: Dimethylsulfate is used to methylate nitrogen 7 of guanine and nitrogen 3 of adenine. In the presence of base, a series of reactions leads to an opening of the guanine ring, displacement of the modified guanine base, and cleavage of the DNA. Even though adenine is methylated by dimethysulfate, the 3-methyl adenine remains attached to ribose, and the DNA is not cleaved.
3. C + T: Hydrazine causes the pyrimidine rings of cytosine and thymine to be chemically modified. The products of this reaction are a mixture of ureido and hydrazone derivatives of ribose which are cleaved by alkali.
4. C: Since the reaction of hydrazine with thymine is suppressed by high salt, specific cleavage of DNA at the cytosine positions only is achieved by including salt in the reaction mixture.

Each of the four reactions is carried out under conditions that allow only partial chemical modification in order to obtain all possible cleavage products of the DNA. The products of the cleavage reactions are separated electrophoretically on polyacrylamide gels. Exposure of the gels to X-ray film visualizes the class of products containing the 5′-end-labeled nucleotide. Since the cleavage products are separated according to the number of nucleotides in the chain, it is possible to read the sequence of the DNA by examining the pattern of radioactive bands in the channels corresponding to the products of each of the four reactions.

B. Sequencing Strategy for Yeast mtDNA

Although there are no theoretical limits on the size of the DNA that can be sequenced, in practice the length should not exceed 3000–4000 base pairs and must have a sufficient number and distribution of restriction sites to yield fragments not more than 400 base pairs long. The amount of DNA available is also an important consideration. Both requirements are dictated by the limitations of the sequencing method. The resolution of sequencing gels is usually good up to 250 nucleotides after which the reading of the gels becomes difficult and inaccurate. By sequencing complementary strands, however, it is possible to obtain the sequence of fragments that are up to 400 base pairs long. Having reasonably large amounts of DNA eliminates the problem of low radioactivity which results in poor-quality autoradiograms.

Most wild-type strains of *S. cerevisiae* contain mtDNA with a molecular weight of 50×10^6 which is roughly equivalent to 70,000 base pairs. From the above comments, it is obvious that in order to sequence a DNA of this size it is necessary to first obtain segments representing different regions of the genome. There are several ways in which this can be done. One is to amplify specific regions by cloning restriction fragments of the wild-type DNA in bacteria using plasmid or phage vectors. An alternative approach is to use as a source of DNA the mitochondrial genomes of genetically marked ρ^- clones. We have already seen that such mutants can serve as a source of small regions of the wild-type genome by virtue of their ability to amplify the nondeleted segment, the amplification factor being inversely proportional to the length of the segment retained. Since the sequencing studies done up to now have employed mtDNA of ρ^- mutants, this approach and a specific example showing how it was used to deduce the sequence of the ATPase proteolipid gene will be described.

The overall strategy to sequence the wild-type mitochondrial genome involves the isolation of ρ^- mutants with segments of mtDNA representing different regions of the wild-type genome. In view of the numerous genetic markers now available, it is possible to select ρ^- mutants enriched for almost all of the known genes. Once appropriate mutants are obtained, detailed restriction maps of their DNAs are constructed. This information is important not only for the actual sequencing but also for localizing the segment of the mutant on the wild-type map. The eventual reconstruction of the wild-type sequence depends on having a sufficient number of different ρ^- mutants with overlapping segments of mtDNA.

C. Nucleotide Sequences of Mit Genes

1. ATPase Genes

The ATPase proteolipid gene was a logical first choice for sequencing studies since it is the only yeast mitochondrial gene product with a known primary structure. Furthermore, the *oli* 1 and *pho* 2 markers serve as useful tools for selecting ρ^- clones enriched for this gene.

The properties of several ρ^- mutants selected for the retention of *oli* 1, *oli* 3, and *pho* 2 were described earlier (Table 11.8). The genome of the simplest clone, DS400/A3, has been completely sequenced. This segment contains the proteolipid gene plus 5′ and 3′ flanking sequences extending from 80.4 to 83 map units. Part of the nucleotide sequence of the DS400/A3 genome is shown in Fig. 11.19. The sequence represents the span of wild-type mtDNA from approximately 80.4 to 82 map units. The structural gene begins with an ATG initiation codon at nucleotide +1 and ends with an ochre terminator (TAA) at nucleotide +221. With the exception of the single amino acid at residue 46, the DNA sequence is in complete agreement with the known primary structure of the protein. The threonine at position 46 of the protein is encoded by CTA (= CUA), one of the standard leucine codons. The four codons of the leucine CUN family have been shown to code for threonine rather than leucine in yeast mitochondria. This is one of several recently discovered deviations in codon usage by mitochondria (see Section VIIID).

In the DS400/A3 clone, the sequence preceding the amino terminal end of the gene is 346 nucleotides long and consists almost exclusively of A + T. A similar (A + T)-rich sequence occurs at the carboxyl terminal end of the gene. This long stretch of DNA, however, contains a number of short (G + C)-rich clusters that are palindromic and in several instances are homologous to each other. The function of these sequences will be discussed later in this chapter.

The (A + T)-rich regions adjoining the proteolipid gene are probably spacers since they do not have any recognizable coding sequences and are punctuated by frequent ochre termination codons in all the possible reading frames.

The smaller mtDNA of DS400/A4 (see Fig. 11.15) has a sequence identical to that of DS400/A3 starting with nucleotide 132 in the ATPase proteolipid gene. The genome of DS400/A4, therefore, was formed by a deletion that started within the structural gene. This accounts for the absence of the *pho* 2 marker (mutation at amino acid residue 42) in this clone.

The DS400/A3 segment was oriented with respect to the wild-type mtDNA by comparative restriction mapping. As mentioned in Section VI.A, the segment of DS400/A3 is contained entirely within the larger genome of DS401. The latter has a segment with a unique HincII and two HhaI sites that have been mapped at 83, 82.3, and 84.1 units of wild-type mtDNA (Figs. 11.16, 11.17). Of these three sites, only one of the two HhaI sites (82.3 units) is present in DS400/A3. This single HhaI and the HpaII sites common with DS401 allow a precise localization and orientation of the DS400/A3 segment on the wild-type map. The orientation of the amino and carboxyl terminal ends of the gene is such that the direction of transcription is clockwise according to the convention of Fig. 11.17.

The second ATPase gene sequenced codes for subunit 6 of the ATPase. This gene contains the three genetic markers *oli* 2, *oli* 4, and *pho* 1. The nucleo-

```
5' ATATATATATGAATT AATATTTAATAATAA ATAATAATATAATTA ATAATATTATTATTA TTATAATTTTTATTT ATAATATTATAAATA

   TTATTATATATATAT TATAATAATATTAAT AAGATATATAAATAA GTCCCTTTTTTTTTA TTTAAAATAAAGAAG ATAATTAATATATTT

   TAATAATTTAATTAA ATGTGTATTAAAAGA ATAATAAAAAGATAA TATTAATATGTTAAT TATATATAATATATT ATATATAATTATATA

                                                                                   fMet-Gln-Leu-Val-
   TATATATATAAATAA TAATAAATATATATA TAATATAAAAATAAG AATAGATTAAATATT TAATAAATAAATATT ATG CAA TTA GTA

    5                   10                  15                  20                  25
   Leu-Ala-Ala-Lys-Tyr-Ile-Gly-Ala-Gly-Ile-Ser-Thr-Ile-Gly-Leu-Leu-Gly-Ala-Gly-Ile-Gly-Ile-Ala-Ile-
   TTA GCA GCT AAA TAT ATT GGA GCA GGT ATC TCA ACA ATT GGT TTA TTA GGA GCA GGT ATT GGT ATT GCT ATC

       30                  35                  40                  45                  50
   Val-Phe-Ala-Ala-Leu-Ile-Asn-Gly-Val-Ser-Arg-Asn-Pro-Ser-Ile-Lys-Asp-Thr-Val-Phe-Pro-Met-Ala-Ile-
   GTA TTC GCA GCT TTA ATT AAT GGT GTA TCA AGA AAC CCA TCA ATT AAA GAC CTA GTA TTC CCT ATG GCT ATT

           55                  60                  65                  70                  75
   Phe-Gly-Phe-Ala-Leu-Ser-Glu-Ala-Thr-Gly-Leu-Phe-Cys-Leu-Met-Val-Ser-Phe-Leu-Leu-Leu-Phe-Gly-Val-
   TTT GGT TTC GCC TTA TCA GAA GCT ACA GGT TTA TTC TGT TTA ATG GTT TCA TTC TTA TTA TTA TTC GGT GTA

   ochre     ochre
   TAA TATATA TAA   TATATTATAAATAA ATAAATAAATAAATA ATGAAATTAATAAAA AAATAAAATAAAATA AAATCTCATTTGATT

   AAATTAATAACATTC TTATAATTATATAAT TATTATAAATATATA AATATTATAATAATA ATAATATATAT... ≃130 nucleotides

   to . . .    TTAT ATATATTATGATATT ATTATGTAACATTAT ATAATAATATAAATT ACCATAATGAAATAT ATTATTTATTAATAA

   TAAAATATTTATTAA TAATAGAATATATAT ATTATGATAATATTT ATTAATAAATAATAA ATTCTTTATATATAA ATATATTAAATATAT

   TTAATTGGACACAAT ATAATTTTTATTATA ATTATGATAATATTT ATTAATAAATAATAA ATTCTTTATATATAA ATATATTAAATATAT

   TTAATTGGACACAAT ATAATTTTTATTATA TTATTCATTTAATAA TATTAATATTAATAT TAATATTAATATAAT ATTGGTGAAACATCT

   CCTTTCGGGGTTCCG G
```

FIG. 11.19. Nucleotide sequence of the ATPase proteolipid (subunit 9) and of the flanking regions. The sequence shown is that of the nontranscribed strand.

tide sequence of the gene has been obtained from three independent ρ^- clones isolated from respiratory-competent strains with different genotypes. The clone DS520 was derived from an oligomycin-sensitive parent (*oli* 2^s *oli* 4^s), DS14 from a resistant parent with a mutation in *oli* 2 (*oli* 2^r*oli* 4^s), and DS500 from another oligomycin-resistant strain with a mutation in *oli* 4 (*oli* 2^s *oli* 4^r). The mtDNA segments of the three clones had slightly different deletion endpoints but in each case arose from a region of wild-type mtDNA where the two oligomycin-resistance loci and the *pho* 1 locus had previously been mapped (60–67 units of the map). The ρ^- genomes had a common sequence with a reading frame coding for a hydrophobic protein consisting of 259 residues. The nucleotide sequences of this gene in the three clones were identical except for two nucleotide positions corresponding to amino acid residues 171 and 232 (Fig. 11.20). The nucleotide substitutions in the *oli* 2 resistant strain leads to a replacement of methionine for isoleucine at residue 171. In the *oli* 4 resistant mutant, the base substitution replaces a phenylalanine for a leucine at residue 232 (Table 11.10). These data indicate that the *oli* 2 and *oli* 4 loci are in the same gene and that oligomycin resistance results from mutations causing single amino acid changes in the protein, a situation analogous to that of the proteolipid gene.

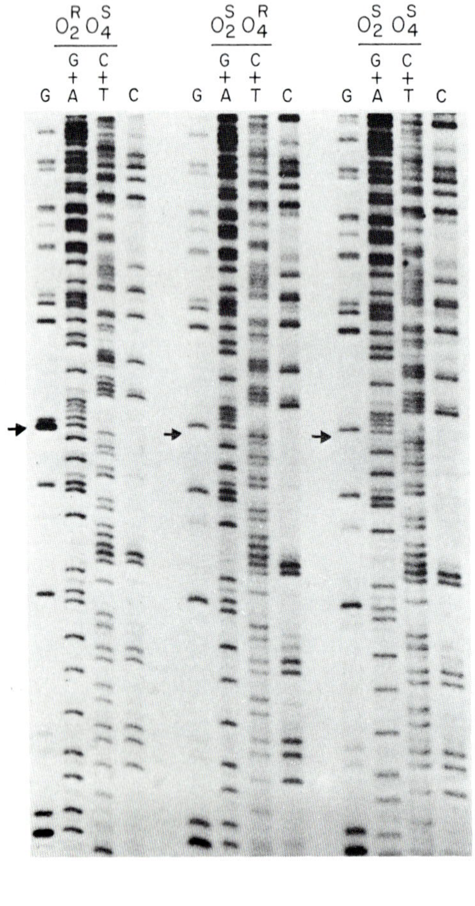

FIG. 11.20. DNA-sequencing gels of part of the ATPase subunit 6 in three different strains of *S. cerevisiae*. The arrow points to nucleotide position 513 where a G has been substituted for a T in the gene of the *oli* 2 resistant strain.

TABLE 11.10. Nucleotide Substitutions in *oli* 2 and *oli* 4 Resistant Mutants[a]

Strain	Genotype	Nucleotide		Amino acid residue	
		513	696	171	232
DS520	$oli2^s oli4^s$	T	A	Ile	Leu
DS14	$oli2^r oli4^s$	G	A	Met	Leu
DS500	$oli2^s oli4^r$	T	T	Ile	Phe

[a]Taken from Macino and Tzagoloff (1981).

Several lines of evidence suggest that the *oli* 2 gene codes for subunit 6 of the ATPase. (1) The amino acid composition and size of the *oli* 2 gene product is in good agreement with the known properties of this protein. (2) Mutations closely linked to *oli* 2 have been shown to give rise to size alterations in subunit 6.

2. Cytochrome Oxidase Genes

Yeast mitochondria code for three subunit polypeptides of cytochrome oxidase. The three genes are encoded in the *oxi* 1, *oxi* 2, and *oxi* 3 loci.

The *oxi* 1 locus has been mapped at approximately 13 units of the wild-type genome. The nucleotide sequence of this gene has been determined by the same approach used to study the ATPase genes. A segment of mtDNA present in a ρ^- clone selected for the retention of *oxi* 1 markers was sequenced and found to contain a reading frame capable of generating a continuous amino acid sequence of 251 residues. The gene starts with an ATG initiator, terminates with an ochre codon, and is flanked by (A + T)-rich sequences (Fig. 11.21). The reading frame has been identified as the gene of subunit 2 of cytochrome oxidase. The gene codes for a protein whose overall homology with the known amino acid sequence of bovine subunit 2 is 49%. The extent of primary structure homology of the two proteins is depicted graphically in Fig. 11.22 and is seen to be especially extensive in the carboxyl half of the proteins. The *oxi* 1 gene contains five UGA codons which normally function as translational stops. Four of the five UGAs occur at positions where the human subunit 2 has tryptophans. This finding indicated that in yeast mitochondria UGA functions as a codon for tryptophan. UGA has been shown to be used as a tryptophan codon not only in yeast but in human, bovine, and *N. crassa* mitochondria. This deviation from the universal code therefore appears to be a general feature of the mitochondrial genetic code.*

The *oxi* 2 locus has been localized in a coding sequence identified as the gene of subunit 3 of cytochrome oxidase. The gene is 810 nucleotides long and is located between 19.8 and 20.9 map units. It is a hydrophobic protein with an amino acid composition matching that of bona fide subunit 2. The first ten

*This statement has to be modified in view of a recent report that UGA has the usual terminator function in plant mitochondria.

amino acids encoded in the yeast gene are homologous to the experimentally determined sequence of bovine subunit 3.

The *oxi* 3 locus of *S. cerevisiae* D273-10B has been mapped between 43 and 58 map units. Prior to any information about the DNA sequence of this region, it was already evident that the *oxi* 3 locus codes for subunit 1 of cytochrome oxidase and that the gene was likely to be split with numerous intervening sequences. The evidence supporting this conclusion is summarized in Table 11.11.

The sequence of the *oxi* 3 gene has been obtained from the genomes of a series of ρ^- clones each having a different span of the *oxi* 3 region. Although the clones used for the sequence analysis had unique segments of mtDNA, there was sufficient overlap to permit a reconstruction of the entire gene. Based on the final DNA sequence, the *oxi* 3 gene was deduced to be 9979 nucleotides in length occupying the region from 43.9 to 58 map units. Since the amino acid sequence of the yeast subunit 1 has not been determined, the identification of

```
                                                                            fMet-Leu-Asp-Leu-Leu-Arg-Leu-Gln-Leu-
AAATAAATTTTAATT AAAAGTAGTATTAAC ATATTATAAATAGAC GAGAGTCAAAGGTTA AGATTTATTAAA ATG TTA GAT TTA TTA AGA TTA CAA TTA

Thr-Thr-Phe-Ile-Met-Asn-Asp-Ala-Pro-Thr-Pro-Tyr-Ala-Cys-Tyr-Phe-Gln-Asp-Ser-Ala-Thr-Pro-Asn-Gln-Glu-Gly-Ile-Leu-
ACA ACA TTC ATT ATG AAT GAT GTA CCA ACA CCT TAT GCA TGT TAT TTT CAG GAT TCA GCA ACA CCA AAT CAA GAA GGT ATT TTA

Glu-Leu-His-Asp-Asn-Ile-Met-Phe-Tyr-Leu-Leu-Val-Ile-Leu-Gly-Leu-Val-Ser-Trp-Met-Leu-Tyr-Thr-Ile-Val-Ile-Thr-Tyr-
GAA TTA CAT GAT AAT ATT ATG TTT TAT TTA TTA GTT ATT TTA GGT TTA GTA TCT TGA ATG TTA TAT ACA ATT GTT ATA ACA TAT

Ser-Lys-Asn-Pro-Ile-Ala-Tyr-Lys-Tyr-Ile-Lys-His-Gly-Gln-Thr-Ile-Glu-Val-Ile-[Trp]-Thr-Ile-Phe-Pro-Ala-Val-Ile-Leu-
TCA AAA AAT CCT ATT GCA TAT AAA TAT ATT AAA CAT GGA CAA ACT ATT GAA GTT ATT [TGA] ACA ATT TTT CCA GCT GTA ATT TTA

Leu-Ile-Ile-Ala-Phe-Pro-Ser-Phe-Ile-Leu-Leu-Tyr-Leu-Cys-Asp-Glu-Val-Ile-Ser-Pro-Ala-Ile-Thr-Ile-Lys-Ala-Ile-Gly-
TTA ATT ATT GCT TTC CCT TCA TTT ATT TTA TTA TAT TTA TGT GAT GAA GTT ATT TCA CCA GCT ATA ACT ATT AAA GCT ATT GGA

Tyr-Gln-[Trp]-Tyr-[Trp]-Lys-Try-Glu-Tyr-Ser-Asp-Phe-Ile-Asn-Asp-Ser-Gly-Glu-Thr-Val-Glu-Phe-Glu-Ser-Tyr-Val-Ile-Pro-
TAT CAA [TGA] TAT [TGA] AAA TAT GAA TAT TCA GAT TTT ATT AAT GAT AGT GGT GAA ACT GTT GAA TTT GAA TCA TAT GTT ATT CCT

Asp-Glu-Leu-Leu-Glu-Glu-Gly-Gln-Leu-Arg-Leu-Leu-Asp-Thr-Asp-Thr-Ser-Ile-Val-Val-Pro-Val-Asp-Thr-His-Ile-Arg-Phe-
GAT GAA TTA TTA GAA GAA GGA CAA TTA AGA TTA TTA GAT ACT GAT ACT TCT ATA GTT GTA CCT GTA GAT ACA CAT ATT AGA TTC

Val-Val-Thr-Ala-Ala-Asp-Val-Ile-His-Asp-Phe-Ala-Ile-Pro-Ser-Leu-Gly-Ile-Lys-Val-Asp-Ala-Thr-Pro-Gly-Arg-Leu-Asn-
GTT GTA ACA GCT GCT GAT GTT ATT CAT GAT TTT GCT ATC CCA AGT TTA GGT ATT AAA GTT GAT GCT ACT CCT GGT AGA TTA AAT

Gln-Val-Ser-Ala-Leu-Ile-Gln-Arg-Glu-Gly-Val-Phe-Tyr-Gly-Ala-Cys-Ser-Glu-Leu-Cys-Gly-Thr-Gly-His-Ala-Asn-Met-Pro-
CAA GTT TCT GCT TTA ATT CAA AGA GAA GGT GTC TTC TAT GGG GCA TGT TCT GAG TTG TGT GGG ACA GGT CAT GCA AAT ATG CCA

Ile-Lys-Ile-Glu-Ala-Val-Ser-Leu-Pro-Lys-Phe-Leu-Glu-[Trp]-Leu-Asn-Glu-Gln-ochre                     ochre        ochre
ATT AAG ATC GAA GCA GTA TCA TTA CCT AAA TTT TTG GAA [TGA] TTA AAT GAA CAA TAA TTAATATTTACTTATTAT TAA TATTTT TAA


TTATTAAAAATAATA ATAATAATAATAATT ATAATAATATTCTTA AATATAATAAAGATA
```

FIG. 11.21. Nucleotide sequence of the *oxi* 1 gene. The amino acid sequence is based on the nucleotide sequence and assumes that UGA codes for tryptophan. The five tryptophan residues conserved in the bovine protein are boxed. The DNA sequence is that of the nontranscribed strand.

the *oxi* 3-coding regions was inferred from a comparison of the human and yeast mitochondrial genes. Barrell and Sanger have shown that the subunit 1 gene of human mitochondria lacks introns. The nucleotide sequence of the human gene can therefore be directly converted to the primary amino acid sequence of the protein. A search for homologous sequences in the yeast mtDNA sequence has revealed the *oxi* 3 gene to be composed of seven or eight exons and eight or nine introns. The structure of the yeast gene is shown in Fig.

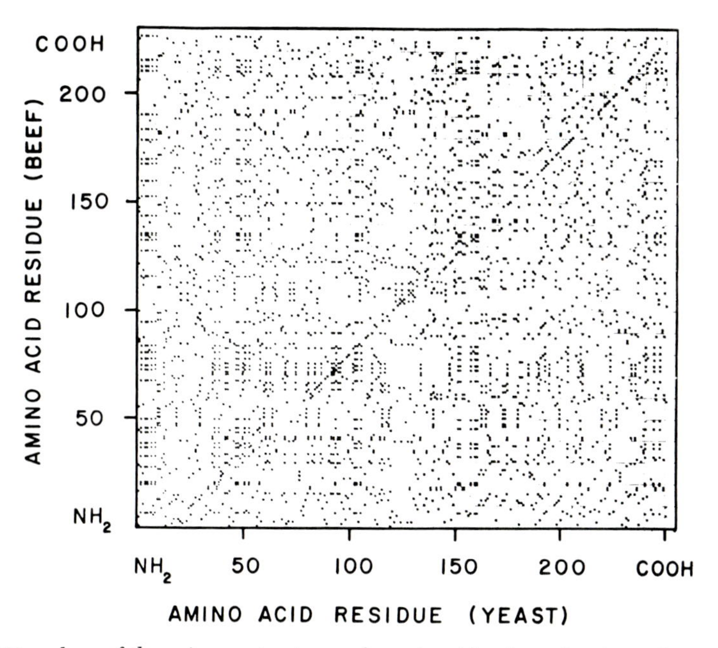

FIG. 11.22. Homology of the primary structures of yeast and bovine subunits 2 of cytochrome oxidase. The abcissa and ordinate represent the amino acid sequences of the two proteins. The residues are numbered 1–251 in the case of the yeast sequence and 1–227 in the case of the bovine sequence. The diagonal line of identity of amino acids in the two sequences intercepts the abcissa at residue 17 because of the presence of an extra 16 amino acids at the amino terminal end of the yeast protein. Insertions and deletions in the yeast protein lead to downward and upward shifts in the diagonal line. Each point in the graph represents an amino acid identity.

TABLE 11.11. Properties of the *oxi* 3 Locus

Genetic
Genetically unlinked alleles
Easily dissected by ρ^- clones
Single complementation group
Physical
Mutational sites scatter over 10 kbp
oxi 3 mutants with deletions of up to 7.5 kbp
Transcripts
Complex pattern of transcripts
Biochemical
oxi 3 mutants with temperature-sensitive cytochrome oxidase
Most *oxi* 3 mutants are deficient in subunit 1 of cytochrome oxidase

11.23. The exons are of variable lengths but make up only 15% of the total sequence.

One of the more remarkable findings to have emerged from the DNA sequence of the *oxi* 3 gene is the existence within the first four intervening sequences (introns) of open reading frames potentially able to code for additional proteins. The intron reading frames are contiguous with and in the same register of the DNA as the exons. Although proteins corresponding to the intron genes have not been detected among the *in vivo* labeled mitochondrial translation products of yeast, their presence at low concentrations cannot be excluded at present. As will be discussed in connection with transcription of mtDNA, the intron genes have been proposed to code for enzymes involved in the splicing of the primary RNA transcripts to form the mature subunit 1 message.

3. Apocytochrome b

Apocytochrome *b* is encoded in a complex gene located between the *oli* 1 and *oli* 2 loci. This region has been shown to contain the *cob*, *box*, and drug-resistance loci *ana*, *muc*, and *diu*. Genetic and physical analysis of the *cob/box* region provided the first evidence that the gene has a mosaic structure similar to that of the subunit 1 gene of cytochrome oxidase. This has been confirmed by the DNA sequence of this region. In *S. cerevisiae* D273-10B, the apocytochrome *b* gene has a relatively simple structure, being made up of three exons and two introns. The sequence of the gene was determined from the genomes of ρ^- clones that spanned the wild-type mtDNA from 70–80 map units. The gene proper is located between 71.5 and 76.2 map units, representing a sequence of 3309 nucleotides. The exons and introns have been deduced from the homology of the amino acid sequences encoded in the yeast and bovine genes—the latter lacks intervening sequences and is therefore colinear with the protein sequence.

The structure of the apocytochrome *b* gene in *S. cerevisiae* D273-10B is shown in Fig. 11.24. The first 252 residues from the amino terminal end of the protein are specified by exon bl. This exon is separated from the next short exon b2 by an approximately 1500-nucleotide-long intervening region. The intron has a reading frame continuous and in frame with the first exon. The entire reading frame starting from the amino terminus of exon bl continues for 1917 nucleotides even though the exon itself comprises only the first 756 nucleotides. The third exon, b3, codes for the last 116 amino acids. This exon is preceded by a shorter intron (730 nucleotides) with an (A + T)-rich character.

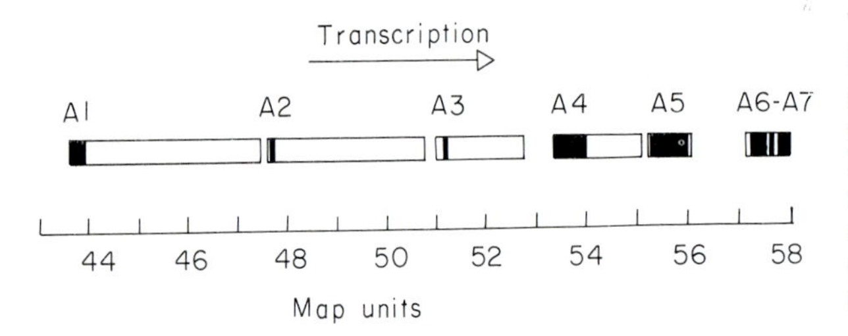

FIG. 11.23. Structure of the *oxi* 3 gene of subunit 1 of cytochrome oxidase. The exons are denoted by the solid bars and are designated by the letters A1–A8. The clear part of each bar corresponds to the reading frames that are in phase with the exons.

This second intron has no reading frames of any significant length. The exon/intron boundaries of the gene have also been determined by sequencing of the messenger RNA. Based on amino acid homology with bovine apocytochrome *b* and the messenger sequence, yeast apocytochrome *b* has the amino acid sequence shown in Fig. 11.25. The molecular weight of the protein is 44,000, a

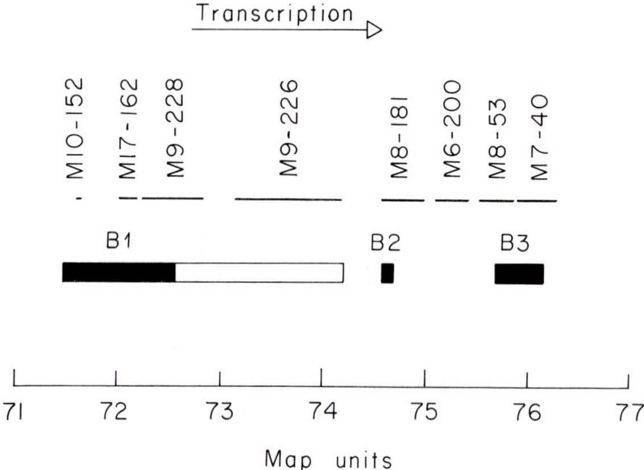

FIG. 11.24. Structure of the apocytochrome *b* gene in *S. cerevisiae* D273-10B. The three exons (B1, B2, B3) are shown as solid bars. The reading frame in the first intron is denoted by the open bar. The physical spans within which cytochrome *b* mutations have been mapped is indicated by the lines in the upper part of the figure.

fMet-Ala-Phe-Arg-Lys-Ser-Asn-Val-Tyr-Leu-Ser-Leu-Val-Asn-Ser-Tyr-Ile-Ile-Asp-Ser-Pro-Gln-Pro-Ser-Ser 25
Ile-Asn-Tyr-Trp-Trp-Asn-Met-Gly-Ser-Leu-Leu-Gly-Leu-Cys-Leu-Val-Ile-Gln-Ile-Val-Thr-Gly-Ile-Phe-Met 50
Ala-Met-His-Tyr-Ser-Ser-Asn-Ile-Glu-Leu-Ala-Phe-Ser-Ser-Val-Glu-His-Ile-Ile-Arg-Asp-Val-His-Asn-Gly 75
Tyr-Ile-Leu-Arg-Tyr-Leu-His-Ala-Asn-Gly-Ala-Ser-Phe-Phe-Phe-Met-Val-Met-Phe-Met-His-Met-Ala-Lys-Gly 100
Leu-Tyr-Tyr-Gly-Ser-Tyr-Arg-Ser-Pro-Arg-Val-Thr-Leu-Trp-Asn-Val-Gly-Val-Ile-Ile-Phe-Ile-Leu-Thr-Ile 125
Ala-Thr-Ala-Phe-Leu-Gly-Tyr-Cys-Cys-Val-Tyr-Gly-Gln-Met-Ser-His-Trp-Gly-Ala-Thr-Val-Ile-Thr-Asn-Leu 150
Phe-Ser-Ala-Ile-Pro-Phe-Val-Gly-Asn-Asp-Ile-Val-Ser-Trp-Leu-Trp-Gly-Gly-Phe-Ser-Val-Ser-Asn-Pro-Thr 175
Ile-Gln-Arg-Phe-Phe-Ala-Leu-His-Tyr-Leu-Val-Pro-Phe-Ile-Ile-Ala-Ala-Met-Val-Ile-Met-His-Leu-Met-Ala 200
Leu-His-Ile-His-Gly-Ser-Ser-Asn-Pro-Leu-Gly-Ile-Thr-Gly-Asn-Leu-Asp-Arg-Ile-Pro-Met-His-Ser-Tyr-Phe 225
Ile-Phe-Lys-Asp-Leu-Val-Thr-Val-Phe-Leu-Phe-Met-Leu-Ile-Leu-Ala-Leu-Phe-Val-Phe-Tyr-Ser-Pro-Asn-Thr 250
Leu-Gly-*His-Pro-Asp-Asn-Tyr-Ile-Pro-Gly-Asn-Pro-Leu-Val-Thr-Pro-Ala-Ser-Ile-*Val-Pro-Glu-Trp-Tyr-Leu 275
Leu-Pro-Phe-Tyr-Ala-Ile-Leu-Arg-Ser-Ile-Pro-Asp-Lys-Leu-Leu-Gly-Val-Ile-Thr-Met-Phe-Ala-Ala-Ile-Leu 300
Val-Leu-Leu-Val-Leu-Pro-Phe-Thr-Asp-Arg-Ser-Val-Val-Arg-Gly-Asn-Thr-Phe-Lys-Val-Leu-Ser-Lys-Phe-Phe 325
Phe-Phe-Ile-Phe-Val-Phe-Asn-Phe-Val-Leu-Leu-Gly-Gln-Ile-Gly-Ala-Cys-His-Val-Glu-Val-Pro-Tyr-Val-Leu 350
Met-Gly-Gln-Ile-Ala-Thr-Phe-Ile-Tyr-Phe-Ala-Tyr-Phe-Leu-Ile-Ile-Val-Pro-Val-Ile-Ser-Thr-Ile-Glu-Asn 375
Val-Leu-Phe-Tyr-Ile-Gly-Arg-Val-Asn-Lys 385

FIG. 11.25. Amino acid sequence of yeast apocytochrome *b*. The sequences encoded in each of the three exons are separated by the asterisks.

value considerably in excess of what has been estimated on SDS polyacrylamide gels.

Both genetic and RNA:DNA hybridization studies indicate that there are at least two different versions of the apocytochrome *b* gene. The one described above in D273-10B is referred to as a "short" gene. In other strains, notably KL14, the gene has a more complex organization because of the presence of additional introns totaling some 3.5 kbp of DNA. Although the "long" gene has not been completely sequenced, restriction mapping and partial sequence data provide a fairly accurate picture of how the two genes differ. As shown in Fig. 11.26, the most substantive difference occurs in the structural sequence coding for the first 252 residues of the protein. Whereas in D273-10B, this part of the protein is specified by a single exon, in KL14 the DNA sequence is fragmented into four shorter exons. The nucleotide sequences of the first two introns of the "long" gene indicate that only the second contains a reading frame.

The DNA sequence has been useful not only in clarifying the structure of the gene but in mapping cytochrome *b* mutations to different exon and intron regions of the gene. Slonimski and co-workers have shown that *box* 3 mutations are confined to the first two introns and *box* 10 to the third intron of the "long" gene. Since these intervening sequences are absent in D273-10B, none of the cytochrome *b* mutants derived from this strain are allelic to the above *box* loci. Mutations in D273-10B designated as *cob* can, however, be related to other *box* loci that have been mapped either in the exons or the last two introns of the "long" gene (Fig. 11.26).

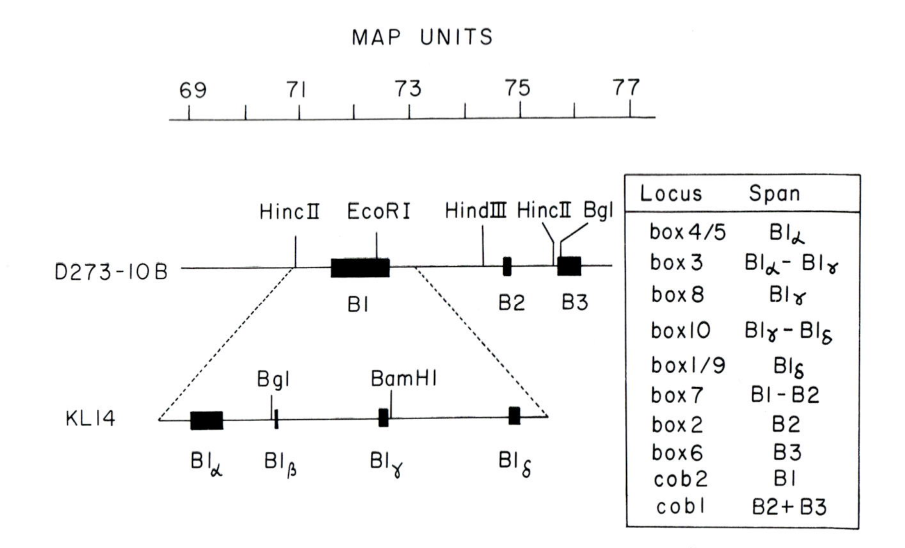

FIG. 11.26. Structures of the "short" and "long" apocytochrome *b* genes. The exons of the two genes are shown as dark bars. The relationship of the *box* and *cob* loci to the different exons and introns are based both on genetic and physical mapping of cytochrome *b* mutations (e.g., *box7* mutations map between exons B1 and B2, *box2* mutations are in the structural sequence corresponding to exon B2, etc.).

D. Sequences of Ribosomal and tRNA Genes of Yeast Mitochondria

1. Ribosomal RNAs

The sequences of the 21S and 15S rRNAs show partial homology to the procaryotic RNAs. The 21S rRNA gene has a 1-kbp intervening sequence in ω^+ but not ω^- strains of *S. cerevisiae*. Strains referred to as ω^+ and ω^- can be distinguished genetically by the recombination of markers in the 21S rRNA gene. In homosexual crosses ($\omega^+ \times \omega^+$ or $\omega^- \times \omega^-$), drug-resistance markers such as *cap* and *ery* recombine with equal frequency; i.e., the output of the two recombinant classes is the same. This is not true of heterosexual crosses ($\omega^+ \times \omega^-$) where there is a strong bias for one of the two recombinant classes. The explanation for the observed unequal transmission of markers in heterosexual crosses is not clear at present.

2. Transfer RNAs

The genes coding for the yeast mitochondrial tRNAs have been sequenced and precisely localized on the physical map. Present evidence indicates the presence of 24 tRNA genes. There is at least one tRNA for each of the 20 different amino acids. In addition, there are two isoacceptors for threonine, arginine, and serine. The initiator tRNA is also a separate gene product.

The limited set of mitochondrial tRNA genes raises the question of how this system is capable of decoding mitochondrial messages. It should be remembered that according to the Wobble hypothesis of base pairing, the minimum number of tRNAs needed to recognize all the codons of the universal code is 32. Since there is no evidence for importation of cytoplasmic tRNAs, it is unlikely that tRNAs other than those encoded in the mitochondrial genome function in translation. The alternative explanation that mitochondrial genes make use of only a subset of the codons is also untenable. The codons of six mitochondrial genes (Table 11.12) show that with the exception of the CGN family of arginine, most of the other codons are used although not with equal frequency. An important clue that has helped to explain how mitochondrial tRNAs function came from the sequences of the tRNAs. As shown in Fig. 11.27, there is a simple but strict rule governing the nucleotide at the Wobble position of the anticodon. With the single exception of the tRNA in the CGN box of arginine, all of the other tRNAs for the four codon families have a U in the Wobble position of the anticodon. Direct sequencing of several of these tRNAs has confirmed that the U is unmodified. Since only one tRNA has been found for each of the eight families coding for single amino acids, the implication is that in mitochondria U:U or U:C pairing at the Wobble position is permitted or that two out of three base pairs are sufficient to give a stable interaction with the messenger. A further examination of the codon table shows that in the eight families with mixed codons there are two tRNAs, one with a G and the other with a U in the Wobble position. The U in the latter set of tRNAs has been shown to be modified. Although the exact mechanism is not known, the modification of the Wobble U probably restricts the tRNA to recognize only the two codons ending with an A or G. The second tRNA in the mixed family boxes has

TABLE 11.12. Codons of Yeast Mitochondrial Genes[a]

Amino acid	Codon	Number	Amino acid	Codon	Number
Ala	GCA	53	Met	AUG	65
	U	66			
	C	5	Phe	UUU	51
	G	2		C	6
Arg	AGA	37	Pro	CCA	35
	G	0		U	39
	CGA	0		C	2
	U	0		G	0
	C	0			
	G	0			
			Ser	UCA	71
Asn	AAU	62		U	34
	C	9		C	1
				G	0
Asp	GAU	41			
	C	2		AGU	15
				C	0
Cys	UGU	12			
	C	0			
			Thr	ACA	51
Gln	CAA	29		U	34
	G	4		C	1
				G	0
Glu	GAA	37		CUA	14
	G	3		U	2
				C	0
Gly	GGA	28		G	2
	U	88			
	C	0	Trp	UGA	36
	G	6		G	0
				G	0
His	CAU	47			
	C	3	Try	UAU	73
				C	13
Ile	AUU	149			
	C	26			
	A	7	Val	GUA	76
				U	43
Leu	UUA	224		C	3
	G	2		G	6
Lys	AAA	32			
	G	1			

[a]The codons have been tabulated from the gene sequences of the ATPase subunits 6 and 9, cytochrome oxidase subunits 1, 2, and 3, and apocytochrome *b*. Anomalous codons have been boxed.

a G in the Wobble position, thereby recognizing the other two codons ending in a U or C. It is significant that in the case of the methionine tRNA, the Wobble base in the anticodon is a C which can only base pair with the terminal G of the methionine codon.

The decoding rules spelled out above allow for an economy of 8 tRNAs. At present, this simple system of tRNAs has only been found to operate in yeast, *N. crassa*, and mammalian mitochondria.

E. General Organizational Features of Yeast mtDNA

Yeast mitochondrial genes including the complex genes with introns account for only 45–50% of the genome length. The rest of the DNA is comprised of gene spacers averaging 95% A + T. The (A + T)-rich regions are interspersed with short sequences that have a very high G + C content (70–90%). The G + C clusters contain most of the HpaII and HaeIII sites. Bernardi and co-workers have estimated some 150–200 HpaII/HaeIII site clusters in the

<table>
<tr><td>UUU
Phe AAG
UUC
UUA
Leu AAU
UUG</td><td>UCU
UCC
Ser AGU
UCA
UCG</td><td>UAU
Tyr AUG
UAC
UAA
Ter
UAG</td><td>UGU
Cys ACG
UGC
UGA
Trp ACU*
UGG</td></tr>
<tr><td>CUU
CUC
Thr GAU
CUA
CUG</td><td>CCU
CCC
Pro GGU
CCA
CCG</td><td>CAU
His GUG
CAC
CAA
Gln GUU
CAG</td><td>CGU
CGC
Arg GCA
CGA
CGG</td></tr>
<tr><td>AUU
AUC Ile UAG
AUA
AUG Met UAC</td><td>ACU
ACC
Thr UGU
ACA
ACG</td><td>AAU
Asn UUG
AAC
AAA
Lys UUU
AAG</td><td>AGU
Ser UCG
AGC
AGA
Arg UCU
AGG</td></tr>
<tr><td>GUU
GUC
Val CAU
GUA
GUG</td><td>GCU
GCC
Ala CGU
GCA
GCG</td><td>GAU
Asp CUG
GAC
GAA
Glu CUU
GAG</td><td>GGU
GGC
Gly CCU
GGA
GGG</td></tr>
</table>

FIG. 11.27. Codons and anticodons of the yeast mitochondrial genetic code. The codons (5′ → 3′) are at the left, and the anticodons (3′ → 5′) are at the right in each box. The Wobble nucleotides of the anticodons are underlined.

yeast genome. Of those sequenced, a significant number appear to be palindromic, and others are either homologous or have sequences that are inverted repeats of other clusters. The function of the A + T spacers and the G + C clusters is still not known. The frequent and random occurrence of the clusters argues against a role in transcription or DNA replication.

F. Mammalian mtDNA

Barrell and Sanger have recently completed the sequences of human and bovine mtDNA. The organization of genes, codon usage, and even informational contents of these mammalian genomes are significantly different from those of yeast. The most conspicuous feature of the mammalian mtDNAs is the contiguity of the coding sequences. Almost all the genes are butt-joined, resulting in a virtually complete saturation of the DNA with genetic information. The economic use of the DNA sequence is also evidenced in the absence of intervening sequences in the genes and the exploitation of both strands for coding purposes.

The DNA sequences of the mammalian genomes have confirmed earlier conclusions based on hybridization data that most of the genes are located in the heavy strand. This strand codes for two ribosomal RNAs, 14 transfer RNAs,

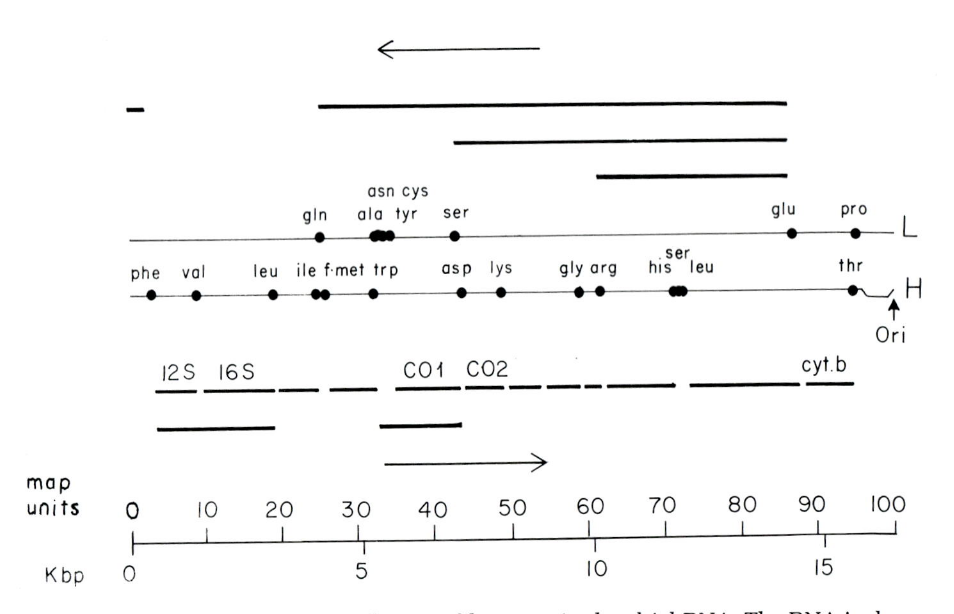

FIG. 11.28. Physical and transcriptional maps of human mitochondrial DNA. The DNA is shown in a linear form. The positions of the tRNA genes on the heavy (H) and light strands (L) are denoted by the solid circles. The transcripts originating from the light (upper) and heavy (lower) strands are shown as heavy lines. The 12S and 16S transcripts correspond to the small and large rRNAs, Co1 and CO2 code for subunits 1 and 2 of cytochrome oxidase, and cyt.b for apocytochrome *b*. The longer transcripts that hybridize to the 12S, 16S and CO1 genes are probably precursor RNAs. Ori refers to the origin of replication. The directions of transcription of the two strands are indicated by the arrows. (Taken from Attardi *et al.*, 1980.)

and in addition contains ten open reading frames, five of which have been shown to be genes for subunits of cytochrome oxidase, ATPase, and apocytochrome *b*. The locations of the identified genes on the human genome are shown in Fig. 11.28. The nature of the proteins encoded in the other five reading frames of the heavy strand is still obscure; they do, however, give rise to stable polyadenylated transcripts with sizes matching the lengths of the reading frames. With one exception, all of the reading frames of the heavy strand are separated by at least one tRNA gene. As will be pointed out subsequently, the tRNA sequences are believed to act as signals for processing of mitochondrial RNA. The light strand of the mammalian mtDNA codes for eight tRNAs. This strand also contains several reading frames potentially capable of coding for proteins. At present, no stable polyadenylated transcripts other than a small RNA mapping near the origin of replication have been found to hybridize to the light strand.

The sequences of mitochondrial DNA have necessitated a reappraisal of the previously held belief that the usual codon assignments of the genetic code are general for all biological systems. In addition to the use of UGA as a tryptophan codon, mammalian mitochondria have modified the code to their own unique needs. Table 11.13 summarizes some of these differences between mammalian and yeast mitochondria. Both human and bovine mitochondria lack the tRNA for the AGA and AGG codons of arginine. Barrell and Sanger have found several instances in which AGA functions as a translational stop. Another deviation in the mammalian systems is the use of AUA either as an initiator or as a methionine codon. In contrast to yeast mitochondria, however, the CUN family is recognized as leucine rather than threonine codons.

VIII. Transcription of mtDNA and Processing of RNA

Transcription and RNA processing in mitochondria are only now beginning to be examined but already there are indications of substantive differences from other known organisms. The discovery of mosaic genes in yeast mitochondria has made this a particularly attractive experimental system for studying the manner in which intervening sequences are removed from precursor ribosomal and messenger RNAs, and a number of laboratories are exploiting the tools of genetics and molecular biology to solve this important problem.

TABLE 11.13. Anomalous Codon Usage in Mitochondria

Codon	Universal Code	Mammalian mitochondria	Yeast mitochondria
UGA	Stop	Tryptophan	Tryptophan
AGA	Arginine	Stop	Arginine
AUA	Isoleucine	Methionine	Isoleucine
CUN	Leucine	Leucine	Threonine

A. Mammalian Mitochondria

In human mitochondria, both strands of the DNA are continuously transcribed, presumably from single promoter sites. The disposition of the tRNA genes in relation to the ribosomal and other genes coding for messenger RNAs has led Barrell and Sanger to propose a simple mechanism by which the mature RNAs are generated from the long primary transcripts. According to this model (Fig. 11.29), the first processing step involves cleavage of the RNA at the 5′ and 3′ terminal nucleotides of the tRNA genes by an endonuclease that recognizes the cloverleaf structure of the tRNAs. Since the genes coding for the ribosomal and messenger RNAs are separated by tRNAs, these cleavage events result in a series of smaller transcripts that can be further modified by polyadenylation (messages), base modification (tRNA, rRNA) and addition of ACC to the acceptor stems of the tRNAs.

This processing scheme is supported by studies on the transcriptional map of human mtDNA (Fig. 11.28). Attardi and co-workers have found that human mitochondrial messages map exactly to the regions of DNA demarcated by the tRNA genes. Furthermore, the 5′ termini of the messages start almost always with the initiation codons, suggesting that the maturation of the transcripts formed after the endonucleolytic excision of the tRNAs involves a simple polyadenylation of the 3′ ends. The uniformity in size and gene organization of all the mammalian genomes studied to date tends to indicate that transcription and RNA processing in the higher eucaryotic systems are likely to involve similar mechanisms.

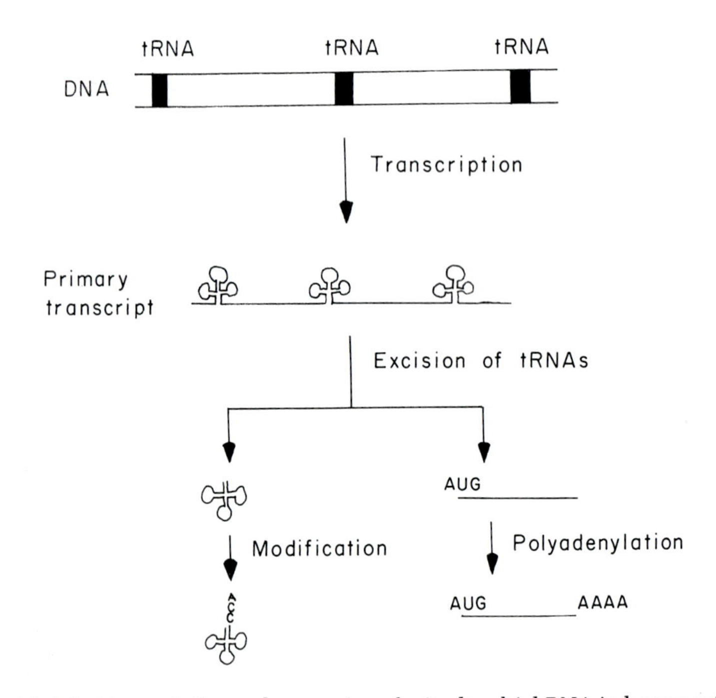

FIG. 11.29. Model of transcription and processing of mitochondrial RNA in human mitochondria.

B. Yeast Mitochondria

The situation in yeast is considerably more complex. There is presently good evidence that the two ribosomal and some of the messenger RNAs are transcribed separately. It is also clear, however, that some of the tRNA genes are cotranscribed with genes coding for subunits of cytochrome oxidase. The mechanism of processing of the primary transcripts is still not understood, although it is likely that the tRNA sequences and perhaps some of the G + C clusters may act as signals for endonucleolytic processing events.

The production of mature RNAs in yeast mitochondria seems to be a rather inefficient process. Numerous transcripts differing in size are found for most of the gene coding regions. For example, a DNA probe from the cytochrome oxidase subunit 2 gene hybridizes to ten different transcripts ranging in size from 800 to 3400 nucleotides. The higher-molecular-weight transcripts have been assumed to represent processing intermediates, although it is not excluded that they may also function as messengers. Regardless of whether they do or not, their presence is indicative of slow processing.

The 5′ leader sequences in the yeast messages vary considerably in length. In the case of cytochrome oxidase subunit 2, the leader has been determined to be 54 nucleotides long and to have a sequence identical to the genomic DNA upstream from the gene. Other messages have considerably longer leaders. It was already pointed out in the previous chapter that three of the messages whose 5′ leaders have been identified contain certain sequences complementary to a 10-nucleotide sequence in the 3′ tail of the 15S rRNA. These sequences probably serve as the ribosomal binding sites. Although functionally equivalent, the putative mRNA binding sequence of the yeast 15S rRNA is different from the Shine and Dalgarno sequence in the 16S rRNA of *E. coli*.

Most of the enzymes involved in processing of mitochondrial RNA in yeast are probably encoded in nuclear DNA. Thus, certain regions of mtDNA are transcribed and correctly processed in ρ^- mutants. This is true of the 21S rRNA and the messages for subunits 2 and 3 of cytochrome oxidase and the ATPase proteolipid. The ribosomal RNA is of special interest since its maturation requires not only trimming of the 5′ and 3′ termini but also removal of the 1-kb intervening sequence. Not all mitochondrial messages, however, are processed in such mutants. For example, the mature messages for subunit 1 of cytochrome oxidase and apocytochrome *b* fail to be formed in ρ^- mutants whose mitochondrial genomes contain the corresponding genes as well as extensive flanking sequences. Both genes are known to have intervening sequences. Their excision therefore probably depends on mitochondrially encoded RNA or protein cofactors (see maturase model, Section VIIID).

C. Processing Pathway of the Yeast Apocytochrome *b* Messenger

The order in which the introns are excised from the primary transcript of the apocytochrome *b* gene has been studied by hybridization of DNA probes from the exon and intron regions to total mitochondrial RNA of wild-type and mutant yeast strains. The analysis of the apocytochrome *b* transcripts is conveniently done by the Northern blot hybridization technique. In essence the procedure involves separation of a mixture of RNAs by electrophoresis on agarose. The RNA bands are then transferred by blotting from the agarose to a suitable paper onto which they are covalently bound, thus providing an exact replica of the agarose gel. The original procedure made use of cellulose paper derivatized with diazobenzyloxymethyl groups (DBM) which react covalently with nucleic acids. Once the RNA is transferred, the DBM paper is hybridized with radioactive DNA fragments (probes) to determine which bands of RNA contain sequences complementary to the probes. The RNA bands which hybridize to the probe are visualized by exposure of the DBM paper to X-ray film.

When total mitochondrial RNA prepared from the wild-type strain of *S. cerevisiae* D273-10B was fractionated on agarose and blotted on DBM paper, four different transcripts were found to hybridize with an exon B1 probe (Fig. 11.30). The apparent sizes of the transcripts were 4, 3.3, 2.6, and 1.9 kb. The same four transcripts were seen when the DBM blot was challenged with a probe containing the exon B3 sequence. The sequence compositions of the transcripts detected by the exon probes were further defined by repeating the hybridizations with intron-1 and intron-2 probes. The 4-kb transcript hybridized to all the exon and intron probes and must therefore be either the primary transcript or an early unspliced intermediate. The 3.3-kb transcript represents a partially processed intermediate lacking the second intron since it is only detected by the exon and the first intron probes (Fig. 11.30). The second-largest RNA species (2.6 kb) is not detected by the first intron probe. Since this tran-

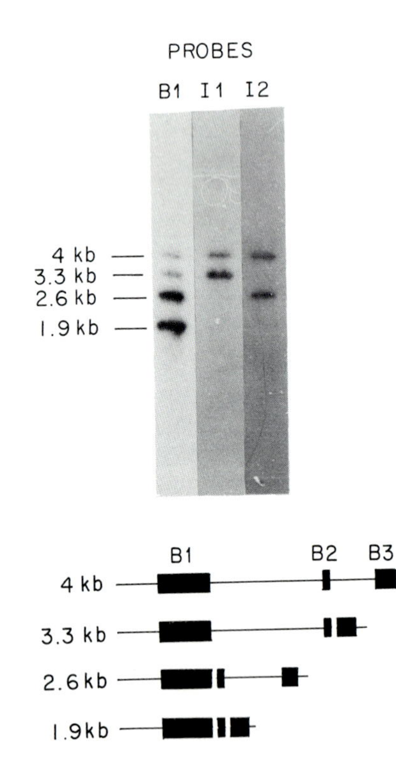

FIG. 11.30. Northern blot analysis of apocytochrome *b* transcripts in the wild-type strain, *S. cerevisiae* D273-10B. Total mitochondrial RNA was separated on a 1% agarose gel, transferred to DBM paper and probed with 5′-end-labeled single-stranded DNA fragments containing exclusively sequences from exon B1 (B1), intron 1 (I1) and intron 2 (I2). The sizes of the transcripts in kilobases are indicated in the left-hand margin. The compositions of the transcripts based on the hybridization patterns are shown in the lower half of the figure.

script hybridizes to a second intron probe, it must be another splicing intermediate containing only the exon and second intron sequences. The identification of the 1.9-kb transcript as the mature message is based on the observation that it is the smallest and most abundant RNA and that it hybridizes exclusively to exon probes. Similar analysis using probes with sequences from the flanking regions of the gene indicate the 5′ leader of the message to be 700–800 nucleotides long and the 3′ extension to consist of less than 100 nucleotides. The leader sequence of the message in strains with the "long" gene has also been found to have a similar length. In D273-10B, none of the processed introns appear to exist as stable transcripts. This is to be contrasted with the situation in strains with the "long" gene where the first intron accumulates as a stable circular RNA.

The presence in wild-type yeast of two splicing intermediates lacking either the first or second intron implies that the message can be generated by two alternate processing pathways. This has been substantiated by Northern blot hybridization analysis of the RNA species in mitochondrial mutants (mit^-) blocked at different steps of message maturation. The mutations in the mit^- strains M9-226 and M6-200 have been localized to introns 1 and 2, respectively, by petite deletion mapping. Both strains fail to synthesize cytochrome *b* and, as shown in Fig. 11.31, are unable to make mature apocytochrome *b* message. The results of hybridizations performed on the mitochondrial RNAs in the two mutants indicate that each mutation prevents excision exclusively of the intron

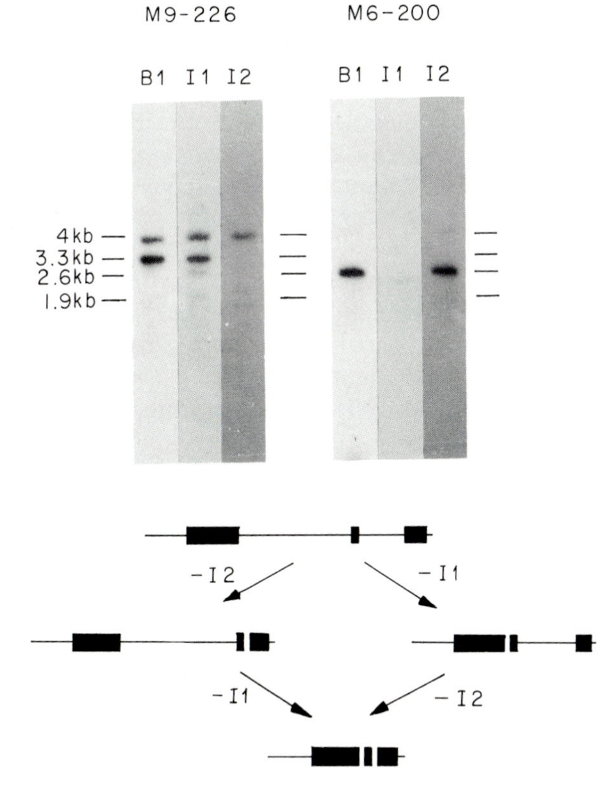

FIG. 11.31. Apocytochrome *b* transcripts present in intron mutants. Mitochondrial RNA from the intron-1 mutant (M9-226) and intron-2 mutant (M6-200) were separated on 1% agarose, transferred to DBM paper and probed with exon B1 (B1), intron 1 (I1) and intron-2 (I2) probes. The absence of the 2.6-kb transcript in M9-226 (lanes 1–3) indicates that the mutation prevents processing of the first intron. Similarly M6-200 cannot excise the second intron as seen by the absence of the 3.3-kb transcript (lanes 4–6). The two processing routes leading to the mature message are shown in the lower part of the figure.

with the mutated allele. Thus, M6-200 is capable of excising the first intron, as evidenced by the presence in the mutant of the 2.6-kb transcript. The absence in M6-200 of the 3.3-kb transcript indicates that the second intron fails to be processed. A similar analysis of the first-intron mutant M9-226 shows that even though excision of intron 1 is blocked, it does not interfere with removal of the second intron. These results confirm the nonobligatory nature of the splicing pathway.

Because of the presence of three additional introns in the "long" gene, the splicing pathway has been more difficult to study. Grivell and co-workers have shown that processing starts with an obligatory excision of a 10S circular RNA corresponding to the first intron (see Fig. 11.26). The 28S transcript resulting from this first splicing event is further processed by the excision of the third, fourth, and fifth introns in any one of six possible orders. The 22S product of these splicing events is then converted to the messenger by the removal of the second intron. Alternatively, a 24S intermediate formed after excision of the third and fourth introns can be further processed by the removal of the second, followed by the fifth, introns.

D. RNA Splicing Mechanism

Several mechanisms have been proposed to account for the precise excision of intervening sequences from precursor RNAs containing split gene sequences. The first invokes the participation of a guide RNA whose function is to align the ends of neighboring exons by virtue of sequence complementarity to the exon–intron junctions. This mechanism has certain attractive features: for one, it dispenses with the need of multiple splicing enzymes with different substrate specificities. At present, however, there is no evidence for the presence of low-molecular-weight RNAs in yeast mitochondria that might fulfill the role of guides. A novel mechanism of splicing based on the occurrence of reading frames in the introns of the apocytochrome *b* and subunit 1 of cytochrome oxidase genes has recently been proposed by Slonimski and co-workers and by Church and Gilbert. In both models, the intron-encoded reading frames are postulated to code for a special class of maturase enzymes that catalyze some enzymatic step in splicing (Fig. 11.32). Slonimski has delineated in some detail how this might occur in the "long" apocytochrome *b* gene. As pointed out earlier, the initial splicing step is the excision of the first intron (10S circular RNA) from the 5′ end of the precursor transcript. This step has been shown to occur in ρ^- mutants and must, therefore, be catalyzed by a nuclear gene product. The DNA sequence of the "long" gene reveals that the excision of the first intron generates a reading frame 1269 nucleotides in length. This reading frame consists of the coding sequences corresponding to the first two exons (429 nucleotides) and part of the second intron (840 nucleotides). The composite reading frame is proposed to code for a maturase protein which catalyzes the splicing reaction leading to the ligation of the second and third exons. Inherent in the maturase model is the destruction of the maturase message. Since a number of introns in the apocytochrome *b* and subunit 1 genes have been found to contain reading frames in register with the upstream exons, in principle each is potentially capable of coding for a different maturase specific for the ligation of the bordering exons.

In support of this model, Slonimski and co-workers have shown that mutations in the intron reading frames cause the appearance of high-molecular-weight translation products not seen in wild-type cells. These novel products have been found in strains with nonsense mutations in the intron reading frames and are presumed to be prematurely terminated maturases. Since the defective maturases cannot catalyze the splicing reaction, the mutants accumulate high levels of the maturase message and hence of the translation product.

IX. Evolution of Mitochondria

This book began with an account of the historical development of mitochondrial studies; it may therefore be appropriate to end with some thoughts on how the organelle itself could have evolved. This question became a popular topic of speculation in the 1960s when mitochondria were first recognized to contain their own genetic system and a special class of ribosomes that synthesize a limited but functionally important set of proteins. It was especially intriguing that the mitochondrial machinery for protein synthesis was strikingly similar to that of bacterial systems. These and other observations formed the basis for a proposal of mitochondrial evolution known as the endosymbiont hypothesis. According to this theory, the modern eucaryotic cell is viewed to have arisen as the result of two evolutionary events. The first was the appearance of a nucleated cell which was devoid of mitochondria and depended entirely on glycolysis for its source of ATP. A second evolutionary line saw the

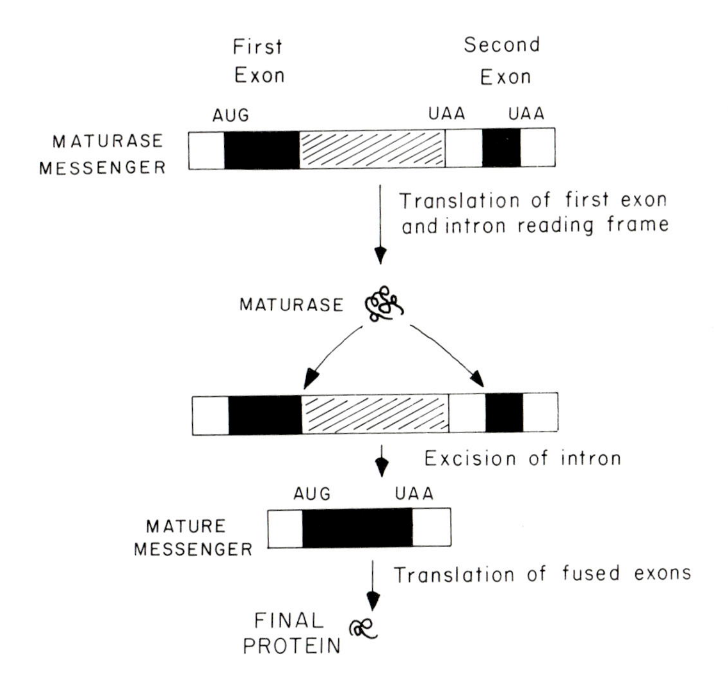

FIG. 11.32. Maturase model for the processing of intervening sequences in yeast mitochondrial genes. (Adapted from Lazowska *et al.*, 1980.)

emergence of oxidative phosphorylation as a means of energy conservation. Cells endowed with this capacity were anucleated and probably gave rise to the present-day eucaryotes. The theory further postulates that at some early time, the nucleated cell was invaded by the procaryote. The invasion resulted in the establishment of a symbiotic relationship of distinct advantage to the host because of the more efficient means of energy production (Fig. 11.33).

Since there may have been a substantial duplication of genetic information in the two different cells, some genes of the procaryote were lost, and others, particularly those involved in the specification of respiratory functions, were transferred to the host genome. A limited number of genes, however, were retained by the procaryote. According to this view, the mitochondrion is a remnant of the original eucaryote stripped of all functions except those related to the oxidative production of ATP.

An alternative but equally plausible interpretation sees all present-day organisms as having been derived from a simple ancestral cell by two divergent evolutionary lines. The starting point of this cluster-clone hypothesis is a cell which already had a full complement of genes needed for the glycolytic and aerobic pathways of energy metabolism. For the sake of argument, it is further

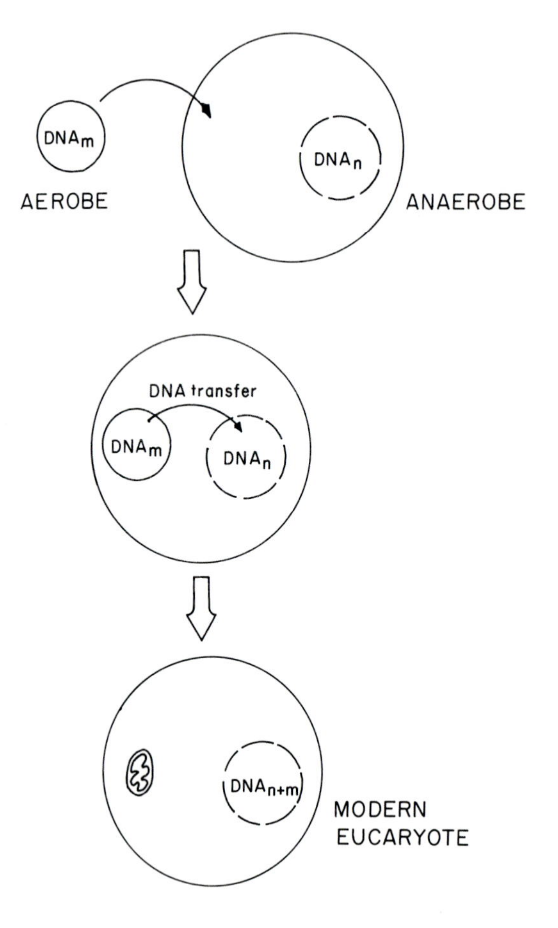

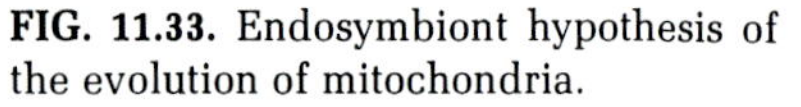
FIG. 11.33. Endosymbiont hypothesis of the evolution of mitochondria.

assumed that the genes of the ancestral cell were dispersed in different genomic elements or plasmids. As depicted in Fig. 11.34, the integration of the plasmid DNAs into a central storehouse of genetic information gave rise to modern day procaryotes. Eucaryotes followed another evolutionary route in which there was a segregation of genes into spatially separate compartments. The assortment of functionally related genes may have been the first step in the development of specialized organelles such as the nucleus, mitochondria, and chloroplasts. The compartmentation probably necessitated an initial duplication of genes involved in the replication, transcription, and translation of the organellar genomes. To account for the similarities of mitochondrial, chloroplast, and bacterial ribosomes, one need only imagine that some organellar genomes were slower to evolve.

Recent data on the fungal and mammalian mitochondrial genomes have already revealed numerous interesting differences in their genetic codes, gene structure, and genetic contents. These new discoveries point to profound evolutionary divergences. The pressures that have led to these adaptations are still

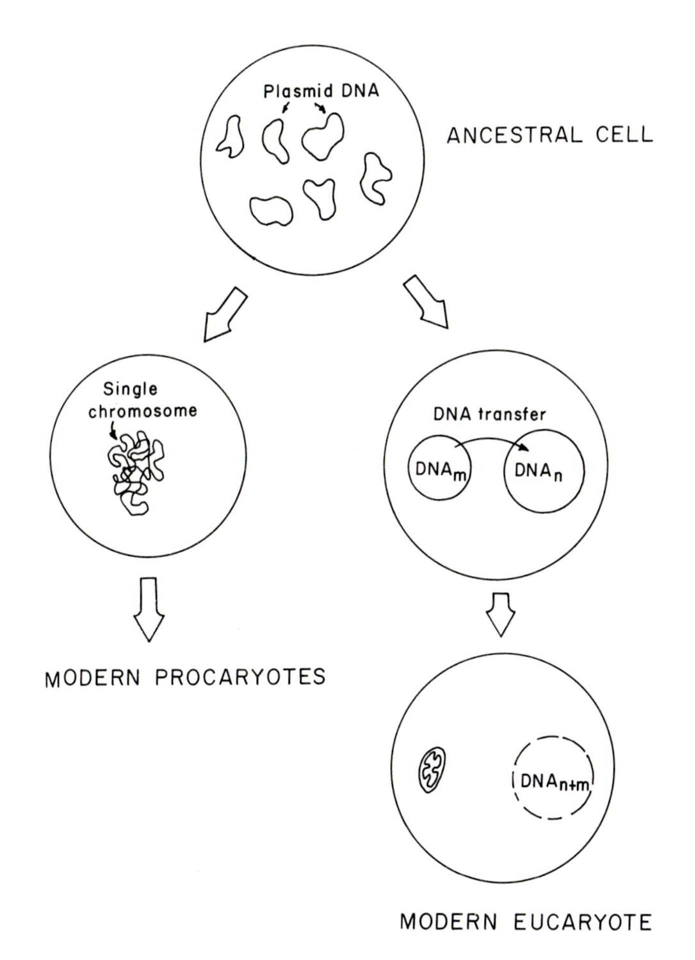

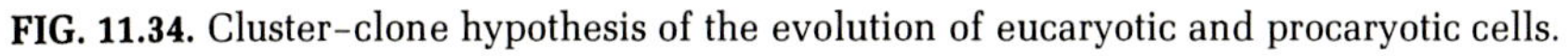
FIG. 11.34. Cluster–clone hypothesis of the evolution of eucaryotic and procaryotic cells.

not understood. It should be evident, however, that mitochondria offer a unique opportunity to study general questions related to the evolution of DNA and the transfer of genetic information. The substantial efforts being invested in the study of other mtDNAs should help not only to answer these questions but also to assess some of the current theories of the origin of this organelle.

Selected Readings

Alwine, J. C., Kemp, D. J., and Stark, G. R. (1977) Method for detection of specific RNAs in agarose gels by transfer to diazobenzyloxymethyl paper and hybridization with DNA probes, *Proc. Natl. Acad. Sci. U.S.A.* **76**:731.

Attardi, G., Cantatore, P., Ching, E., Crews, S., Gelfand, R., Merkel, C., Montoya, J., And Ojala, D. (1980) The remarkable features of gene organization and expression of human mitochondrial DNA, in *The Organization and Expression of the Mitochondrial Genome* (A. M. Kroon and C. Saccone, eds.), North-Holland, Amsterdam, pp. 103–119.

Barrell, B. G., Bankier, A. T., and Drouin, J. (1979) A different genetic code in human mitochondria, *Nature* **282**:189.

Barrell, B. G., Anderson, S., Bankier, A. T., deBruijn, M. H. L., Chen, E., Coulson, A. R., Drouin, J., Eperon, I. C., Nierlich, D., Staden, R., and Young, I. G. (1980) Different pattern of codon recognition by mammalian mitochondrial tRNAs, *Proc. Natl. Acad. Sci. U.S.A.* **77**:3164.

Bernardi, G., Peperno, G., and Fonty, G. (1972) The mitochondrial genome of wild type yeast cells. I. Preparation and heterogeneity of mitochondrial DNA, *J. Mol. Biol.* **65**:173.

Bogorad, L. (1972) Evolution of organelles and eucaryotic genomes, *Science* **188**:891.

Bonitz, S. G., Coruzzi, G., Thalenfeld, B. E., Tzagoloff, A., and Macino, G. (1980) Assembly of the mitochondrial membrane system. Structure and nucleotide sequence of the gene coding for subunit 1 of yeast cytochrome oxidase, *J. Biol. Chem.* **255**:11927.

Bonitz, S. G., Berlani, R., Coruzzi, G., Li, M., Macino, G., Nobrega, F. G., Nobrega, M. P., Thalenfeld, B., and Tzagoloff, A. (1980) Codon recognition rules in yeast mitochondria, *Proc. Natl. Acad. Sci. U.S.A.* **77**:3167.

Borst, P. (1972) Mitochondrial nucleic acids, *Annu. Rev. Biochem.* **41**:333.

Borst, P., and Grivell, L. A. (1978) The mitochondrial genome of yeast, *Cell* **15**:705.

Cabral, F., and Schatz, G. (1978) Identification of the structural gene for yeast cytochrome oxidase subunit II on mitochondrial DNA, *J. Biol. Chem.* **253**:297.

Casey, J., Cohen, M., Rabinowitz, M., Fukuhara, H., and Getz, G. S. (1972) Hybridization of mitochondrial transfer RNA's with mitochondrial and nuclear DNA of Grande (wild type) yeast, *J. Mol. Biol.* **63**:431.

Claisse, M. L., Spyridakis, A., Wambier-Kluppel, M. L., Pajot, P., and Slonimski, P. P. (1978) Mosaic organization and expression of the mitochondrial DNA region controlling cytochrome *b* reductase and oxidase. Analysis of proteins translated from the *box* region, in *Biochemistry and Genetics of Yeast* (M. Bacila, B. L. Horecker and A. O. M. Stoppani, eds.), Academic Press, New York, pp. 369–390.

Coen, D., Deutsch, J., Netter, P., Petrochilo, E., and Slonimski, P. P. (1970) Mitochondrial genetics. I. Methodology and phenomenology, *Symp. Soc. Exp. Biol.* **24**:449.

Dawid, I. B. (1972) Mitochondrial RNA in *Xenopus laevis*. The expression of the mitochondrial genome, *J. Mol. Biol.* **63**:201.

Dujon, B. (1980) Nucleotide sequences of the intron and flanking exons of the mitochondrial 21S rRNA gene of yeast strains with different alleles at the *omega* and *ribl* genetic loci, *Cell* **20**:185.

Dujon, B., Slonimski, P. P., and Weill, L. (1974) Mitochondrial genetics IX. A model for recombination and segregation of mitochondrial genomes in *Saccharomyces cerevisiae*, *Genetics* **78**:415.

Dujon, B., Colson, A. M., and Slonimski, P. P. (1978) The mitochondrial genetic map of *Saccharomyces cerevisiae:* Compilation of mutations, genes, genetic and physical maps, in *Mitochondria 1977: Genetics and Biogenesis of Mitochondria* (W. Bandlow, R. J. Schweyen, K. Wolf and F. Kaudewitz, eds.), Walter de Gruyter, Berlin, pp. 579–669.

Ephrussi, B., Hottinguer, H., and Chimenes, A. M. (1949) Action de l'acriflavine sur les levures. I. La mutation "petite" colonie, *Ann. Inst. Pasteur* **76**:351.

Fox, T. D. (1979) Five TGA "stop" codons occur within the translated sequence of the yeast mitochondrial gene for cytochrome *c* oxidase subunit II, *Proc. Natl. Acad. Sci. U.S.A.* **76**:6534.

Gillham, N. W. (1974) Genetic analysis of chloroplast and mitochondrial genomes, *Annu. Rev. Genet.* **8**:347.

Gillham, N. W. (1978) *Organelle Heredity*, Raven Press, New York.

Goldring, E. S. L., Grossman, L. J., Drupnick, D., Cryer, D. R., and Marmur, J. (1970) The petite mutation in yeast. Loss of mitochondrial deoxyribonucleic acid during induction of petites with ethidium bromide, *J. Mol. Biol.* **52**:323.

Heckman, J. E., Hecker, L. I., Schwartzbach, S. D., Barnett, W. E., Baumstark, B., and RajBhandary, U. L. (1978) Structure and function of initiator methionine tRNA from the mitochondria of *Neurospora crassa*, *Cell* **13**:83.

Heckman, J. E., Sarnoff, J., Alzner-Deweerd, B., Yin, S., and RajBhandary, U. L. (1980) Novel features in the genetic code and codon reading patterns in *Neuropora crassa* mitochondria based on sequences of six mitochondrial tRNAs, *Proc. Natl. Acad. Sci. U.S.A.* **77**:3159.

Hollenberg, C. P., Borst, P., and Van Bruggen, E. F. J. (1970) Mitochondrial DNA. V. A 25-μ closed circular duplex DNA molecule in wild-type yeast mitochondria. Structure and genetic complexity, *Biochim. Biophys. Acta* **209**:1.

Lazowska, J., Jacq, C., and Slonimski, P. P. (1980) Sequence of introns and flanking exons in wild type and *box* 3 mutants of cytochrome *b* reveals an interlaced splicing protein coded by an intron, *Cell* **22**:333.

Li, M., and Tzagoloff, A. (1979) Assembly of the mitochondrial membrane system. Sequences of yeast mitochondrial valine and an unusual threonine tRNA gene, *Cell* **18**:47.

Lynch, D. C., and Attardi, G. (1976) Amino acid specificity of the transfer RNA species coded for by HeLa cell mitochondrial DNA, *J. Mol. Biol.* **102**:125.

Macino, G., and Tzagoloff, A. (1979) Assembly of the mitochondrial membrane system. Two separate genes coding for threonyl-tRNA in the mitochondrial DNA of *Saccharomyces cerevisiae*, *Mol. Gen. Genet.* **169**:183.

Macino, G., and Tzagoloff, A. (1981) Assembly of the mitochondrial membrane system. Sequence analysis of a yeast mitochondrial ATPase gene containing the *oli2* and *oli4* loci, *Cell* **20**:507.

Macino, G., Coruzzi, G., Nobrega, F. G., Li, M., and Tzagoloff, A. (1979) Use of the UGA terminator as a tryptophan codon in yeast mitochondria, *Proc. Natl. Acad. Sci. U.S.A.* **76**:3784.

Mahler, H. R., Hanson, D., Miller, D., Bilinski, T., Ellis, D. M., Alexander, N. J., and Perlman, P. S. (1977) Structural and regulatory mutations affecting mitochondrial gene products, in *Mitochondria 1977: Genetics and Biogenesis of Mitochondria* (W. Bandlow, R. H. Schweyen, K. Wolf and F. Kaudewitz, eds.), Walter de Gruyter, Berlin, pp. 345–370.

Margulis, L. (1970) *Origin of Eukaryotic Cell*, Yale University Press, New Haven.

Maxam, A. M., and Gilbert, W. (1977) A new method for sequencing DNA, *Proc. Natl. Acad. Sci. U.S.A.* **74**:560.

Morimoto, R., and Rabinowitz, M. (1979) Physical mapping of the yeast mitochondrial genome. Derivation of the fine structure and gene map of strain D273-10B and comparison with a strain MH41-7B differing in genome size, *Mol. Gen. Genet.* **170**:25.

Morimoto, R., Merten, S., Lewin, A., Martin, N. C., and Rabinowitz, M. (1978) Physical mapping of genes on yeast mitochondrial DNA. Localization of antibiotic resistance loci, and rRNA and tRNA genes, *Mol. Gen. Genet.* **163**:241.

Nagley, P., Sriprakash, K. S., Rytka, J., Choo, K. B., Trembath, M. K., Lukins, H. B., and Linnane, A. W. (1976) Physical mapping of genetic markers in the yeast mitochondrial genome, in *The Genetic Function of Mitochondrial DNA* (C. Saccone and A. M. Kroon, eds.), North-Holland, Amsterdam, pp. 231–242.

Nass, M. M. K. (1969) Mitochondrial DNA. II. Structure and physiochemical properties of isolated DNA, *J. Mol. Biol.* **42**:529.

Nass, M. M. K., Nass, S., and Afzelius, B. A. (1965) The general occurrence of mitochondrial DNA, *Exp. Cell Res.* **37**:190.

Nobrega, F. G., and Tzagoloff, A. (1980) Assembly of the mitochondrial membrane system. DNA sequence and organization of the cytochrome *b* gene in *Saccharomyces cerevisiae* D273-10B, *J. Biol. Chem.* **255**:9828.

Prunell, A., and Bernardi, G. (1977) The mitochondrial genome of wild type yeast cells. VI. Genome organization, *J. Mol. Biol.* **110**:53.

Sager, R. (1972) *Cytoplasmic Genes and Organelles*, Academic Press, New York.

Schatz, G., and Mason, T. L. (1974) The biosynthesis of mitochondrial proteins, *Annu. Rev. Biochem.* **43**:51.

Schatz, G., Halsbrunner, E., and Tuppy, H. (1964) Deoxyribonucleic acid associated with yeast mitochondria, *Biochem. Biophys. Res. Commun.* **15**:127.

Schweyen, R. J., Weiss-Brummer, B., Backhaus, B., and Kaudewitz, F. (1978) The genetic map of the mitochondrial genome in yeast. Map positions of $drug^r$ and mit^- markers as revealed from population analysis of ρ^- clones in *Saccharomyces cerevisiae*, *Mol. Gen. Genet.* **159**:151.

Sebald, W., Wachter, E., and Tzagoloff, A. (1979) Identification of amino acid substitutions in the DCCD-binding subunit of the mitochondrial ATPase complex from oligomycin-resistant mutants of *S. cerevisiae*, *Eur. J. Biochem.* **100**:599.

Slonimski, P. P. (1953) *Formation des Enzymes Respiratoire chez la Levure*, Masson, Paris.

Slonimski, P. P., and Tzagoloff, A. (1976) Localization in yeast mitochondrial DNA of mutations expressed in a deficiency of cytochrome oxidase and/or coenzyme QH_2-cytochrome *c* reductase, *Eur. J. Biochem.* **61**:27.

Strausberg, R. L., Vincent, R. D., Perlman, P. S., and Butow, R. A. (1978) Asymmetric gene conversion at inserted segments on yeast mitochondrial DNA, *Nature* **276**:577.

Thalenfeld, B. E., and Tzagoloff, A. (1980) Assembly of the mitochondrial membrane system. Sequence of the *oxi2* gene of yeast mitochondrial DNA, *J. Biol. Chem.* **255**:6173.

Tzagoloff, A., Akai, A., Needleman, R. B., and Zulch, G. (1975) Assembly of the mitochondrial membrane system. Cytoplasmic mutants of *S. cerevisiae* with lesions in enzymes of the respiratory chain and in the mitochondrial ATPase, *J. Biol. Chem.* **250**:8236.

Van Ommen, G. J. B., Groot, G. S. P., and Grivell, L. A. (1979) Transcription maps of mtDNAs of two strains of *Saccharomyces*: Transcription of strain specific insertions; complex RNA maturation and splicing, *Cell* **18**:511.

Wesolowski, M., and Fukuhara, H. (1979) The genetic map of transfer RNA genes of yeast mitochondria. Correction and extension, *Mol. Gen. Genet.* **170**:261.

Glossary

Active transport The transfer of solutes against a concentration gradient. This mode of transport requires a net input of energy.

ADP/O A measure of the extra oxygen reduced in state 3 respiration when a given amount of ADP is added to mitochondria. This simple polarographic measurement was introduced by Britten Chance as an alternative to P/O determinations.

Amobarbital An inhibitor of NADH-coenzyme Q reductase. Also amytal.

Amphipathic Term used to denote the hydrophobic and hydrophilic properties of certain compounds, viz. phospholipids, detergents.

Ant^R Abbreviation meaning antibiotic resistance.

Antimycin A A specific inhibitor of coenzyme QH_2–cytochrome *c* reductase. This compound probably binds to cytochrome *b*, thereby preventing its oxidation by cytochrome c_1.

Antiport Carrier-mediated transport in which the transfer of one solute is obligatorily coupled to the counter-transfer of another solute.

Apocytochrome b The protein portion of cytochrome *b* minus the heme prosthetic group.

Ascus Structure consisting of a resistant wall that encloses the four spores of yeast resulting from meiosis and mitosis.

Atmungsferment Original term used by Warburg for cytochrome oxidase or the CO binding pigment of the respiratory chain.

ATP synthetase An inner membrane enzyme that uses oxidatively derived energy for ATP synthesis. This complex enzyme is also referred to as the OS-ATPase, CF_0-F_1, and complex V.

ATPase Abbreviation for adenosine triphosphatase. General term used for cellular enzymes that catalyze the hydrolysis of ATP to ADP and inorganic phosphate. The mitochondrial ATPase is located in the inner membrane, its primary function being to synthesize ATP during oxidative phosphorylation. See also ATP synthetase, CF_0-F_1, F_1.

ATPase inhibitor A low-molecular-weight protein naturally occurring in mitochondria. This protein binds to F_1 and inhibits its ATPase activity.

ATPase proteolipid A low-molecular-weight subunit of the oligomycin sensitive ATPase. This protein, also known as subunit 9 or DCCD-binding protein, has been proposed to form the proton channel of the ATPase.

Atractyloside Plant steroid that binds to the adenine nucleotide carrier and inhibits ATP/ADP exchange across the inner mitochondrial membrane.

Aurovertin An inhibitor of oxidative phosphorylation and of the mitochondrial ATPase. Aurovertin binds to the β subunit of F_1.

Beta oxidation An important oxidative pathway of mitochondria. The enzymes of beta oxidation metabolize saturated and unsaturated fatty acids to acetyl CoA.

Bonkrekic acid Inhibitor of ATP/ADP exchange in mitochondria.

Box Term designating genetic loci in the yeast apocytochrome *b* gene. See also *Cob*.

Cardiolipin A phospholipid present exclusively in the inner membrane of mitochondria.

Carnitine A quaternary amine derivative of butyric acid involved in the transport of fatty acids through the inner mitochondrial membrane.

m-CCCP Abbreviation for carbonyl cyanide m-chlorophenyl hydrazone, a potent uncoupler.

CF_0 A hydrophobic protein fraction of the OS-ATPase capable of binding to and conferring oligomycin sensitivity on F_1.

CF_0-F_1 Term introduced by Racker and co-workers to describe the oligomycin-sensitive ATPase.

Chaotropes Reagents such as urea, guanidine salts, and thiocyanate that promote solubilization of proteins. They are thought to weaken hydrogen bonds.

Chemical hypothesis Oldest interpretation of the mechanism of oxidative phosphorylation, which invoked the existence of nonphosphorylated and phosphorylated high-energy compounds as intermediates in the coupling process.

Chloramphenicol A bacterial antibiotic used as a specific inhibitor of mitochondrial protein synthesis.

Cholate A bile acid frequently used to solubilize hydrophobic membrane proteins. Cholate is a slightly less efficient detergent than deoxycholate.

Cholate dialysis A procedure developed by Racker and co-workers to reconstitute proton-impermeable membrane vesicles from purified electron transfer complexes and from the oligomycin-sensitive ATPase.

Chondriosome Term used to denote the mitochondrial organelle in mature sperm cells.

Cluster-clone hypothesis An evolutionary theory of how the present-day eucaryotic cell arose. The central idea of this hypothesis, formulated by Lawrence Bogorad, is that the ancestral cell contained many different copies of plasmid type of DNA whose genetic information became centralized into a few organelles such as the nucleus, mitochondria, and chloroplasts.

Cob Term used to designate genetic loci in the apocytochrome *b* gene of yeast mtDNA. *Box* and *cyb* are synonymous terms used by Slonimski and co-workers and Linnane and co-workers, respectively.

Codon A three-nucleotide sequence in mRNA specifying a unique amino acid.

Coenzyme Q A lipid-soluble quinone, also referred to as *ubiquinone*, that functions as both electron and proton acceptor and donor in the main pathway of the mitochondrial respiratory chain.

Complex I NADH–coenzyme Q reductase. One of the four electron transfer complexes.

Complex II Succinate–coenzyme Q reductase.

Complex III Coenzyme QH_2–cytochrome *c* reductase.

Complex IV Cytochrome *c* oxidase.

Complex V Another term for the oligomycin-sensitive ATPase.

Condensed conformation In this state the mitochondrial matrix is reduced in volume and the inner membrane forms tubelike structures. The condensed conformation is seen in actively respiring and phosphorylating mitochondria.

Conformational hypothesis A mechanistic model of oxidative phosphorylation. According to this view the energy of oxidation is first converted to an energetically unstable conformational state of a protein.

Core proteins The two highest-molecular-weight polypeptides of coenzyme QH_2–cytochrome *c* reductase. The function of these proteins is not known at present.

Cotranslational transport Mechanism of protein transport according to which the synthesis of the polypeptide chain is coupled to its transfer across the membrane.

Coupling Term used in connection with the ability of mitochondria and certain types of submitochondrial particles to use the energy of oxidation of NADH or succinate for ATP synthesis. The efficiency with which the energy is conserved reflects the degree of coupling of the two processes.

Coupling factor Term used to describe soluble protein fractions of mitochondria capable of increasing the efficiency of oxidative phosphorylation or energy-dependent reactions in submitochondrial particles.

Cristae Invaginations of the mitochondrial inner membrane giving rise to one of the most characteristic morphological features of the organelle.

Cross-over point The slowest step in a linear sequence of enzymatic reactions. The cross-over points in the electron transfer chain were used by Chance and co-workers to identify the three phosphorylation sites.

Cycloheximide A specific inhibitor of cytoplasmic protein synthesis in eucaryotic cells.

Cytochrome Present in all known electron transfer chains, cytochromes are identified by their absorption spectra with well-defined maxima in the visible spectrum. The absorption properties of cytochromes are due to the heme prosthetic groups associated with the apoproteins.

Davson and Danielli model An early model of membrane structure in which the protein constituents were proposed to be bound to the two surfaces of the lipid bilayer through electrostatic interactions with the polar groups of the phospholipids.

DCCD Abbreviation for dicyclohexylcarbodiimide, a potent inhibitor of the mitochondrial oligomycin-sensitive ATPase. DCCD binds covalently to the proteolipid of the enzyme.

DCCD-binding protein See *ATPase proteolipid.*

Deoxycholate A bile acid frequently used to solubilize hydrophobic membrane enzymes and proteins.

Dichlorophenolindophenol Commonly used as an artificial electron acceptor in reactions catalyzed by electron transfer enzymes.

Digitonin Plant steroid with detergent properties used to selectively disaggregate the outer mitochondrial membrane.

Dio-9 An inhibitor of the F_1 ATPase.

DNP Abbreviation for dinitrophenol, one of the oldest known uncouplers of oxidative phosphorylation.

Electrogenic exchange Exchange of two solutes with different charges across a membrane.

Electron transfer chain A mitochondrial pathway that catalyzes the transfer of reducing equivalents from NADH or succinate to oxygen. A universal device to reoxidize NADH in aerobic cells.

Electron transfer complex Also *respiratory complex* in this text. This term refers to the four enzymes that jointly catalyze the oxidation of NADH or succinate by oxygen.

Endosymbiont hypothesis A hypothesis proposed by L. Margulis to explain how the modern eucaryotic cell arose during evolution. The main tenet of the hypothesis is the colonization of an early nucleated cell by a procaryote with a more efficient means of energy metabolism (oxidative phosphorylation, photosynthesis). Accordingly mitochondria and chloroplasts are viewed as being the vestiges of the original invaders stripped of all functions except energy metabolism.

EPR spectroscopy Abbreviation for electron magnetic resonance spectroscopy. A method used to measure the oxidation reduction states of metals and organic compounds with unpaired electrons.

Erythromycin A bacterial antibiotic used to specifically inhibit mitochondrial protein synthesis.

ETP Abbreviation for electron transfer particle. This is an early term used for vesicle preparations obtained after ultrasonic irradiation of mitochondria.

ETP_H Submitochondrial vesicle preparations obtained from bovine mitochondria. Characterized by their high P/O.

Exon That part of a gene sequence that eventually appears in the mature mRNA, rRNA, or tRNA.

External proteins A class of membrane proteins peripherally associated with the lipid bilayer. External proteins can usually be separated from the integral membrane proteins by extraction of membranes with salt or relatively weak chaotropes such as urea or guanidine hydrochloride.

F_1 Name originally given to a mitochondrial ATPase preparation capable of increasing the phosphorylation capacity of submitochondrial particles. The F_1 ATPase is a water-soluble protein composed of 5 different subunit polypeptides and is a part of the oligomycin sensitive ATPase.

$F_1 \cdot X$ A preparation of F_1 complexed to OSCP.

F_6 A soluble protein with coupling factor activity. This inner membrane protein has been shown to be a part of the oligomycin-sensitive ATPase and to be required for the binding of F_1 to the hydrophobic CF_0 units.

Facilitated diffusion Mechanism of solute transport involving the participation of a porter or carrier protein.

Factor B A coupling factor of oxidative phosphorylation. Probably a subunit of the oligomycin-sensitive ATPase.

FAD–$FADH_2$ Oxidized and reduced forms of flavin adenine dinucleotide. Common coenzyme of dehydrogenases. In the mitochondrial respiratory chain, FAD is covalently linked to succinate dehydrogenase.

Flavoenzymes Dehyrogenases with FAD or FMN as prosthetic groups.

Fluid mosaic model A general model of membrane structure proposed by Singer and Nicolson. The fluid mosaic model departed from the earlier view of membrane structure by permitting certain proteins to span the entire width of the lipid bilayer. This model also ascribed more dynamic properties to the lipid and protein constituents, which could move laterally through the bilayer.

FMN–$FMNH_2$ Oxidized and reduced forms of flavin mononucleotide. Common coenzyme of dehydrogenases. In the respiratory chain of mitochondria FMN is a prosthetic group of NADH dehydrogenase.

Genetic code The assignment of codons to the 21 amino acids.

Genetic map The order and location of genes and loci as deduced from genetic analysis of mutants. The map distances are usually expressed in units of recombination.

Glucogenic amino acids Amino acids such as alanine, threonine, etc., that can be converted to pyruvate and in turn used for glucose synthesis.

Gluconeogenesis Metabolic pathway responsible for the biosynthesis of glucose from phosphoenolpyruvate. Gluconeogenesis occurs partly in mitochondria.

Glycolysis A metabolic pathway present in all known cells. The enzymes of this pathway are located in the cytoplasm. Jointly they catalyze the anaerobic conversion of glucose to pyruvic acid.

Gramicidin Linear pentadecapeptide with ionophoric activity for K^+ and Na^+.

Heme a Heme prosthetic group of "a" type cytochromes.

Hemin The ferroprotoporphyrin prosthetic group of "b" and "c" type cytochromes.

Initiation codon The first codon of mRNA translated into an amino acid.

Inside out vesicles Vesicles whose membrane orientation with respect to the external medium is opposite to the orientation *in situ*. Submitochondrial particles prepared by sonic disruption of mitochondria are an example of inside out vesicles.

Integral proteins Water-insoluble membrane proteins that are embedded in and anchored to the membrane through hydrophobic interactions with the lipid bilayer. Operationally this class of proteins can be extracted with detergents or strong chaotropes such as guanidine thiocyanate.

Intermembrane space Mitochondrial compartment located between the outer and inner membrane. This compartment is also referred to as the *intracristal space*.

Intervening sequence See *intron*.

Intron Another term for *intervening sequence*. A region inside a gene sequence that does not appear in the mature tRNA, rRNA, or mRNA.

Ionophore A compound that promotes the transfer of ions through the membrane bilayer.

Isosbestic wavelength The wavelength at which a compound (electron transport carrier in the present context) has the same absorbance when oxidized or reduced.

Ketogenic amino acids Amino acids, such as leucine or phenylalanine, that are metabolized to acetyl CoA.

Ketone-bodies Term used for the three compounds, acetone, acetoacetate, and β-hydroxybutyrate. Ketone bodies are made in liver mitochondria during periods of fat utilization. They are transported to other tissues, where they can serve as a source of oxidizable substrates.

5′ leader The sequence preceeding the initiation codon in a messenger RNA.

Locus Literally, a location in DNA. In genetics this term implies a closely linked set of alleles expressing a certain phenotype.

Map units Length of DNA as expressed in percent. Each map unit represents 1% of the total length of DNA.

Matrix Mitochondrial compartment enclosed by the inner membrane. Most of the soluble enzymes of mitochondria are present in the matrix.

Maturase Term coined by Slonimski and co-workers to designate a protein encoded in an intervening sequence. As the term implies, the function of such a protein is to excise the intervening sequence from the precursor RNA.

Membrane factor CF_0 from which OSCP has been extracted. Probably similar to the hydrophobic protein fraction of the ATPase complex.

Midpoint potential The potential at which the oxidized and reduced forms of a compound or metal are equal in concentration.

Mit genes Mitochondrial genes coding for certain subunit polypeptides of cytochrome oxidase, coenzyme QH_2–cytochrome *c* reductase, and the oligomycin-sensitive ATPase.

Mit⁻ mutants Strains of *S. cerevisiae* with genetic lesions in mit genes.

Mitochondria Subject of this book.

Mitoplast A fraction of mitochondria consisting of the inner membrane with the full complement of matrix proteins.

Mitotic segregation The sorting out of genetic markers during vegetative growth of diploid or haploid cells.

Mobile carrier Term reserved for coenzyme Q and cytochrome *c*.

mtDNA Abbreviation for mitochondrial DNA.

N particles Submitochondrial particles prepared by disruption of mitochondria in a Nossal shaker.

NAD^+-NADH Abbreviation for the oxidized and reduced forms of nicotinamide adenine dinucleotide, an essential coenzyme of many dehydrogenases. In the earlier literature this coenzyme is referred to as *diphosphopyridine nucleotide* (DPN).

Negative staining A technique used to stain biological materials for electron microscopic examination. The sample is covered with a solution of an electron-opaque polymer (phosphotungstate) and viewed directly in the electron microscope. This rapid method reveals the contours of the biological structure with a high degree of resolution.

Nigericin An ionophoric compound capable of effecting the exchange of protons for monovalent cations.

Nonheme iron proteins A class of proteins that mediate electron transport through the oxidation and reduction of iron complexed to elemental sulfur and the sulfhydryl groups of cysteine residues.

Northern blot A technique by which RNA separated electrophoretically on agarose is transferred and covalently linked to appropriately derivatized paper.

Oligomycin An inhibitor of oxidative phosphorylation. This antibiotic binds to the CF_0 component of the oligomycin-sensitive ATPase.

O/R loop A term used by Peter Mitchell for the proton and electron carrier systems of the respiratory chain. The O/R loop consists of two coupled transport systems, one effecting the transfer of electrons between two carriers and the other, that of protons across the inner membrane.

Orthodox conformation Morphological state of mitochondria defined by the presence of cristae. This conformation is commonly observed in thin sections of tissue and is associated with metabolically resting mitochondria.

OS-ATPase Abbreviation for oligomycin-sensitive adenosine triphosphatase. See also *CF_0-F_1, ATP synthetase, Complex V.*

OSCP Abbreviation for oligomycin-sensitivity-conferring protein. A constituent of the oligomycin-sensitive ATPase believed to be part of the stalk seen in negatively stained preparations of the inner membrane. OSCP has coupling factor activity.

Oxi Term designating genetic loci in mitochondrial genes coding for subunits of cytochrome oxidase. *Cox* is a synonymous term used by Linnane and coworkers.

Oxidative phosphorylation Process by which ATP is formed from ADP and inorganic phosphate during electron transport.

Palindromic sequence A symmetrical nucleotide sequence such that the two halves of the sequence can base pair to form a hairpin.

Partial reactions of oxidative phosphorylation Exchange reactions (ATP-P_i, ATP-H_2O, etc.) believed to reflect catalytic steps of the overall mechanism of oxidative phosphorylation.

Passive diffusion Diffusion of solutes across membranes owing solely to differences in their concentration inside and outside.

Pet Mutants of *S. cerevisiae* with genetic lesions in nuclear genes required for the morphogenesis of respiratory functional mitochondria.

Petite deletion mapping A method used to map genetic markers in mitochondrial DNA by analyzing their loss or retention in the genomes of ρ^- mutants of *S. cerevisiae*.

Petite mutants Respiratory deficient mutants of *S. cerevisiae* resulting from long deletions in mitochondrial DNA. Cytoplasmic petite mutants are also known as ρ^-.

Phenazine methosulfate Compound used as an artificial electron acceptor or donor.

Pho Term designating genetic loci in mitochondrial genes coding for subunits of the oligomycin-sensitive ATPase.

Phosphorylation sites Spans of the electron transport chain that are coupled to ATP synthesis.

Physical map The order and locations of genes or loci as deduced from the analysis of DNA by methods capable of assigning distances in physical units of length of DNA or number of nucleotides.

Piericidin An inhibitor of NADH–coenzyme Q reductase.

P/O The molar ratio of inorganic phosphate esterified and oxygen reduced during oxidative phosphorylation. The P/O value has been and continues to be an important measurement of the efficiency of energy coupling in mitochondria.

Porter Term used for membrane proteins that bind low-molecular weight compounds and effect their transfer across the membrane. Synonymous with *carrier*.

Post-translational transport Mechanism in which a protein is transported through a membrane following completion of the polypeptide chain.

Primary structure The order of building units (e.g., amino acids, nucleotides) in a polymer.

Primary transcript The initial RNA transcribed from a given region of DNA.

Promitochondria Mitochondrial organelles lacking a functional respiratory chain.

Proton gradient The difference in the chemical activity of protons inside and outside of the mitochondrial inner membrane. The establishment of a proton gradient during electron transport is considered to be the primary energy conservation event of oxidative phosphorylation in the chemiosmotic hypothesis.

Protonmotive force Term used by Peter Mitchell to describe the energy of the proton gradient formed as a result of electron transport. The protonmotive force is equal to the energy stored in the membrane potential and the osmotic force of the proton gradient.

Quaternary structure The topological arrangement of subunit polypeptides in an oligomeric protein.

Q cycle A hypothetical electron transfer scheme proposed by Peter Mitchell to account for the transfer of two protons across the inner membrane for every electron traversing the coenzyme Q–cytochrome *c* segment of the chain.

R.C. Abbreviation for respiratory control. This value is calculated by dividing the rate of state 3 by that of state 4 respiration. The respiratory control value provides a measure of how well phosphorylation is coupled to respiration.

Respiratory carriers Catalytic components of the electron transfer chain that have been shown to undergo oxidation–reduction reactions during substrate oxidation.

Respiratory chain Another term for the *electron transfer chain.*

Respiratory complex See *electron transfer complex.*

Restriction endonucleases Enzymes that cleave double-stranded DNA at specific nucleotide sequences.

Restriction fragment Piece of linear DNA produced as a result of cleavage of DNA by a restriction endonuclease.

Restriction map A map of DNA showing the location of sites for restriction enzymes in physical units of distance.

Rho^0 (ρ^0) A mutant of *S. cerevisiae* completely devoid of mitochondrial DNA.

Rho$^-$ (ρ^-) Cytoplasmic petite mutants of *S. cerevisiae* with long deletions in mitochondrial DNA.

Rhodamine 123 A fluorescent dye used to stain mitochondria in living cells.

Rieske nonheme iron protein An electron carrier of coenzyme QH_2-cytochrome *c* reductase.

Right side out vesicles Membrane vesicles whose membranes have the same orientation with respect to the external medium as the *in situ* membrane has with respect to the cytoplasm.

RNA maturation The steps leading to the conversion of the primary RNA transcript to a functional rRNA, tRNA, or mRNA.

Rotenone An inhibitor of NADH–coenzyme Q reductase.

Rutamycin Oligomycin D.

SDS Abbreviation for sodium dodecyl sulfate. A powerful ionic detergent widely used for the total dissociation of protein–protein and protein–lipid interactions. Also causes loss of quaternary and tertiary structure in a protein.

SDS gel electrophoresis A technique used to separate proteins on polyacrylamide gels purely on the basis of their sizes.

Secondary structure Structural features of a protein having to do with turns in the polypeptide chain, regions of β sheets and α helix. In nucleic acids this term refers to the folding of the molecule owing to base pairing.

Signal hypothesis A cotranslational mechanism of protein transport. An essential feature of this mechanism is the synthesis of the protein as a precursor with an amino terminal signal sequence.

Signal peptide That portion of the polypeptide chain that contains the binding sequence for attachment to a membrane receptor. In most proteins this peptide constitutes the beginning amino terminal sequence that is cleaved during translocation of the protein.

Silent copper The copper of cytochrome oxidase that does not have an EPR signal.

SMP Abbreviation for *submitochondrial particles.* An alternative name for ETP.

Soret The γ-absorption band of a cytochrome.

Splicing Process by which intervening sequences are removed from RNA.

State 3 respiration The rate of oxygen utilization by mitochondria when inorganic phosphate and ADP are plentiful.

State 4 respiration The rate of oxygen utilization by mitochondria when either phosphate or ADP is limiting.

Syn genes tRNA and rRNA genes in yeast mitochondrial DNA.

Syn⁻ mutants Yeast strains with genetic lesions in syn genes.

TCA cycle Abbreviation for tricarboxylic acid cycle. Also known as *citric acid* or *Krebs cycle*. The enzymes of this metabolic pathway are located in mitochondria. They catalyze the oxidative decarboxylation of acetylCoA to CO_2 and water.

Termination codon A three-nucleotide sequence immediately following the last amino acid codon in mRNA. The termination codons signal the end of translation.

Tertiary structure The manner in which a polypeptide chain is folded in three dimensions.

Tetrad analysis Term used to describe the analysis of the genotypes of the four spores arising from meiosis of a diploid yeast cell.

2-Thenoyltrifluoroacetone Inhibitor of succinate–coenzyme Q reductase.

TMPD Abbreviation for tetramethylphenylene diamine. A dye commonly used as an artificial electron acceptor or donor.

Triton A neutral synthetic detergent used in the fractionation of membrane proteins.

U particle SMP extracted with urea.

Ubiquinone Another name for *coenzyme Q*.

Uncouplers Compounds that lower the efficiency of oxidative phosphorylation. According to the chemiosmotic hypothesis uncouplers are proton carriers.

Uniport Transport mechanism involving a unidirectional transfer of a solute by a carrier molecule.

Urea cycle An important metabolic pathway in ureotelic animals (land vertebrates). The reactions of the urea cycle are used for arginine biosynthesis and for the removal of ammonia in the form of the endproduct urea. Two enzymes of this cycle are located in the matrix compartment of mitochondria.

Valinomycin A cyclic polypeptide that acts as a potassium ionophore.

Varl Genetic locus in mtDNA of *S. cerevisiae*. This locus is believed to be in the gene coding for a subunit of mitochondrial ribosomes. Also name of the ribosomal subunit synthesized in yeast mitochondria.

Wobble The ability of guanine and uridine to base pair with one another.

Wobble hypothesis Base pairing rules governing the interaction of tRNA anticodons with the mRNA codons first proposed by Francis Crick.

Index